The Process of Animal Domestication

The Process of Animal Domestication

MARCELO R.
SÁNCHEZ-VILLAGRA

PRINCETON UNIVERSITY PRESS
Princeton and Oxford

Published by Princeton University Press
41 William Street, Princeton, New Jersey 08540
6 Oxford Street, Woodstock, Oxfordshire OX20 1TR

press.princeton.edu

All Rights Reserved
ISBN 9780691217666
ISBN (pbk.) 9780691217673
ISBN (e-book) 9780691217680

British Library Cataloging-in-Publication Data is available

Editorial: Alison Kalett & Whitney Rauenhorst
Production Editorial: Ali Parrington
Jacket/Cover Design: Wanda España
Production: Jacqueline Poirier
Publicity: Matthew Taylor & Amy Stewart
Copyeditor: Lucinda Treadwell

Jacket/Cover Credit: Pablo Picasso, "Le Taureau" (series of 11 lithographs), 1945–46. © 2021 Esate of Pablo Picasso / Artists Rights Society (ARS), New York. Digital images: National Gallery of Art, Washington

This book has been composed in Minion Pro

Printed on acid-free paper. ∞

Printed in the United States of America

10 9 8 7 6 5 4 3 2 1

Contents

Preface   vii

CHAPTER 1   Pathways in time and space   1

CHAPTER 2   Domesticated mammals and birds: Species accounts   36

CHAPTER 3   The genetics of domestication   67

CHAPTER 4   Evolutionary development   90

CHAPTER 5   Ontogenetic change   122

CHAPTER 6   Life history and growth   144

CHAPTER 7   Morphological diversification   169

CHAPTER 8   Feralization and experimental domestication   203

CHAPTER 9   Fish domestication   220

CHAPTER 10   Insect domestication   236

             Epilogue   247

             Acknowledgments   249

             References   251

             Index   311

             Image Credits   321

Preface

Domestic animals are ubiquitous. In contrast to the biodiversity crisis currently impacting many kinds of animals and plants, not a single domesticated species is endangered. Instead, like humans, domesticated species show wide geographic distributions, in some cases even becoming invasive species. The estimated biomass of livestock, led by cattle, is twice as much as that of all wild mammals combined.

Domestication is important. And yet there is a kind of snobbery against domestic animals on the part of biologists. For example, domesticates are largely neglected or ignored by those building or maintaining zoological collections. While the modest radiation of Galapagos finches is celebrated as a paramount of evolution, domesticated pigeons are many orders of magnitude more diverse morphologically. Cichlid fishes from African lakes have received attention for their impressive and fast evolution, and as such are always included in textbooks on evolution. In contrast, the cranial diversity of goldfish, which may surpass that of cichlids in many parameters, is understudied.

Maybe disdain for domestic animals is an example of the "nature fallacy" pervasive in many people, the notion that nature and wildness are good, or at least better than human-created things, which are bad. This may have led to the perception of domesticated animals as "not natural." A biologist is interested in the natural world, and domestic animals are not quite part of that. In addition, boundaries of scholarly disciplines play a role. Domesticated animals are a subject tied to veterinary medicine. While this discipline has generated huge amounts of knowledge, traditionally it has lacked the evolutionary view.

But things are changing. Some of the most innovative work on genetics of morphological diversification is conducted in domestic pigeons, rabbits, and a myriad of fishes as subjects of aquaculture. Canaries, zebra finches, and Bengalese finches provide insights into the study of language. Ideas on domestication have inspired work on the neural crest, a fundamental group of embryonic cells of vertebrates traditionally studied only in model organisms. The subject of domestication is becoming more prominent in controversial but meaningful comparisons between the origin of modern *Homo sapiens* and some aspects of the domestication process.

The most influential work on evolution, Darwin's (1859) *On the Origin of Species*, starts with a chapter about domestication. The conceptual and methodological developments in studies of domestication since Darwin have been enormous. The study of the origins of, or rather transitions to domestication, and of the multidimensional consequences of domestication, require contributions from different disciplines, including zooarchaeology, ethnology, molecular biology and evolutionary morphology. I summarize salient aspects that come from these perspectives. My view, biased and incomplete as any other, is that of an organismal zoologist. The emphasis of this book is on the natural history of the phenotype and its developmental origin. The patterns of wonderful variation concern different levels of organization.

It pains me to know there are many great examples of subjects I treat that I do not mention, and relevant papers I have not included. I have tried to use my exposure to German Academia and my multicultural background to strengthen the scholarship of this book, but I have surely failed in many ways. The subjects that domestication touches upon are so many and disparate, and the information so vast, that it is impossible to embrace them all.

I have provided drafts of this book to several colleagues, and without exception their critical reading resulted in useful feedback and sometimes strong comments about attempts to define domestication. Defining domestication is a thankless task, and it is just impossible to please everybody. I did my best to provide several views and convey the difficulties in finding a universal concept for so many species and differences in interactions. One thing we all agree on is that domestication is an ongoing process rather than an invention or an event. Another point of general agreement is the importance of the animal perspective in understanding the transition to domestication. And that is where the commonality of domestication across species ends. The main message of this book is that there are few if any universals in the patterns that result from the domestication process.

Domestication encompasses many processes and mechanisms and many degrees of interaction between humans and the domesticated organisms.

Domestication is just evolution, with human influence to some degree or another. As such, it includes many mechanisms—for example, hybridization—and if we discuss all animals as I do, then the historical contingencies are diverse. Thus, the main message of the book I will emphasize is that domestication produces, as it happens in evolution, wonderful variation and a lack of universals. There are common principles, and these are explained in the book, but the resulting patterns are diverse depending on local conditions. This explains, for example, why universal features in a "domestication syndrome" are not to be expected.

The diversity of patterns and kinds of domestication make the subject extremely rich and challenging. This is not some hopeless and almost naive "biology is complicated and it is not physics" perspective of some biologists of a few generations ago. Physics itself has changed in the last decades, with ideas of emergence and local laws. Biology like physics is complicated, surely, but tractable and even predictable. However, we need to look at local conditions and not necessarily expect universal patterns, even if these are our null hypothesis.

The famous maxim that "more is different" of condensed matter physics by Philip W. Anderson in 1972 applies well to our current understanding of evolution. The emergent patterns that result during domestication, an example of developmental evolution, mean that it is not just the finding of individual genes and their interactions that will reveal how diversity originates.

The same biological principles that apply to mammals and birds apply to fishes and insects, but their different evolutionary histories mean that comparisons of some organ systems are too broad to be practical. An example is brain size. More importantly, the literature available on fishes and insects has different histories and contexts, and the number of "classical" or old domesticated species are few. The chapters on fishes and insects follow the same topic sequence as the rest of the book, so the reader finds the same story line.

Throughout the book I present short discussions on fundamental aspects of the environmental crisis, animal welfare issues, and food consumption. These are complex subjects that require a more extensive discussion, but I did not want to leave these matters unmentioned. There is no doubt that selective breeding has brought great benefits to humans, in some cases without compromising the welfare of animals in significant ways. But there is a tipping point. Those that claim that genetic manipulation is just an example of what humans have been doing for millennia with selective breeding are right, but they overlook the fact that there is a nonlinear relationship between the extent or kind of manipulation and the biological consequences or the welfare of the animals in question.

The Process of Animal Domestication

CHAPTER 1

Pathways in time and space

Animal domestication encompasses many kinds of interactions between humans and other species. It is a continuum of stages of a gradually intensifying relationship. This relationship ranges from anthropophily to commensalism, from control in the wild to control of captive animals, from extensive to intensive breeding, and in some cases it extends to owning of pets (e.g., Vigne 2011, Zeder 2012a, b; Larson and Fuller 2014). A fundamental and primary aspect of domesticated animals is their tameness, meaning that they tolerate and are unafraid of human presence and handling. The genetics and the physiological and morphological correlates of tameness have thus been a central focus of studies of domestication. However, tameness alone does not imply domestication, as exemplified by tamed elephants living in close association with humans. Keeping an animal as a pet does not make it domestic. Examples from the Amazon region abound. Changes in reproduction can be seen at the core of domestication (Vigne 2011).

Domestic animals emerged from small groups of individuals of their respective wild form that became increasingly reproductively isolated from the stem forms as a result of the influence of humans. They adapted to the peculiar ecological conditions imposed by an anthropogenic environment and in some cases developed considerable population sizes. Domesticated animals are subject to environmental conditions and selective pressures different from those faced by their wild counterparts. Furthermore, the conditions to which populations of domestic animals are exposed vary greatly (e.g, culling patterns, availability of food, protection from predators). Altered natural selection and continual targeted and non-targeted selection by humans led to divergence from the wild norm in morphology, physiology, and behavior. Domestic animals are increasingly used for economic and leisure purposes in diverse ways. The variety of perspectives by which to characterize domestication (e.g., symbiotic interactions: Budiansky 1992; resulting domesticated phenotype: Price 1984; Kohane and Parsons 1988) make a unique and universal definition a challenging and unrealistic goal (Ladizinky 1998; Balasse et al. 2018).

Traditionally, domestication has been defined and conceptualized from the human perspective, with our species as the domesticator. This view is no longer universally accepted, and in fact different perspectives have contributed to this change. A new look at naturalistic observations demonstrates the active role played by animals in approaching humans and in looking for benefits resulting from human proximity and interaction. It is thus relevant to examine the reciprocal impact of animals in shaping the trajectory of human biological and cultural evolution (Zeder 2017). Animal-human interactions have been discussed in terms of niche construction, a subject often treated in discussions of an expanded evolutionary synthesis (Smith 2011a; Zeder 2018). Niche construction refers to the evolutionary impact of ecosystem engineering activities that create new or modify existing selection pressures acting on present and future generations (Odling-Smee et al. 2003). Humans have been characterized as the ultimate niche constructors, and cultural niche construction has been discussed in the context of the initial phase of domestication (Smith 2011b). Domestic animals are also niche constructors. Independent of the discussion around the repetitive nature of the subject of niche construction in the literature (Gupta et al. 2017), its relevance to conceptualizing and describing ecological interactions is uncontested.

Another perspective that questions the traditional and human-centered conceptualization of domestication (e.g., Zeuner 1963) is a philosophical/sociological one. People tend to create narratives (Diogo 2017), and we have done so with domestication, in which we present ourselves as central and the makers of destinies of organisms. This notion ignores the active role of the "domesticated" and is a traditional Western European view of our place in nature that is not universal among humans (Ingold 2000; Descola 2013; Figure 1.1). The argument has been made for abandoning the notion of domestication in favor of a continuum of human-nonhuman animal relationships (Russell 2002). Although there is merit in this idea, it does not solve the issue of defining the complex phenomenon we call domestication. It is more productive to discuss the pathways to domestication and the different kinds of interactions entailed by domestication. These reflections should not obviate the recorded cases in which humans have played and directed a one-sided role in domestication, as in the case of canaries native to the Canary Islands brought to Europe and domesticated simply because of their singing (Birkhead 2003).

When the focus is on intense, selective breeding and animal management, the conceptualization of domestication leads to a view in which humans are the sole agents (Fig. 1.1, "Ego"). This view also sees domestication as an intentional and goal-oriented interaction. An alternative view arises if one concentrates on the first steps of the domestication continuum. At this point, people did not have long-term domestication plans, and interactions between humans and other animals were voluntary on both sides; therefore, from this perspective, the agent is not as obvious. The argument has been made that, based on some parameters, some domesticated animals and plants have benefited more from the interaction than humans themselves (Budiansky 1992). The increased distribution and multiplication of species that became domesticated contrast with the many challenges and disadvantages faced by humans following the Neolithic transition. The idea of human demise following the Neolithic transition has an element of retro-romantic thinking. What is needed is a multivariate evaluation and quantification of human prosperity across time, so that a nuanced evaluation of how human life has changed can be attained. Surely the result will show nonlinear changes, geographic variation, and a lack of universals.

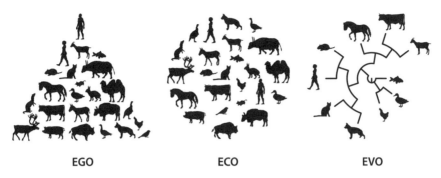

EGO   ECO   EVO

Fig 1.1. Ego, Eco, and Evo views of the human-animal interactions. Only domesticated animals are shown.

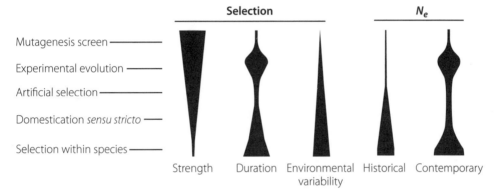

Fig 1.2. Comparison of domestication *sensu stricto* versus artificial selection and other kinds of selection regimes. The geometric shapes represent the relative magnitude of variables shown along the top of the figure. N_e is the effective population size. The historical population size influences the amount of variation present in the population.

It is fundamental to differentiate the intense "artificial selection" typical of the creation and preservation of breeds ("intensive breeding") from the domestication pathways described below, associated with the initial phase of interaction, in which a dependence of the domesticated form on humans has not yet been established. Mutagenesis screens, experimental evolution, artificial selection, domestication, and selection within species differ in important parameters in space and time (Stern 2011; Figure 1.2). A mutagenesis or genetic screen is an experimental approach used in research to generate a mutated population to identify and select for individuals with a specific target phenotype, providing information on gene function. The difference between domestication *sensu stricto* versus selection for "improvement" traits or artificial selection, as well as with other kinds of evolutionary and human-induced phenomena, becomes evident when comparing degree of selection and population sizes.

There are different pathways to domestication. Likewise, the kinds of interactions at the other end of the domestication continuum (Figure 1.3) are not all the

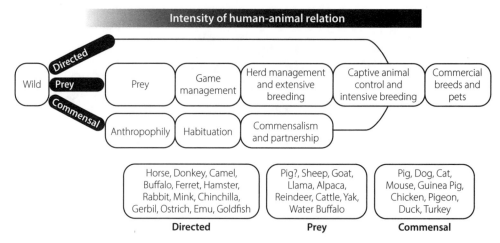

Fig 1.3. Pathways to domestication. Changes in intensity of the human-animal relationship are indicated.

same. Selective breeding, aimed at preserving specific breeds or features, is different from management, which involves manipulation of growth conditions, or the environment that sustains it. The aim of management is to increase the relative abundance and predictability of a population and to reduce the time and energy required to harvest it (Zeder 2015).

Pathways to domestication

Traditionally, domestication has been seen as resulting from goal-driven human action, with narratives about selection for traits that differentiated wild and domestic forms. In reality, domestication of different species has involved different kinds of interactions. Zeder (2012a, b) formally recognized and described three separate pathways followed by animals into a domesticated relationship with humans: a commensal pathway, a prey pathway, and a direct pathway (Figure 1.3).

There is usually no intentionality in the commensal pathway, which involves a coevolutionary process in which a population uses a novel niche that includes another species. That niche could involve human food waste or refuge, which is then taken advantage of by a subset of individuals of another animal species (e.g., wolves) that were less aggressive (i.e., tamer) than the rest. In the absence of human instigation, an interaction could arise, and only later would the human-directed selection that we associate with modern domestic populations have been possible.

The prey pathway involved a human intention to increase the efficiency of resource management. Medium to large herbivores were targeted as prey, including perhaps the case of horses. Although not originally planned as such, domestication resulted from humans altering their hunting strategies toward herd management, eventually leading to control over the animals' diet and reproduction (Zeder 2012a, b). The prey pathway probably took place in human communities that cultivated plants and did not lead a hunter-gatherer life. The directed pathway involved the deliberate use of a species and its incorporation into human life for uses such as transport, although the species in question were sometimes hunted as prey. A classic example of a directed pathway involving consumption is aquaculture (chapter 9).

The domestication pathway followed by different species is in some cases clear and in some cases debatable (Larson and Fuller 2014). There can be mixed cases, as in pigs perhaps having been domesticated via both a commensal and a prey pathway. The zooarchaeological and molecular evidence used to establish domestication pathways is mostly inconclusive, but mortality profiles may provide clues (Payne 1973) if performed with proper sampling and approach (Bartosiewicz 2015; Bartosiewicz and Bonsall 2018). Although recourse to comparative, ethnological data from hunter-gatherers is important, such ethnological data only provide hints on the plausibility of an explanation by analogy, and never a direct test of what happened.

Domestication in other species?

Different kinds of interactions occur among animal species, and some of these have been compared with domestication (Zeuner 1963). Prominent among these interactions is symbiosis, in which both partners benefit. Certainly common aspects are shared by some interactions and domestication, but by definition and considering the cognitive and social aspects associated with humans, it seems reasonable to see those commonalities as superficial.

The sharing of resources and of defense against predators recorded for baboons interacting with feral dogs and cats in Saudi Arabia are a remarkable case recorded in numerous videos and popular accounts. Indeed, mixed-species associations are known to occur and benefit those involved by increasing foraging success, and by aiding in the detection and deterrence of predators (Venkataraman et al. 2015).

In the case of agriculture, some authors have called the case of humans and crops and the "agriculture" practiced by leaf cutter and other ants a convergence (Conway Morris 2003), but there are profound evolutionary differences between the two (Sterelny 2005; Jablonsky 2017). Agriculture has reportedly evolved in three groups of insects: once in ants, once in termites, and seven times in ambrosia beetles. All three groups produce clonal monocultures within their nests and for generations, with monitoring of gardens and additionally managing of microbes that provide disease suppression (Mueller et al. 2005). Other reports include those of fungus farming by a snail in the marine environment (Silliman and Newell 2003), bacterial husbandry in social amoebas (Brock et al. 2011), and a damselfish (*Stegastes nigricans*) and algae (*Polysiphonia* sp.) in a coral reef ecosystem. But the "agriculture" of these animals is neither associated with cultural changes in the domesticator, nor has it led to major geographic expansions and use of natural resources. Furthermore, the associated cognitive, physiological, and developmental aspects of the organisms involved are different from those of humans.

The diversity of domesticated mammals and birds: patterns in time and space

Roughly 70,000 species of vertebrates have been recognized in the world, of which about 5,500 are mammals and 10,900 are birds. Of these, only a few dozen species have been domesticated. The number of species with populations being managed or kept in captivity is much larger, and many of these have been described as "semi-domesticated" (Mason 1984). Distinguishing wild from domestic forms—to use the simple and not always proper dichotomy—in both the zooarchaeological record and even when considering extant population samples, is not an easy task. One aspect to consider is that lifestyle under domestication is quite variable. A wild population may be more similar in its life conditions to a domesticated one than to another wild population, for example.

When Darwin (1868) published his major work on domestication, hypotheses about which ancestral species led to domesticated ones were being first postulated. Darwin wrote that the diversity of dogs was such that origin from a single species would be highly unlikely. He was wrong—although not quite, if we consider the fact that introgression (gene flow resulting from hybridization) has occurred between wolves and coyotes (Lehman et al. 1991), and probably between some groups of dogs and other canids (Norton 2019). On the other hand, Darwin suggested that pigeons have a single ancestor, a surprising (and correct) hypothesis given how remarkably diverse pigeons are (Hansell 1998; Price 2002b). But things are complicated, as some traits of pigeons have been introgressed from other species (Vickrey et al. 2018). A similar case is known for the many breeds of chickens, originating mainly from the red junglefowl but with some degree of introgression from two other species of *Gallus*, at least in some regions, explaining some of the traits of chickens (Eriksson et al. 2008; Wang et al. 2020). Molecular and archaeological studies have hypothesized with great certainty which wild species were the ancestors for domesticated ones, as well as helped to test hypotheses on when and where major domestication phases occurred (Shapiro and Hofreiter 2014).

Mammalian domesticates

Among mammals, more than 25 species of placentals have been domesticated (Table 1.1). I follow Gentry et al.'s (2004) nomenclature, the one more universally used, in spite of the idiosyncratic nature of this decision given the known history of the animals involved, including hybridization (Zeller and Göttert 2019). Most domestic species are herbivores and, of those, most belong to the artiodactyls, which tend to live in herds and are nonterritorial. These features surely contributed to lend themselves to herding and managing by humans. The pig, although usually characterized as an "omnivore," also eats mostly plant material, calculated in one study as around 90% of its diet (Ballari and Barrios-García 2014). Some domesticated artiodactyls such as the yak have remained confined to their original areas of domestication, but others, including cattle, sheep, goats and camels, dispersed widely through their association with humans. Pastoralism spread throughout semidesert lands, steppes, and savannas of Eurasia and Africa.

Table 1.1. A selection of domesticated mammals and their wild ancestors

Domestic form	Wild form
Dog, *Canis familiaris*	Grey wolf, *Canis lupus*
Ferret, *Mustela furo*	European polecat, *Mustela putorius*
American mink, *Neovison vison*	Wild mink, *Neovison vison*
Cat, *Felis catus*	Wildcat, *Felis silvestris lybica*
Horse, *Equus caballus*	Extinct lineage of *Equus ferus*
Ass (plus hybrids mule and onager), *Equus asinus*	North African wild ass, *Equus africanus*
Domestic goat, *Capra hircus*	Bezoar, *Capra aegagrus*
Domestic sheep, *Ovis aries*	Mouflon, *Ovis orientalis*
Pig, *Sus domesticus*	Wild boar, *Sus scrofa scrofa*
Bactrian camel, *Camelus bactrianus*	Bactrian camel, *Camelus ferus*
Dromedary, *Camelus dromedarius*	Dromedary, *Camelus dromedarius*
Llama, *Lama glama*	Guanaco, *Lama guanicoe*
Alpaca, *Vicugna pacos*	Vicuña, *Vicugna vicugna*
Common cattle, *Bos taurus*	Auroch, *Bos primigenius*
Indicine cattle, *Bos indicus*	Auroch, *Bos primigenius*
Bali cattle, *Bos javanicus*	Banteng, *Bos javanicus*
Gayal or mithan, *Bos frontalis*	Gaur, *Bos gaurus*
Domestic yak, *Bos grunniens*	Wild yak, *Bos mutus*
Water buffalo, *Bubalus bubalis*	Asian water buffalo, *Bubalus* spp.
Reindeer, *Rangifer tarandus*	Reindeer, *Rangifer tarandus*
Domestic rabbit, *Oryctolagus cuniculus*	Wild European rabbit, *Oryctolagus cuniculus*
Domestic cavy, *Cavia porcellus*	*Cavia tschudii* and/or *C. anolaimae*
Chinchilla, *Chinchilla brevicaudata* and *C. laniger*	Chinchilla, *Chinchilla brevicaudata* and *C. laniger*
Syrian or golden hamster, *Mesocricetus auratus*	Syrian or golden hamster, *Mesocricetus auratus*
Mongolian gerbil, *Meriones unguiculatus*	Mongolian gerbil, *Meriones unguiculatus*
House mouse (W Europe) and laboratory mouse, *Mus musculus domesticus*	Mouse, *Mus musculus* "group"

Note: Scientific names largely follow Gentry et al. (2004), given the widespread use of that nomenclature (but see Zeller and Göttert 2019). The taxa included follow the review of Larson and Fuller (2014) for the most part, with some modifications, such as adding the gerbil (Stuermer et al. 2003).

The domesticated carnivorans are the dog, the ferret, and the cat, and, more recently, the domesticated mink. Almost half the species of mammals are rodents, but few of them became domesticated. The laboratory rat can be considered domesticated, and together with the mouse and the domestic cavy or guinea pig, they are important in biomedical research.

Some species not listed in Table 1.1 are considered "domesticated" in a most general way, including many species that are simply kept in captivity or managed for diverse economic purposes but were never tamed over generations resulting in the genetic or morphological changes characteristic of domestication, nor were their reproductive patterns significantly changed (Vigne 2011). In a compendium of do-mesticated animals, Mason (1984) listed among others the following species: mus-kox (*Ovibos moschatus*), American (*Bison bison*) and European (*Bison bonasus*) bison, silver fox (*Vulpes vulpes*), raccoon dog (*Nyctereutes procyonoides*), Egyptian mongoose, Indian grey mongoose, and the small Asian mongoose (*Herpestes ich-neumon, H. edwardsi,* and *H. javanicus,* respectively), some civets (*Viverra* spp. and *Viverricula indica*), coypu or nutria (*Myocastor coypus*), capybara (*Hydrocho-eris hydrochaeris*), muskrat (*Ondatra zibethicus*), giant pouched rat and greater cane rat (*Cricetomys* spp. and *Thryonomys swinderianus*), and Arctic or white fox (*Vulpes lagopus*).

There are no domesticated marsupials, even though opossums, possums, and kangaroos and their relatives both in the Americas and in Australia have played a role in the culture and traditions of humans (e.g., Smith and Litchfield 2009). Fur-thermore, no domesticates are included among two of the four large clades of pla-centals, the xenarthrans (armadillos, sloths, and anteaters) and the afrotherians (el-ephants, tenrecs, golden moles, sirenians, hyraxes). However, in these groups many species have been important as pets or are being or have been managed in different cultures (e.g., kangaroos), in some cases for centuries.

The evolutionary relationships among the domesticated mammalian species are solidly supported by comprehensive analyses of placental mammals (Francis 2015; Figure 1.4). This phylogenetic framework is fundamental to understanding the commonalities and differences among species of domesticates regarding changes in morphology and life history that result from domestication, as the evolvability and modularity of traits are usually clade-specific. For example, the "domestication syn-drome" is not a universal and uniform set of characters, as different clades exhibit different sets of modifications arising from selection for tameness (chapter 3). Like-wise, an understanding of the evolutionary relationships and distances among spe-cies is important for predicting the likelihood of transmission of infectious diseases between them (Farrell and Davies 2019). It has been speculated that infections from parasites outside their normal phylogenetic host range are more likely to result in death. In fact, the odds of lethality were estimated to double for each additional 10 million years of evolutionary distance (Farrell and Davies 2019).

Avian domesticates

Poultry are birds kept by humans for their eggs, meat, or feathers. Most of these birds are members of the Galloanserae (fowl), especially the Galliformes, including chickens, guinea fowls, quails, and turkeys, which are a sister group to the Anseri-formes, which include ducks, Muscovy ducks, and geese. All these constitute the sister group to the Neoaves, including the pigeons in the Columbiformes and the great radiation of Passeriformes, examples of which are the Bengalese or society finch and the canary among domesticates (Table 1.2, Figure 1.5).

In addition to the species listed in Table 1.2, the budgerigar (*Melopsittacus undu-lates*), the zebra finch (*Taeniopygia guttata*), and the ostrich (*Struthio camelus*) are considered domesticates by many authors. Several bird species are usually kept in

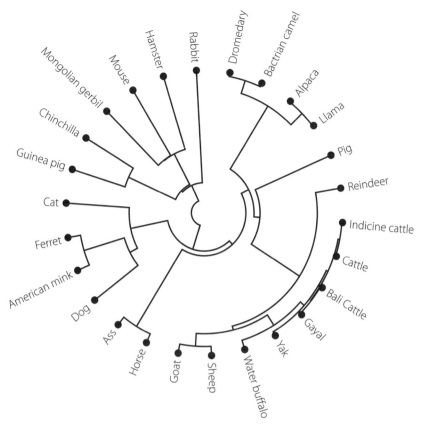

Fig 1.4. Evolutionary relationships among species of domesticated mammals. In some works, the yak is hypothesized as a sister group to the cattle/indicine cattle. Branch lengths are proportional to the estimated distance in time among species.

captivity or managed for diverse economic purposes. Mason (1984) listed the following species as semidomesticated, or routinely captive-bred: Barbary dove (*Streptopelia risoria*) domesticated as the African collared dove (*Streptopelia roseogrisea*), African lovebirds (*Agapornis*), cockatiel (*Nymphicus hollandicus*), mute swan (*Cygnus olor*), peafowl (*Pavo cristatus*), including Indian peafowl (*Pavo muticus*), green peafowl (*P. m. spicifer*), and Burmese form (*P. m. muticus* and *P. m. imperator*), pheasants (*Phasianus colchicus*, common pheasant, and *P. versicolor*, green pheasant), as well as partridges: grey (*Perdix perdix*), red-legged (*Alectoris rufa*), rock (*A. graeca*), and chukar (*A. chukar*).

One peculiar human-bird interaction involves the great cormorant (*Phalacrocorax carbo*) and fishermen in rivers in many countries in Asia, a few countries in Europe, and perhaps Peru in the fifth century of the current era (Leight 1960). In a traditional method now disappearing, fishers tie a snare near the base of the bird's throat, preventing the swallowing of large fish. When a cormorant has caught a fish

Table 1.2. A selection of domesticated birds and their wild ancestors. Additional species are discussed in the text.

Domestic form, common and scientific name	Wild form, common and scientific name
Domestic fowl / chicken, *Gallus gallus domesticus*	Red junglefowl, *Gallus gallus*
Domestic guinea fowl, *Numida meleagris*	Guinea fowl, *Numida meleagris*
Domestic turkey, *Meleagris gallopavo*	Mexican wild turkey, *Meleagris gallopavo gallopavo*
Japanese quail, *Coturnix coturnix japonica*	Japanese quail, *Coturnix coturnix*
Domestic duck, *Anas platyrhynchos*	Green-headed mallard, *Anas platyrhynchos platyrhynchos*
Muscovy duck, *Cairina moschata*	Muscovy duck, *Cairina moschata*
Goose, *Anser anser* and *Anser cygnoides*	Greylag goose, *Anser anser anser*, and Swan goose, *A. cygnoides*
Pigeon, *Columba livia*	Rock dove / rock pigeon, *Columba livia*
Canary, *Serinus canarius*	Canary, *Serinus canarius*
Bengalese or Society finch, Munia, Uroloncha, *Lonchura striata*	Striated, white-rumped, white-backed or sharp-tailed finch, manikin or munia, *Lonchura striata*

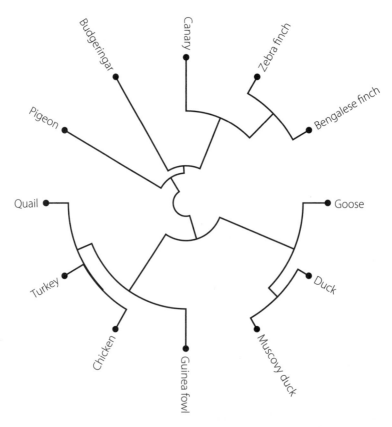

Fig 1.5. Evolutionary relationships among main species of domesticated birds. Branch lengths are proportional to the estimated distance in time among species.

in its throat, the fisher brings the bird back to the boat and has the bird spit up the fish. This peculiar and close interaction requires careful management, but to my knowledge it has not led to any generational changes in reproduction. There may have been rapid evolution of morphological features (Schilthuizen 2018), but longitudinal studies of skeletons or other organ systems are unlikely to be feasible.

The beginnings and antiquity of domestication and transitions from wild to domesticated

Concerning the antiquity of the (complex, continuous, ongoing) domestication process, it is important to avoid the term "event," as domestication is complex and entails multiple and parallel events and population admixtures (Larson and Fuller 2014). It may be more appropriate to ask questions in terms of "transitions," as in matters with a strong historical dimension such as those in evolutionary biology and developmental biology. Fixing a specific time and place for the origin of domestication of a species is not possible. What is possible is to provide a general framework of minimal ages, an approximation of reliable documentation of domestication in diverse species, as has been done for many mammals and birds (Figures 1.6, 1.7).

The search for and excessive focus on oldest occurrences as a leitmotif in archaeological research, tied to a progressivist rhetoric, endure in the mass media, but zooarchaeology and related fields dealing with domestication are better off having other foci (Gifford-Gonzalez and Hanotte 2011; Sykes 2014). The archaeological record is fundamental but of limited assistance in providing definitive earliest dates of domestication. This record will never be complete, as the first domesticated individual (actually, if there were such a thing, which is quite questionable, as discussed above) is unlikely to be recorded archaeologically (Perreault 2019). The oldest record of domesticated forms fails to represent the first domestication phase, but instead an approximation of that, and a minimum date. Paleontologists are faced with an analogous situation, what has been coined the Signor–Lipps effect. Given that the fossil record of organisms is incomplete, it is very unlikely that the first or the last organism of a given taxon will be recorded as a fossil (Signor and Lipps 1982).

Our knowledge of the earliest phase of domesticated animals consisted, until recently, of educated guesses based on reasonable but in many cases untested assumptions about morphological changes and mortality profiles suggested by zooarchaeological studies and limited studies of a few genes. Advances have been made over the years, both methodological and conceptual (Vigne et al. 2005a, b). As quantification and more data have become available, it is now more evident how little we know for sure. Furthermore, recognition of varying degrees of intensity in the animal-human relationship—as opposed to an oversimplistic dichotomous categorization of wild versus domestic—has also been a major step forward (Balasse et al. 2016). These categorizations also vary depending on the geographic region and the species in question.

Several years ago most genetic data sets were restricted to mitochondrial sequences, a non-recombining maternally inherited DNA, which by itself cannot be used to identify or quantify hybridization between wild and domestic populations

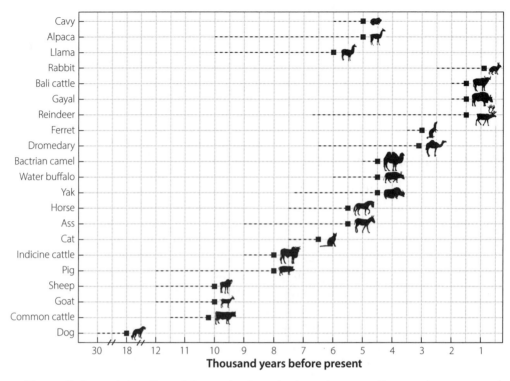

Fig 1.6. Estimated time line of domestication of selected mammalian species.

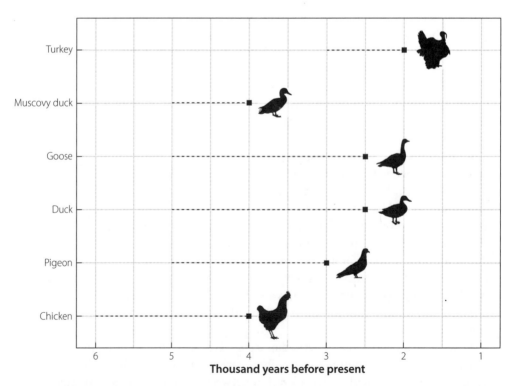

Fig 1.7. Estimated time line of domestication of selected bird species.

or among geographically differentiated domestic populations. This lack of discriminatory power led to false claims of independent and multiple events of domestication for pigs (Larson et al. 2005), goats (Luikart et al. 2001), sheep (Pedrosa et al. 2005), horses (Vilà et al. 2001), and cattle (Hanotte et al. 2002), based on the presence of divergent mitochondrial haplotypes in domestic populations. With current genomic data, including population genetic studies of nuclear DNA sequences, it is possible to determine whether those haplotypes result from an independent domestication process involving genetically divergent wild populations or from introgression of a wild population into domestic stock (Larson and Burger 2013; Gerbault et al. 2014). Gene flow is common not only between domestic and wild populations, but also among geographically diverse domestic populations of the same species (Larson and Fuller 2014; Frantz et al. 2019). Genetic data as currently analyzed can also be used to date domestication phases, migrations, and mixing of populations, in spite of the caveats and challenges these estimates involve (Sykes et al. 2019; Frantz et al. 2020).

Zooarchaeologists have documented changes in the management strategies of hunted sheep, goats, pigs, and cows in the Fertile Crescent by measuring the size, sex ratios, and mortality profiles of assemblages of animal remains (Zeder 2012b). By 10,000 ybp, some people were preferentially killing young males of a variety of species and allowing the females to live to produce more offspring.

Traditionally, two alternative explanations have been offered for the beginning of domestication—here meaning management and some kind of selective breeding. One hypothesis suggests that domestication started independently of any population–resource imbalances (Smith 2011a, b; Zeder 2012c) and was driven by intentional management of wild resources and experimentation. The intentionality aspect of this hypothesis is often questioned. An alternative hypothesis is based on predictions from foraging theory models and behavioral ecology, and assumes that domestication arose at times of need, of Malthusian population–resource imbalance, which led people to try to acquire more food from the environment (Hawkes and O'Connell 1992). It is possible that different mechanisms operated in the many places where domestication occurred, without a universal driving force in all of them. Thus, empirical studies of specific areas and a broad and pluralistic framework seem justified, one firmly based on our knowledge of human behavior and evolutionary biology (Gremillion et al. 2014). The domestication of plants that occurred in Eastern North America approximately 5,000 ybp was associated with population–resource imbalances, as inferred based on changes in radiocarbon date density and site counts as proxies for human population data (Weitzel and Codding 2016). Larger populations, along with decreased resource abundance, may have led to domestication in this area of the world. For other regions, other conditions and dynamics were probably involved.

A different kind of question concerns the earliest domestication phase, one of close human-animal interactions. A combination of ethnographic and anthropological data, and a refreshed view of the zooarchaeological record are needed to address this (Sykes 2014). It seems that nonutilitarian aspects drove those interactions, and this is reflected in the archaeological record of many species that had a close contact with humans, some of which were never domesticated later. This could be the case for the monk seal (*Monachus monachus*), of which a burial on the island of Rhodes in Greece is known (Masseti 2012). Francis Galton (1865) related a story of a tamed seal from the Shetland islands (it must have been either the common seal *Phoca vitulina*, the grey seal *Halichoerus grypus*, or the fur seal *Arctocephalus gazella*) and speculated on the possibility of populations of this species becoming domesticated.

The translocation of a species outside its native range can be used as circumstantial evidence for domestication. Morphological traits of domestication are not detectable in the archaeological record of sheep, goats, cattle, pigs, and cats before 10,000 ybp, but populations of these species were translocated to Cyprus at least 10,600 ybp, suggesting that management of some kind occurred back then (Vigne et al. 2012).

Identifying morphological changes associated with domestication has been a major interest among zooarchaeologists, aiming at finding signs in isolated and often fragmentary bones and teeth that can be used for this purpose. The artifacts of preservation and the impossibility of separating the many factors involved make the search for universal or even species-specific markers of the first phase of domestication an almost hopeless task. Experimental studies are one approach (Harbers et al. 2020a, b); improvements in the zooarchaeological record will surely help as well.

Given the diverse evolutionary (phylogenetic) background of the different groups of species of domestics and therefore different evolvability of skull modules, kinds of tissue, and organs, it is unrealistic to expect universal features (e.g., changes in size) or simple markers of a clear wild-versus-domestic dichotomy. More importantly, there are fundamental if not insurmountable challenges in such a search.

Complete skeletons of the earliest domesticates will never become available. A standard approach to identifying morphological changes associated with domestication has been to compare wild and domestic modern forms of the same species, assuming that the current populations accurately reflect both the ancestral wild form and the domesticated counterpart in its first phases of differentiation (Price 2002a). We can arrive at approximations by looking at populations of domesticated forms that have not diverged much from the wild ones, as done in a study comparing growth series of skulls (Sánchez-Villagra et al. 2017). However, hybridization, feralization, bottlenecks, and the complex interactions between natural and artificial selection pressures can introduce considerable noise to such a standard approach. In fact, no living population is any group's ancestral population. Furthermore, in many cases, wild populations of a domesticated form no longer exist, as in the case of cattle and camels, and likely also the horse.

Many genetic studies compare the wild form with current domestic ones, sometimes a specific breed, or even a wide array of them, and discuss the finding as revealing selection for that gene and the associated trait in domestication, even for the early phases. Given the antiquity of domestication and the different intensity of the interaction, it is clear that there will be biases in such studies. This was discussed and demonstrated for genetic features of chickens (Girdland Flink et al. 2014). Over the past 2,000 years there has been variation in two genes in ancient European chickens: the *BCDO2* gene, which underlies yellow skin, and the thyroid stimulating hormone receptor gene *TSHR*, related to the control of development of the thyroid gland and its functions, affecting the regulation of growth, brain development, and metabolic rate. The study of these genes showed that a mutation thought to be associated with domestication was not subjected to strong human-mediated selection until much later in time than what all experts agree was the start of chicken domestication. This is an example of the challenge of addressing any issue concerning the process of (early) domestication with existing and consequently "derived" forms, known as breeds. Studies of ancient DNA—combined with sound morphological studies from zooarchaeological studies (Evin et al. 2017b; Evin 2020)—may help in making more meaningful wild-domestic comparisons, if the goal is to address domestication *per se* and not some aspect of selective breeding.

The study of ancient DNA recovered from remains of different time periods can be used to reconstruct patterns of genetic variation and admixture at earlier stages of the domestication process, get better estimates of the time when initial stages took place, or more specifically provide insights into whether specific variants were already present in past populations, for example coat color mutations (Frantz et al. 2020).

In the case of forms from which milk is consumed, there is another approach to domestication research: detecting milk residue in pottery (Evershed et al. 2008). This approach has provided evidence of early horse domestication, studying organic residue analysis using δ13C and δD values of fatty acids (Outram et al. 2009). Furthermore, it is possible to use proteomics to differentiate yak, cattle, and goat milk (Yang et al. 2013). Another approach has been to detect residues in the dental calculus of humans, thus directly showing consumption from dairy livestock. The protein β-lactoglobulin (BLG) is a species-specific biomarker of dairy consumption of cattle, sheep, and goat milk products and is preserved for example in human dental calculus from the Bronze Age, circa 3000 BCE (Warinner et al. 2015).

The anatomy of hair can be informative about taxonomic allocation and domesticated status (De Marinis and Asprea 2006). Through the identification of hairs, the oldest evidence for domestic goat in Neolithic Finland was reported, from a pastoral herding economy, the Corded Ware Culture, dated ca. 2800–2300 BCE (Ahola et al. 2018). The study consisted of microscopic analyses of soil samples collected during the 1930s from a grave.

Material culture associated with domestication can also be used to provide evidence of the latter in the archaeological record. An example is provided by artifacts that were parts of headgear worn by transport reindeer, remains of these dating back to around 2,000 ybp (Losey et al. 2020).

Dental signs of domestication in pigs and dogs

Given the high preservation potential of teeth and that teeth are taxonomically informative and they consist of the most mineralized tissues of the body, efforts have been made to find signs of domestication in them. Here I discuss pigs and dogs, which have been intensively studied.

A study of molar teeth of current wild and domestic West Palearctic pigs compared maximum length, size, and shape variables from 2D geometric morphometrics

(Evin et al. 2013). Size was a poor indicator of wild and domestic status, whereas shape provided a high degree of confidence distinguishing the two. The authors concluded that geometric morphometrics is a better alternative to traditional biometric techniques. This and most other similar studies described differences between modern wild and domestic forms. It is a major improvement when additional categories are used—that is, feral, hybrid, captive, and insular, as in a study of diversity of dental size, shape, and allometry (Evin et al. 2015a). Although the amount of variation among domestic pigs does not exceed that of their wild counterpart, domestication has produced new dental phenotypes not found in wild boar. Domestic breeds can be distinguished by distinct dental phenotypes, and captive and insular pigs are also distinctive in dental shape (Evin et al. 2015a).

Tooth crowding in domestic dogs in contrast to wolves has been proposed as a criterium to infer the process of domestication in the zooarchaeological record (Benecke 1994a). A comprehensive test of this hypothesis using landmark-based metrics examined 750 modern dogs versus 205 modern wolves from across the modern geographic range of the latter and 66 Late Pleistocene wolves from Alaska (Ameen et al. 2017). This study found a higher than expected frequency of crowding in both modern (~18%) and ancient (~36%) wolves, thus questioning assumptions linking tooth crowding with the process of early dog domestication. The strength of this study is supported by its examination of alternative approaches to quantify and compare tooth crowding, which show that the results are reliable.

Osteological signs of domestication in dogs, pigs, goat, and sheep

Geometric morphometrics of wild and domestic pig crania reveal strong discrimination among wild, domestic, and hybrid pigs that applies to both the complete and the subsections of the crania (Owen et al. 2014). Based on a study of adults of 42 modern domestic pigs representing six European domestic breeds, 10 wild/domestic first-generation hybrid pigs, and 55 wild boars, it was possible to discriminate among the breeds on the basis of cranial morphology (Figure 1.8). The skull of first-generation hybrid wild/domestic pig more closely resembles that of wild pigs than domestic. As with dental metrics, it was concluded that geometric morphometrics can provide a quantifiable separation between wild and domestic pigs, even when considering partial cranial remains. As in other wild-domestic pairs (chapter 7), the variation in skull shape in domestic pigs is larger than that of wild boar.

A potential osteological marker of domestication in mammals is the petrosal or periotic bone, a small, compact bone at the base of the skull. Even though goats and sheep are closely related taxa that may not be obviously distinguished based on fragmentary osteological material, an anatomical and metric study of the petrosal bone showed that it was possible to discriminate between these two species (Mallet et al. 2019). Distinguishing wild from domestic forms within the same species pair, or from other categories such as feral or hybrids, can be more complicated, as small differences could also result from phenotypic plasticity.

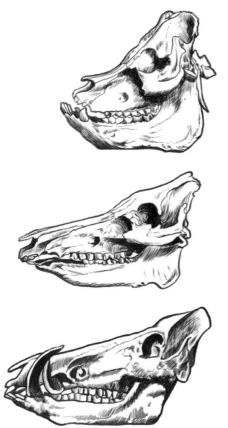

Fig 1.8. Domestic pig skulls (top and middle) and wild boar (bottom). The skull of domestic forms is easy to distinguish from the wild counterpart even in less derived forms. But distinguishing the two based on fragmentary skull parts in individuals from the early phases of domestication, in which hybridization occurred as it still does today in many regions, is a challenge.

Studies of dogs using microcomputed tomography and 3D geometric morphometrics have attempted to discern differences in the organs of hearing and balance located within the petrosal bone (Figure 1.9). The interspecific variation of the inner ear is well documented for many groups of mammals. The proportions among canals and their shape, and the number of cochlear coils are variable features, and this variation is correlated with locomotory habits and hearing frequencies (Ekdale 2016). An investigation of a sample of 24 wolves, 8 dingoes, 39 modern domesticated dogs from 20 different breeds, and 21 prehistoric domestic dogs reported that shape variance is slightly higher for the different parts of the inner ear in domestic dogs than in wolves, but these differences are not significant (Schweizer et al. 2017). The variation detected in inner ear shape was size-related, and this work did not identify criteria by which to differentiate between domestic dog and wolf inner ear. Although wolves have smaller levels of size variation than dogs, they show a greater level of variance in the angle between the lateral and the posterior canal than domestic dog breeds (Schweizer et al. 2017).

The study of a different sample of inner ears using an alternative 3D geometric morphometric method led to contrasting conclusions (Janssens et al. 2019b). This study measured 20 modern Eurasian wolves and 20 modern dogs of comparable skull length and reported that dogs had on average a significantly smaller bony

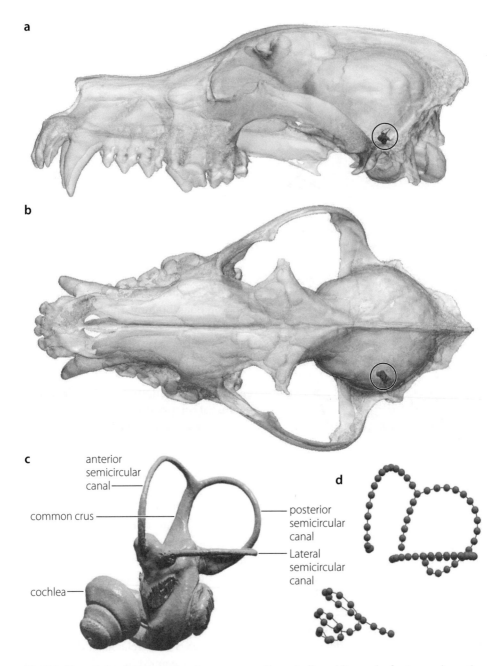

Fig 1.9. Bony labyrinth location in a mammalian skull and its main features. Location in a wolf skull of the bony labyrinth, (a) lateral, and (b) dorsal views based on 3D reconstruction of computer tomographic images. Also illustrated are (c) the bony labyrinth and (d) the location of semilandmarks along the cochlea and the three semicircular canals.

labyrinth than wolves. Furthermore, the shape space of the two groups forms two nonoverlapping clusters, with dogs having a smaller relative size of the vertical canals and oval window, larger relative size of the lateral semicircular canal, and shorter relative cochlea streamline length, with a more anteroventrally tilted modiolus. The authors discussed how these shape differences are not related to allometric effects and could therefore potentially lead to recognition of dog-specific features useful in the identification of samples in the zooarchaeological record. However, the significance of this important study needs confirmation in view of the limited sample evaluated, one that may reflect phenotypic differentiation following selective breeding and not the first phase of domestication.

A quantification of the variation of the wild form is paramount in studies aiming at finding osteological markers of domestication. This was shown for wolves versus dogs in a comprehensive review of several skull and dental parameters (Janssens et al. 2019a). Clearly, further work needs to be conducted in this area, as the implementation of geometric morphometrics and anatomical characterizations of additional species with appropriate samples has only just started. The question arises as to what functional, biomechanical, or other kinds of processes might underlie morphological differences between wild and domestic forms. This is a challenging area of research, as it requires samples of great quality in order to discern subtle differences among populations.

Perhaps as gross morphological features of the skeleton are affected by domestication, so are bone microstructure features. Examination of wild and domesticated sheep bones in petrographic thin sections seemed to identify characteristics that distinguish the two: increase in lacunar size, higher preferential alignment of hydroxyapatite crystals, thicker trabeculae, and a sharp transition between compact and spongy bone (Drew et al. 1971). However, later examinations of the subject showed that aspects first described as differences were linked to diagenesis and to individual variation. Furthermore, the original work assumed naively a dichotomy between wild and domestic unsuitable for the studied samples (Watson 1975; Gilbert 1989). The older and the more recent literature on comparative mammal and bird bone microstructure includes much information on domestic forms (Enlow and Brown 1956; Zedda et al. 2008), but no systematic comparisons with wild forms have been attempted. Studies of bone microstructure may provide information on life history variation (chapter 6).

Isotopic markers of domestication

Whether zooarchaeological specimens are from individuals living in the wild in close proximity to humans or from fully domesticated ones, their status cannot always be assessed based on anatomical or morphometric comparisons alone; this often requires the use of additional approaches. Stable isotope biochemical analysis of animal remains can be used to investigate ancient human-animal relationships. However, the methods involved are plagued with issues of sampling and interpretation; thus establishing uniform protocols and terminology is essential (Roberts et al. 2018).

Potential domesticates differ isotopically from other wild taxa because of different diets, water consumption, and properties of the soil on which they live. Isotopic values represent an average of dietary patterns over many years, given that bone collagen is replaced slowly throughout the life of an organism. In some localities the consumers of the potential domesticates will also be isotopically distinct from consumers of other taxa. The isotopic approach, based on C4 and C3 values in different plants, can be used to study human remains and establish whether diets included domesticated crops (Barton et al. 2009).

The broad dietary plasticity of pigs makes them an excellent subject for isotopic analysis, as it is possible to distinguish among wild, feral, extensively herded, and household pigs, the last being more dependent on domestic sources of animal protein (human consumption leftovers). Various dietary sources can be traced through analysis of stable nitrogen isotope ratio (δ 15 N) of bone, which significantly increases with each trophic level in a food web (Ervynck et al. 2007). This approach was used in a detailed study of pig husbandry in the city of York, England, with a long and more or less continuous zooarchaeological record in handling of pigs from the foundation of the city at the end of the first century AD until post-medieval times. Further examples of studies of this kind are those on Neolithic China (Cucchi et al. 2016), a Celtic village in France (Frémondeau et al. 2015), and Chalcolithic Romania (Balasse et al. 2016). They all documented varying degrees of intensity in the pig-human relationship.

It is possible to combine studies of dental microwear and oxygen isotopes, as in the examination of herding practices of sheep in Çatalhöyük East, in central Anatolia, one of the largest Neolithic sites in southwest Asia (Henton 2013). The two data sets situated an individual sheep in its environment at different points of its life with good resolution. Different models of herding were associated with different shapes, ranges, and summer isotopic values. Those models were characterized as follows: "Sheep raised year-round near settlement on the plain—Marl steppe, alluvial fan, sand-ridges"; "Sheep raised year-round in perennial stream valleys, cutting through terraces and lower hill-slopes"; and finally "Vertical transhumance to higher hill-slopes in summer or pasturing near springs fed by averaged groundwater." Consideration of seasonality in dental wear and reconstructed diet and in the life history of these animals provided a look at mismatches with the natural environment (and the cycles recorded by wild forms living in it) that can characterize conditions of managing of domesticated forms (Henton 2013).

By studying carbon, nitrogen, and oxygen isotopes, it was hypothesized that animal trade and possible captive animal rearing occurred in the Maya region of Ceibal, Guatemala (Sharpe et al. 2018). This study examined animal specimens across almost 2,000 years (1000 BC to AD 950), and the strontium isotope analysis revealed that the Maya brought dogs to Ceibal from distant highlands. Contextual evidence indicated that domesticated and possibly wild animals were deposited in the ceremonial core, showing an association of these managed animals with special events, activities interpreted to have been important in the development of state society (Sharpe et al. 2018).

Cultural evolution and reconstructing the history of domestication

An underexplored and potentially useful tool for examining congruence between cultural and biological data in the history of domestication is the use of statistical techniques to study myths and folktales. A database of 23 myths concerning dogs and 22 geographic areas was analyzed with a neighbor-joining tree based on Jaccard distances (d'Huy 2015). The application of phylogenetic methods showed a correlation between history and geography, and the approach made it possible to reconstruct the paleolithic mythology around dogs.

Another rich area of research is the study of the biological information stored in parchment documents made of animal skin (Ryder 1958). Information about book production can provide data on livestock economies and handling, and provenance of the animals. Advances in molecular methods can make this biological approach a highly relevant discipline in manuscript studies (Fiddyment et al. 2019).

The Neolithic transition

Homo sapiens originated approximately 300,000 ybp, but the domestication process of several species started only around 10,000 ybp. Why then? This is a fundamental issue in human history and biology, so it is justified to propose reasonable hypotheses and even speculations.

The standard account goes that at the start of the Neolithic a transition from a life of hunting and gathering nomadism to farming sedentism occurred in many populations. It was not universal (everywhere, everybody) and, as in domestication ("wild" versus "domestic"), human ways of living cannot be encased in dichotomies (Sykes 2014). Hunting for semi-domesticated forms occurred. The coupling of farming and long-term sedentism is not straightforward, as each of the two shows degrees, and diverse and simultaneous lifestyles most likely occurred in early Neolithic times, with the first cities located in wetlands in which productivity and mobility were both high. The development of multiple cities postdated the first records of domestication by several thousand years (Scott 2017).

For decades the Near East was considered the major center of domestication of plants and animals around 10,000 ybp in what used to be called the "Neolithic revolution" (Childe 1936). We now refer to this as the "Neolithic phase or transition." Many geographic areas of early domestication of plants or animals, or both, were areas of analogous and independent places of transitions to new modes of human life, including many in the Americas (Smith 1998; Piperno and Pearsall 1998; Zeder et al. 2006).

The Neolithic transition happened at the end of the Pleistocene, when the onset of warmer and wetter climates was accompanied by an increase in CO_2

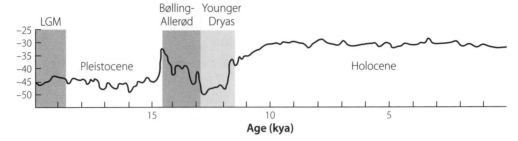

Fig 1.10. Climatic curve, reconstructed from the GISP2 ice core in Greenland. The vertical axis (*left*) refers to temperature (°C). LGM = Last glacial maximum.

levels and the establishment of more stable weather patterns (Figure 1.10). In the temperate latitude zones, productive ecosystems emerged and human societies developed in areas of high human-carrying capacity, especially along river floodplain corridors and lake and marsh/estuary margins. A wide spectrum of plant and animal species were utilized in these early Holocene environments, subject to varying degrees and forms of manipulation and life-cycle changes (Zeder 2012c). Vegetation clearing and the presence of water-management features are recorded in the archaeological record from these regions (Smith 2011a). There were new forms of social learning (Laland and Brown 2011) and interactions associated with sustained economic utilization, including ownership, with a shift by small-scale societies to territorial defense of resources (Dyson-Hudson and Smith 1978). The archaeological record presents evidence of multigenerational corporate ownership of resources, including group burial features and other ceremonial structures.

The Neolithic transition and environmental crisis

In 1928, the renowned archaeologist V. Gordon Childe (1892–1957) stated that domestication provided one of the great moments in prehistory, "that revolution whereby man ceased to be purely parasitic and, with the adoption of agriculture and stock-raising, became a creator emancipated from the whims of his environment." (Childe 1928, p. 2). This view dominated for decades in Western Europe and the United States. After almost 100 years our views have changed dramatically—from celebration of domination to well-founded concern. The age in which we live is arguably one of environmental crisis, and domestication is at the center of it (Ceballos et al. 2015). It all started with the Neolithic transition some 10,000 ybp, but the tipping point in the exponential acceleration of a process of habitat degradation came about when selective breeding became industrialized, tied to excessive consumption and exponential population growth.

Changing patterns of food consumption and human resource use in Western societies have produced an unprecedented reconfiguration of the Earth's biosphere and in many cases of populations of individual species. The domesticated broiler chicken is a classic example of this reconfiguration (Bennett et al. 2018). Human selective breeding has led to a doubling in body size of domesticated chickens from the late medieval period to the present. In the case of the broiler, there has been a fivefold increase in body mass since only the mid-twentieth century, besides the changes in skeletal morphology, pathology, bone geochemistry, and genetics discussed in this book. Broilers cannot survive without human intervention. The huge increases in population sizes mean that broiler chickens have a combined mass exceeding that of all other birds on Earth (Bennett et al. 2018).

A detailed census of the overall biomass composition of the biosphere among all kingdoms of life showed that the mass of humans is an order of magnitude higher than that of all wild mammals combined. Furthermore, there has been an enormous impact of humanity on the global biomass of prominent taxa, including mammals, fishes, and plants (Figure 1.11).

Domestication and agricultural practices impact the evolution and ecology of not only domestic animals themselves but also the wild forms (Turcotte et al. 2017). Special traits of domesticated forms can alter the selective environment of wild species. Domesticated animals can become invasive species; aided by humans, these move to places where they are non-native and can have a detrimental effect on the environment, including effects on populations of other species (Doherty et al. 2016).

Furthermore, the introduction of domesticated animals into new environments can produce a cascade of effects on plants. For example, farming of domestic animals can lead to megafaunal local extinction, which can in turn decrease the dispersal of seeds of plants and consequently their survival

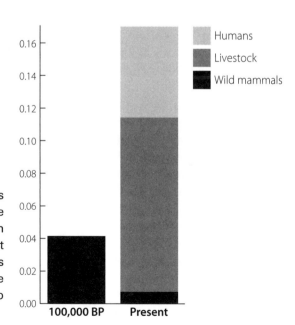

Fig 1.11. Estimated biomass of wild mammals before the rise of domestication and at present. The great proportion of livestock is dominated by cattle. The vertical axis (*left*) refers to gigatons of carbon.

(Onstein et al. 2018). Islands provide many examples, as in the case of diverse native grasses in New Zealand. These evolved in the absence of grazing mammals, and exhibit the strategy of old leaf abscission, a rare characteristic in grasses (Antonelli et al. 2010). This feature increases plant productivity but reduces protection against mammal grazing. Following the introduction of sheep, cattle, and rabbits in the nineteenth century, these endemic grasses were highly affected (Rose and Platt 1992).

The Neolithic transition: mismatches and human health

Intensive selective breeding can and has led to mismatches of past adaptations of humans with the environment, particularly since the industrial revolution. These mismatches concern epidemiological, nutritional, and demographic matters. They have greatly affected human ecology and biology, leading to changes in life history traits including fertility, life span, and age and size at maturity (Corbett et al. 2018), and may lead to gene variants linked to higher fitness in the past now predisposing us to disease. This change in fitness can occur through pleiotropic effects and can predispose humans for example to cancer and coronary artery disease.

It is usually claimed and generalized that the Neolithic transition brought major changes in time management by humans, with mostly an increase in working hours in agricultural societies (Harari 2011). However, a dichotomy of hunter-gatherer versus agricultural is too simplistic to capture the diversity of subsistence styles, and generalizations on their consequences are also multivariate (Higgs and Jarman 1969). The dimensions of "hunter-gatherers" or foragers vary depending on geography and cultural history (Kelly 2013). Notwithstanding this variation, it is highly likely that the activity patterns of the first farmers changed from those of pre-farmers, and their skeletons changed as a result (MacIntosh et al. 2016). Being rather sedentary, farmers' lives strongly contrast with those of foragers, for whom walking, running over long distances, and carrying heavy loads (Carrier 2002; Stock and Pfeiffer 2004) are common activities. The rise of the farming lifestyle has traditionally been associated with a decline in health, and even in physical prowess (Larsen 2006). Greater population densities, as well as greater reliance on domesticated crops and the vagaries of farming contributed to the spread of infectious diseases (Verano and Ubelaker 1992; Steckel and Rose 2002; Scott 2017). The first farmers may have suffered from hunger (Bowles 2011). The relationship between sedentism and the global obesity pandemic is however not simply related to a change in activity and exercise, as cross-cultural studies of metabolic rates have shown (Pontzer et al. 2012).

The patterns of change among the people who experienced the Neolithic transition may not have been universal. The variations among populations and sites and the relations between lifestyle and disease are complex (Carlson and

Marchi 2014; Ash et al. 2016; Ruff 2017). Mortality risks and food shortages can also be substantial among mobile hunter-gatherers.

Nutrition, disease, hormones, and mechanical loading can influence bone development (Hall 2015). It follows that studies of skeletons of past populations can provide clues to their lives. Studying markers of developmental instability during childhood and embryogenesis could offer insights into health and the effects of disease and famine (ten Broek et al. 2012).

The geography of domestication

Not only the old notion of a domestication "event" has been abandoned, but also that of a "center" of domestication. Domestication is a process, a transition, without clear boundaries in place and time. Because of the biases of the archaeological record and the nature of domestication, centers of domestication can refer only to general areas in which domestication was practiced to such an extent as to leave demographic or morphological markers.

The Fertile Crescent was surely an important area for domestication of mammals, but other regions of the world need to be studied to gain a balanced record of the earliest livestock and crops (Figure 1.12). In fact, crop cultivation probably began independently in as many as 20 regions worldwide (Fuller 2010), whereas early animal domestication is associated mainly with just the Middle East, central China, and the Andes. Later cases of animal domestication in antiquity took place in many areas, decoupled from centers of plant domestication. The exponential growth of aquaculture in the last decades is a worldwide phenomenon (Duarte et al. 2007).

The geographic range of wild forms was wide for some species, as in the case of the wild boar, ranging from southeast Asia (where genetic evidence shows it first originated) to Western Europe, or restricted, as in the case of the original distribution of wild sheep and goat in the Middle East. Some of the wild forms are now extinct, such as the aurochs (von Lengerken 1955; Van Vuure 2005).

Concerning the domestication of birds, any geographic pattern is of course conditional on what is assumed to be a domesticated bird, and here there is no consensus. However, a general pattern exists. The subtropical and warm-temperate regions of Asia contributed most, Africa and the middle and high latitudes of Eurasia comparatively less (Donkin 1989). The guinea fowl, *Numida meleagris*, and for some authors also the ostrich, *Struthio camelus*, are the two domesticated bird species from Africa. In the New World, the Muscovy duck, *Cairina moschata*, and the turkey, *Meleagris gallopavo*, were domesticated. The canary, *Serinus canaria*, is a particular case; this song bird is originally from the Macaronesian islands in the Atlantic Ocean, from where it was taken to Europe.

The Americas were the last continents, aside from the polar regions, to be occupied by humans. After the discovery—and at last general acceptance—of several archaeological sites precluding the Clovis phase, the dates of the earliest occurrence of the peopling of the Americas have been moved back in time to at least 15,000 ybp

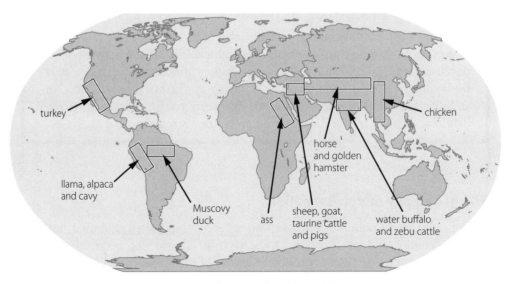

Fig 1.12. Sites of domestication of the main domesticated mammals and birds.

(Waters and Stafford 2013). The human-animal interactions that occurred since then include domestication. Only a few endemic mammals and birds were and still are domesticated in South America. Several New World mammal species had intense interactions with humans in the past, such as the hutia rodent in the Caribbean (Colten et al. 2017). The dog was already domesticated when it arrived in the Americas with human populations in pre-Columbian times (Leonard et al. 2002).

There is no quantitative geography of breeds of domesticated species and their origins. A survey could serve to factor the surface of the regions and human population density in relation to the diversity of local breeds and landraces. It might reveal that areas thought to be poor for domestication, for example some in Africa, are not so. In plants, there are analogous examples of how the domestication process may have started in some region, but the process extended to other areas of even more prolific diversification. The classic example is the tomato, originating in the Andean region, becoming domesticated and selected in Mexico, and after colonial times finding its way into southern Italy, and to all corners of the world, including Ukraine, where delicious varieties are produced (Mann 2011).

On domesticated species, breeds, and landraces—nomenclatural issues

It is regrettable that there is no uniform nomenclature across biological disciplines to refer to domesticated animals, with some exceptions. In both the domestication and the zoological literature, domestic rabbits have the same name, *Oryctolagus cuniculus*. On the other hand, the most widely accepted standard of mammalian taxonomy, *Mammal Species of the World* (Wilson and Reeder 2005), considers the dog a subspecies of the wolf, whereas the literature on domestication assigns different species names to the two, following Linnaeus (1758). Linnaeus (1758) also gave different names to other wild–domestic pairs, as in the aurochs (*Bos primigenius*) and domesticated cattle (*Bos taurus*). The most widely accepted nomenclature for many domestic animals in the literature on domestication is one in which, for many cases, the domesticates constitute different species from the wild form (Tables 1.1, 1.2).

The type species designations of several mammals are based on domestic animals (Gentry et al. 2004). For 16 mammals the name of the domestic form antedates or is contemporary with that of the wild ancestor, as in the case of the wolf and the dog. The contrary case does also occur, as for *Sus scrofa* Linnaeus 1758 for the wild boar pre-dating that for the domestic form, *Sus domesticus* Erxleben 1777 (Gentry et al. 2004).

Even though the nomenclatural separation of wild and domestic for many pairs does not make much sense biologically (Zeller and Göttert 2019), I follow it here for the sake of stability (Tables 1.1 and 1.2). Some authors have argued in favor of naming the domestic forms the same as the wild forms and adding "forma do-mestica" (Herre and Röhrs 1990), others have suggested categorizations below the species level to differentiate domestic from wild forms (Bohlken 1961). The long, tedious history of discussions on the proper nomenclature to use when referring to wild and domesticated forms (Gautier 1993) teaches us that attempts to make the nomenclature uniform across disciplines and research traditions have consis-tently failed.

A different name for the domesticated form could be interpreted as implying reproductive isolation, but hybridization is rampant among breeds within a domestic form and between wild and domestic individuals. Thus, a biological concept (based on reproductive isolation) is problematic at best when used for domestic forms. Many concepts are used for species (Zachos 2016), and the pragmatic decision to give new names to populations that have significantly and measurably diverged morphologically from the wild forms has had influential advocates (Gentry et al. 2004). A species can be considered a distinct cluster of individuals that are morpho-logically and ecologically similar. This "vernacular" species concept is used, for ex-ample, when considering the diversity of cichlids from African lakes (Salzburger 2018). In the case of domesticates, the historical distinction and morphological di-vergence from the wild form have been given much weight. Ecologically, domesti-cated forms can be also seen as distinct.

Addressing the multiplicity of definitions and in search of consilience, one can define a species as a divergent lineage, with a morphological or ecological disconti-nuity, and/or a reproductive isolation having arisen during speciation (de Queiroz 1998). These features could have been attained sequentially. Populations in the pro-cess of speciation (whatever that may mean!) may have different morphological features and yet not be isolated reproductively. On the other hand, cryptic species may be different ecologically or reproductively and yet be morphologically similar. As Darwin (1859, p. 469) stated, "no line of demarcation can be drawn between species . . . and varieties."

Since wild and domestic forms are in most cases recognizable entities by virtue of their phenotypes, and in some cases and for many populations there is imposed reproductive isolation from each other, it is practical to separate them by nomenclature when distinct names exist, which is also justified for the sake of stability.

The categories of landraces and breeds are used for some domestic animals. Landraces result from the culling or disposal of unwanted individuals; they are thus the product of natural selection and postzygotic selection, without direct control over the individual's reproduction. To create breeds, in contrast, there is deliberate mating of preferred animals to perpetuate an observable phenotype that is therefore the result of pre-zygotic selection. A definition of breed was provided by J. A. Clutton-Brock (1999, p. 40): "A breed is a group of animals that has been selected by humans to possess a uniform appearance that is inheritable and distinguishes it from other groups of animals within the same species." But determining what a breed is and what is not is a matter, to a great extent, of convention.

Conventions for defining breeds are particularly established in dogs. Some recognized dog breeds were created when individuals from regional landrace populations were removed and sexually isolated, and then registered as breeds with international kennel clubs; examples include the Italian Maremmas, the French Great Pyrenees, and the Turkish Anatolian Shepherds (Lord et al. 2016). Breeds are recognized by kennel clubs (the British one dates from 1873, the American Kennel Club from 1884, the Fédération Cynologique Internationale from 1911) and have been subject to selective breeding regimes with strict requirements and closed bloodlines.

In some regions of the world, breeding of a particular kind of domesticated animal has been ongoing and has led to a certain singularity, which can then lead to an official request for formal recognition. This is the case of the Ovejero Patagónico or Patagonian Sheepdog from southern Chile's Magallanes region (Barrios et al. 2019). It originated from European breeds, including the extinct British breed Old Welsh Grey, and several varieties of Collies (Fuenzalida 2006). It is highly distinctive in its phenotype (Tafra et al. 2014) even when some of its variation is considered (Figure 1.13). Another example of differentiation leading to a singular and recognized breed is the Colombian Paso horses, whose gait types, genetics, and phenotype have been thoroughly documented (Novoa-Bravo 2021).

Cattle exemplify a case of historical significance concerning the matter of breed recognition. Darwin (1845) wrote quite a bit on a short-snouted cow he encountered in his travels in Argentina and Uruguay, the *vaca ñata*, deemed since then brachycephalic (Veitschegger et al. 2018). The name *ñata* refers to nose in a variation of the local Spanish (used in several tango songs). This breed was included in the writings of Richard Owen (1853), the influential anatomist and contemporary of Darwin, and of the Swiss zoologist Rütimeyer (1866) in his classic treatises on cattle (Figure 1.14). A DNA analysis based on 2,205 single-nucleotide polymorphism (SNP) loci genotyped five ñata samples (Veitschegger et al. 2018) and included 134 other cattle breeds (Decker et al. 2014). The analysis showed clustering of all the ñata samples within taurine breeds, indicating a unity there, as opposed to the independent generation of the peculiar anatomy in diverse individuals (Veitschegger et al. 2018). Furthermore, the anatomical study and that of 3D geometric morphometrics of cattle and aurochs skulls showed all these ñata individuals to be singular and sharing several unique features.

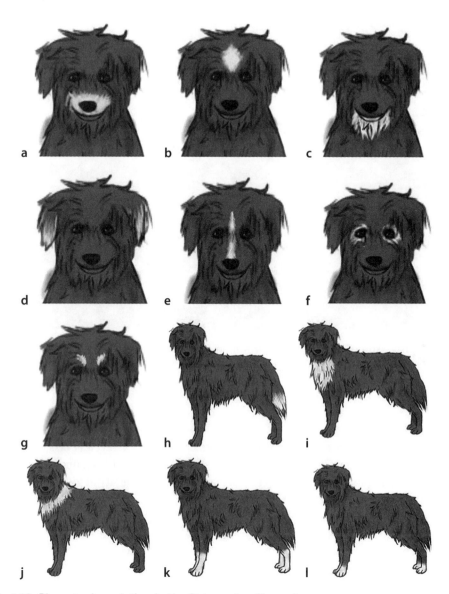

Fig 1.13. Phenotypic variation in the Patagonian Sheepdog.

On the extinction of local breeds

"Men have forgotten this truth," said the fox. "But you must not forget it. You become responsible, forever, for what you have tamed."
—Antoine de Saint-Exupéry, *The Little Prince*

Many domesticated animal species show great diversity in local breeds, and this diversity is usually tied to cultural identity, in addition to the biological properties that match local conditions, or production goals in the case of livestock. Works

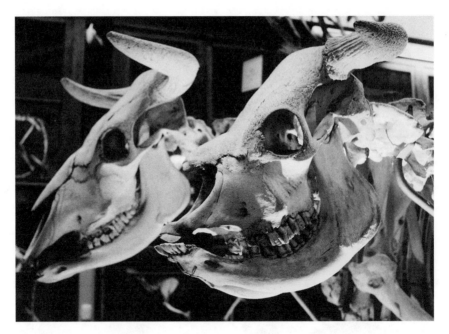

Fig 1.14. The skull of the *vaca ñata* (*front*), in contrast to that of standard cattle (*back*). Skulls displayed in the exhibits of the Museo de La Plata, Argentina.

documenting local breeds of several species come from Greece (Plassará 2005), Japan (Hongo 2017), China (Epstein 1969), Nepal (Epstein 1977), and Africa (Epstein 1971), to name a few examples. Treatises on breed diversity of given species are also available (e.g., Porter 2002).

Industrialization and globalization, with associated urbanization and migration to cities, led to the extinction of many local breeds. There are many examples in cattle, for which, as with many other domesticated forms, phenotypic and genetic diversity are becoming reduced, driven by selection for some productive breeds, with extinction of many regional forms. An example is the extinction of the "Freiburgerkuh" in Kanton Freiburg in Switzerland, a cattle breed well adapted to winter conditions, meager pastures, and unaided birthing in low alpine landscapes. This breed has been replaced by larger and superficially similarly colored Holstein cattle, which are high-production dairy animals. The "Freiburgerkuh" was transported many decades ago to southern Chile around Punta Arenas, with current populations of free-ranging cattle in that southern Patagonian landscape encountering physiological challenges similar to those of Freiburg. Therefore, the "Freiburgerkuh" may either still be around, or may have left significant genetic legacy. Studies by the ProSpecieRara organization of Switzerland have sadly confirmed the extinction of the local Swiss breed also in Chile.

Local movements across the world aim to preserve local breeds and along with them, important cultural and genetic heritage. Examples are the organizations in the

Greek islands of Skyros and Rhodos preserving their unique kinds of small horses (Dimitriadis 1937).

On a more curious note, there have been historical attempts—some also in current times—to "revive" an extinct wild form, the aurochs (Stokstad 2015), perhaps an example of the "taste for the primitive" so characteristic of diverse human endeavors (Gombrich 2002). The case of the aurochs is tied to the Heck cattle and with that to the history of Nazi Germany. It would seem there are better ways to act to restore natural environments.

Landraces and breeds in the Americas

Colonization and exchanges led to new landraces and breeds of many domesticated forms, including, for example, dogs, cattle, horses, and chickens. At present, there are several endemic breeds in the Americas (some extinct) that come mainly from crossbreeding of mostly European lineages (Larson et al. 2012; Leathlobhair et al. 2018). Regardless of their origin, many American breeds are currently recognized as such by organizations such as kennel clubs or the American Standard of Perfection of the American Poultry Association (APA 2015).

Many Indigenous populations adopted domesticated animals brought by Europeans, as is the case in the southwestern United States of the Navajo and sheep herding, which started between 1650 and 1700 CE (Diamond and Bellwood 2003), leading to the celebrated weaving of woolen blankets for which the Navajo have become renowned.

Many are called, but few are chosen? On cultural influences in domestication

Common arguments on the propensity or suitability of populations of a species to be domesticated concern biological aspects of that species and assume a disposition of humans to domesticate. This disposition may not be universal, and this may explain in part the distribution of domesticated forms.

The biological features considered domestication criteria include flexible diet, appropriate growth rate, capacity to be bred in captivity, pleasant disposition toward humans, steady temperament, and appropriate and adaptable social hierarchy. These features were presented by Diamond (1997), who argued that a deficiency in any one of them renders a species "undomesticable." He referred to the "Anna Karenina Principle" of domestication, alluding to Tolstoy's novel with its list of features necessary for a good marriage, whereby a couple missing any one of them is doomed to fail. Francis Galton (1865), a half-cousin of Charles Darwin who wrote much on domestication (Clutton-Brock 2012), presented his own list of requisites for domestication, which included hardiness, fondness for man, desire for comfort, usefulness to man, breeding freely, and easy to tend.

Biological reasons must be considered to explain the few numbers of domesticates in pre-Columbian America. The idea goes that South America was deficient in large land mammals that could have ever been domesticated by humans (Diamond 2002, p. 702). Diamond (1997) argued that the adoption of European domesticates by Native Americans after 1500 is proof that there had been no cultural impediment to domestication among these people. For example, European horses were adopted by several groups of Native Americans across the continent. A large number of American species were not domesticated while having Eurasian close relatives or counterparts that were domesticated (Diamond 1997). These include, for example, the American peccary, similar to the pig; the American bison, similar to the water buffalo, yak, gaur, and banteng (and to the European bison, neither domesticated, though); and the North American bighorn sheep. Diamond also mentioned the widespread practice of keeping wild animals as pets among Native Americans, seen (also by Galton 1865) as a potential initial stage of domestication.

Although biological arguments are sound, other factors can play a role in determining whether a species becomes domesticated in a place and time. As stated by J. Clutton-Brock (1992, p. 79–80), "it was probably for cultural, as much as for many other complicated reasons, that the native Americans did not domesticate the Bighorn Sheep, nor the Australian Aborigines the Kangaroo, nor the 'Kung San the Eland in southern Africa."

The argument of pet keeping (Galton 1865) is not valid, as it wrongly implies this activity as a first step in domestication. A rich body of ethnological literature concerning people in the Amazon region makes clear that keeping of pets, at least in that region, should not be seen as a precursor of domestication, as the conception of human-animal interactions in those people is different from that of Europeans (Descola 2013). As stated by Stahl (2014, p. 227), "Pets cannot be seen as 'mobile protein reserves' because to the hunter taming should be seen as a form of sacrifice, for it is not killing" (Erikson 2000, p. 21). Stahl (2014) argued that the reason that Amazonians "fail to domesticate" animals is not the lack of raw material or opportunity, but instead the different worldview or logic guiding the relations of those people with other animals. According to Stahl (2014, p. 228), "indigenous Amazonians don't domesticate animals because it doesn't make any sense to them." It is thus likely that fewer native domesticates in the Americas has resulted both from biological features of the species in question as well as from cultural aspects of the humans who cohabit with it, surely itself connected to the ecology of the region. Advances in our understanding of Amerindian cultures and their conception of the world will be fundamental to comprehend the history and consequences of the human-animal relations in South America (Fausto 2007), and to realize the singular as opposed to universal nature of the Western world view.

The people in the Americas who adopted European domesticates were not distributed equally across the continent. The environmental and demographic changes brought by the encounter (Mann 2005) were immense. The large biogeographic diversity of the Americas is translated into a large cultural diversity, and influences on the existence and mode of domestication across the continent are perhaps not universal. Concerning the Amazon region, it has been argued that animal protein is easily secured in aquatic environments, leading to no "need" to domesticate ani-

Fig 1.15. Gambian giant rat, *Cricetomys gambianus*, used to detect land mines based on its highly developed sense of smell.

mals. Only the Muscovy duck was domesticated in that area (Sauer 1952). The secure diet accessible at the edges of rivers, shores, and savannas (Harris 1972, p. 188) through hunting and fishing promoted increased sedentism, and an opportunity for plant experimentation (Stahl 2014).

Potential domesticates

Many studies of feasibility of management and selective breeding of species of potential economic use by humans exist. Several bovines as well as four pig species from Asia were examined in terms of their behavior, diet, and reproduction, and deemed suitable (Ruskin 1983). Mason (1984) also listed as "potential domesticants among the Bovidae" species of Oryx (*Oryx*), eland (*Taurotragus*), and some buffalo (presumably *Bubalus*). These are just a few of many examples.

The Gambian giant rat (*Cricetomys gambianus*) has year-round reproduction, no seasonal peak in breeding, and a relatively short estrus cycle. It has been argued that these features make it a potential domesticate and source of meat (Ajayi 1975). Mason (1984) listed the giant rat and cane rats (*Cricetomys* spp., and *Thryonomys swinderianus*) as domesticated, but noted that attempts at domesticating the giant rat by farmers were mostly limited to trapping the juveniles alive during hunting and rearing these for slaughter. Gambian giant rats are used to detect land mines and also tuberculosis, apparently via their highly developed sense of smell (Figure 1.15). This is probably one of the most paradoxical cases of human sophistication—using special skills of other living beings to deal with our own species' proneness to proactive aggression.

Table 1.3. Contemporary vertebrate candidates hypothesized as suitable for domestication and semi-domestication in South America

Turtle: Arrau or South American river turtle (*Podocnemis expansa*)

Birds: rhea (*Rhea americana*), fulvous whistling duck (*Dendrocygna bicolor*), white-faced whistling duck (*Dendrocygna viduata*), blackbellied whistling duck (*Dendrocygna autumnalis*), chachalaca (*Ortalis motmot*), curassow (*Crax* spp.), gray-winged trumpeter (*Psophia crepitans*), diverse parrot species (Psittaciformes)

Mammals: kinkajou (*Potos flavus*), grison (*Galictis vittata*), otter (*Lontra longicaudis*), peccary (*Pecari tajacu*), capybara (*Hydrochoeris hydrochaeris*), Amazon bamboo rat (*Dactylomys dactylinus*), paca (*Cuniculus paca*)

Note: As summarized by Stahl (2014), based on several sources (Gilmore 1950; Morton 1984; Smith 1999). It is not specified what is meant by "semi-domestication," and this notion reflects the continuum of human-animal interactions associated with tameness and domestication referred to in the text. Given that no turtle species is widely considered "domesticated," placing the Arrau or South American river turtle (*Podocnemis expansa*) in this list reflects the wide scope of this proposal.

A long list of potential animals to be domesticated in South America has been postulated by diverse authors (Table 1.3). For decades capybaras have been managed intensively in many countries (Moreira et al. 2013), but this intense interaction has not led to domestication.

Humans in the Americas interact with 19 species of wild canids in different ways, but not involving domestication (Segura and Sánchez-Villagra 2021). Coyote (*Canis latrans*) and bush dog (*Speothos venaticus*) may have domestication potential based on their life history, social system, and diet (Segura and Sánchez-Villagra 2021). In China, an example of a past close relationship is that of the leopard cat (*Prionailurus bengalensis*), as revealed by geometric morphometric analyses of small felid bones dating between 5,500 and 4,900 ybp (Vigne et al. 2016).

No extant turtle is considered domesticated, but several are currently intensively managed, as is the case of the Chinese softshell turtle *Pelodiscus sinensis*. Genetic studies have revealed demographic and geographic changes suggesting that some turtle species may have been subject to intense management in the past. An example is the Colombian river turtle *Podocnemis lewyana* (Vargas-Ramírez et al. 2012), although it is difficult if not impossible to determine the nature of the interaction to refer to it as domestication. Another example is provided by an extensive study of mitochondrial DNA of the Central American river turtle *Dermatemys mawii*, which found a large-scale pattern of haplotype sharing even across hydrological barriers, drastic demographic size fluctuations, and the contemporary absence of a pronounced phylogeographic pattern (González-Porter et al. 2011). The authors hypothesized that these patterns resulted from extensive human-mediated movements and recent bottlenecks produced by overharvesting. The archaeological record suggests that *D. mawii* was consumed by people from Middle American cultures for centuries if not millennia (Jones O'Day et al. 2004), whereby Mayans transported animals far away from their natural distribution range (Lee 1996).

Summary: Variation in the process of domestication

There is no simple and universal way to define domestication that applies to all species we call domesticated. Different kinds of human-animal interactions are used to categorize populations beyond the standard and in many cases misleading wild-versus-domestic dichotomy. Three separate pathways to domestication have been recognized: a commensal, a prey, and a direct pathway. It is best to examine a species domestication process in concordance with that of other species of both plants and animals and not in isolation. For example, the spread of one species may be coupled with another's, as may have happened with chicken and rice.

Reconstructing the earliest phases of the domestication process in the archaeological record is challenging. In the absence of morphological traits associated with domestication, translocation of a species outside its native range may or may not indicate domestication. The domestication process of several species in the Near East probably started about 10,000 ybp, during the Neolithic transition in several human populations. This is described, in an oversimplification, as one from a life of hunting and gathering nomadism to farming sedentism. A similar transition occurred at other times and places as well.

Much has been speculated on how domestication started. It has been proposed that it started independently of population–resource imbalances as a result of intentional management of wild resources. An alternative hypothesis, based on foraging theory models and behavioral ecology, states that domestication of livestock arose in times of need, leading people to try to acquire more food from the environment. However, with domestication occurring in numerous locations, by so many different people and of many species, to expect a universal explanation is unreasonable.

Different hypotheses have been offered about how the process of dog domestication occurred, including considering it a self-domestication example (with wolves gaining access to human food and waste), a result of adoption of young animals (as reported for many hunter-gatherer societies), or following the capture and nurturing of wolf puppies as pets. Maybe all these mechanisms, as well as others, took place either simultaneously or at different times and places. Simple and universal narratives of a domestication event and mode are unjustified.

There is no uniform taxonomic nomenclature across biological disciplines to refer to many domesticated animals. In the domestication literature, domesticates in most but not all cases constitute different species from the corresponding wild forms.

Wild animal species from different areas of the world, including South America, Australia, and Africa, are similar in their biology to Eurasian domesticated animals. The fact that these ecological counterparts were not domesticated may be due to both biological features of the species as well as cultural aspects of their cohabiting humans, or perhaps just to chance. It is difficult if not impossible to generalize on the causes of the many cases that illustrate this contrasting pattern.

The tempo and mode of domestication are unique for each species. Straightforward and simple narratives of a place and time of domestication involving human agency and goals with a path toward use and exploitation have been discredited repeatedly by diverse kinds of evidence, with literature going back decades and growing exponentially as ancient DNA and zooarchaeological studies expand.

CHAPTER 2

Domesticated mammals and birds: Species accounts

This chapter presents overviews of a broad selection of the domesticated forms of mammals and birds and their wild counterparts.

Dogs

Dogs are the most ubiquitous animals and the closest to humans in most geographic areas and cultures around the world, and our understanding of domestication in general has been disproportionately dominated by our knowledge of dogs.

Dogs are descended from ancient wolves (Vilà et al. 1997). Identifying dog remains in archaeological sites is a complex process, as discriminating between dogs and wolves is not straightforward because incipient domesticated dogs likely were morphologically wolf-like (Larson et al. 2012; Janssens et al. 2019a). The dearth of archaeological excavations in many parts of the world also limits our interpretation of early dog domestication. But wolves and dogs are and have been intensively studied, and with increasing access to specimens in collections from around the world, more studies will shed light on their occurrence and biology.

Dog remains dating 12,500 and 8,000 ybp have been found in East and central Asia, respectively. Some skull remains have been interpreted as suggesting an antiquity of the human-dog relationship going back approximately 30,000 ybp (e.g., Belgium: Germonpré et al. 2009, 2015; Siberia: Ovodov et al. 2011), but other studies have contested all or some of these hypotheses (Crockford and Kuzmin 2012; Morey 2014; Drake et al. 2015). Perhaps other reports of early dogs, such as that from Předmostí in the Czech Republic, 28,500 ybp based on dental microwear (Prassack et al. 2020), can be confirmed by further tests. It seems plausible that there were early dog domestication processes that due to climatic and demographic changes left no descendants, and little if any archaeological trace.

Ancient DNA studies, following the pioneering studies of mitochondrial DNA and other markers (Vilà et al. 1997), are refining the estimates on the tempo and mode of dog domestication. The history of dog domestication in Europe and Asia is complex (e.g., Sablin and Kchlopachev 2002; Ovodov et al. 2011; Marinov et al. 2018). Some studies of ancient DNA suggest a dual origin of dog domestication in these two continents (Frantz et al. 2016): one in southern China by 15,000 ybp (Wang et al. 2016) and an earlier one in Europe between 32,000 and 18,000 ybp (Thalmann et al. 2013) and 40,000–20,000 ybp (Botigué et al. 2017), but these datings become refined as new studies are reported. Earliest remains, ~12,000 ybp or older, are present in Europe, the Levant, Iraq, Northern China, and the Kamchatka Peninsula in the Russian Far East. Dogs older than 8,000 ybp are widely distributed and suggest independent domestications of local populations of wolves or human migrations possessing dogs (Frantz et al. 2016).

Joint hunting, which requires highly integrated behavior with human life, seems to be an ancient practice in the human-dog relationship, as shown by pre-Neolithic images of dogs hunting with humans in the Arabian Peninsula dating from the seventh and possibly eighth millennium BCE (Guagnin et al. 2018).

The range of size and shape among domestic dogs, with about 400 currently recognized dog breeds, extends beyond that of their ancestors. Chow chows and salukis are "ancient breeds" dating back a few thousand years, whereas most other breeds are younger than a few hundred years. Some breeds arose by intercrossing breeds while others were selected for a particular role from existing breeds.

It is usually stated that dogs are among the most morphologically diverse terrestrial mammalian species (Stockard 1941). This is an intriguing notion and may well be the case, but a systematic and quantitative assessment of this idea has not been conducted.

For all the diversity among dog breeds, variation in biogeographic origin of village dogs is higher (Shannon et al. 2015). This variation may have resulted from backcrossing and hybridization with other canid species and with domestic breeds, in addition to selection, gene flow, and genetic isolation and drift occurring with coevolution with humans for thousands of years (reviewed in Norton et al. 2019).

As with other domesticated species, genetic analyses of dogs have revealed that admixture across breeds and populations has been a source of morphological diversity (Cadieu et al. 2009). A tree structure does not properly describe the relationships among breeds, but a network does; however, such complicated relationships do not prevent the identification of clades. There are single mutations that produce recognizable traits that are shared across breeds, and these serve to identify clades. Twenty-three clades of breed-types existed before the advent of breed clubs and registries (Parker et al. 2017). During the nineteenth century, the diversity of breeds greatly increased (Figure 2.1). Study of the distribution of haplotypes across breeds has revealed that some common traits have arisen more than once in the evolution of the dog, with transfer of trait and disease alleles among breeds (Parker et al. 2017).

Fig 2.1. Wolf (*Canis lupus*) and a sample of dog breeds (*Canis familaris*). From left to right and top to bottom: Wolf, Bearded Collie, Whippet, Staffordshire Bull Terrier, Airdale Terrier, Australian Terrier, Basset Hound, Pug, Golden Retriever. Not to scale.

The beginnings of domestication: the case of dogs

How domestication of wolves began, eventually leading to domestic dogs, has been long debated (Werth 1944; Scott 1968; Olsen 1985; Serpell 1995). According to the self-domestication hypothesis, adult wolves gained access to food and waste available at campsites (Tanner 1979; Bocherens et al. 2015; Sharp and Sharp 2015). This hypothesis has the merit of emphasizing the niche construction aspect of the interaction and the lack of directionality of humans, both of which seem well founded. A valid critique, however, is summarized by Germonpré et al. (2018), who cite various ethnographic and archaeological studies suggesting that large predators were prevented from accessing food and waste at campsites, although they could scavenge on the remains left at kill sites. The authors also cast doubt on the assumption that consumption of human waste ("familiarization") leads to "tameness," as there is evidence that even then wolves could behave aggressively toward humans and engage in predatory attacks on children.

Germonpré et al. (2018, p. 53) speculate that in the initial phases of domestication, captive wolves were fed by humans. Wolf breeding was likely controlled by selection for behavior and "probably installed a mutual cycle in which humans . . . tended to look after the cute pups of the captive canids." Adopting of young animals, as reported for many hunter-gatherer societies, may have led to the domestication of the wolf. Evidence recorded from ethnographic studies of human populations and the archaeological record show that people obtained diverse products (fur, meat and brain, long bones) from the wolf. To what extent the emotional bond of humans with dogs today (Hare and Woods 2013)—which involves oxytocin secretion as well as other physiological responses (Jeannin 2018)—can be traced back to the origins of domestication is a relevant question that is difficult to address. Germonpré et al. (2018) emphasized how cultural traditions (ritual practices) and an animistic cosmology of Upper Paleolithic societies may have been important in the keeping of captive wolves and subsequent (non-goal oriented) domestication leading to the present-day dog. They hypothesized that the domestication of the wolf was unique and not a model that was later transferred to other animal species (Uerpmann and Uerpmann 2017).

Another aspect of prehistoric dogs and their domestication is their use as hunting partners (Davis 1982). Rock art from northwestern Saudi Arabia dating as far back as the seventh and possibly eighth millennia BCE depict dog-assisted hunting scenes and are direct documentation of this activity (Guagnin et al. 2018). The depiction of leashes on some dogs suggests a great degree of control and association in prehistoric times.

Francis Galton (1865) suggested that wolves were domesticated by humans following the capture and nurturing of wolf puppies, basing this idea on ethnographic observations that pet keeping is not unusual among hunter-gatherer groups across the globe. More recent ethnographic work suggests that in some Amazon populations, and this may apply to other human groups worldwide, pet keeping involves a worldview that is not necessarily conducive to domestication and cannot be seen as a precondition to it (chapter 1).

The dog in the Americas

More than 40 domestic dog breeds from the Americas are currently recognized by kennel clubs. It has been clearly established that these originated in very recent historical times, as did the breeds of European origin. But a diversity of dog forms and kinds has existed in the Americas for millennia at different places and times. This diversity has little or no genealogical continuity with the current one, and much molecular work has been focused on deciphering the contribution, if any, of ancient dogs to their current diversity. Information on this lost diversity of dogs is also retrieved from zooarchaeological records of burials, abundance in sites, diet changes, and ethnologic information (Segura et al. 2022). Hybridization of dogs with wolves and even with coyotes (e.g., Adams et al. 2003, Valadez et al. 2002) may have been an old practice contributing genetic diversity to pre- and post-Columbian dogs.

In several migrations from Asia to America predating the current era, dogs accompanied humans, as revealed by different dog haplotypes reported from ancient DNA studies (Witt et al. 2015; Bergström et al. 2020; Perri et al. 2021). The long history of dogs in the Americas is documented in records of specimens from North American sites dating as far back as 10,000 to 8,400 ybp and from the Andean region of South America from 5,600 to 5,000 ybp (MacNeish and Vierra 1983; Morey and Wiant 1992; Stahl 2013). Pre-Columbian dog diversity in North America is first recorded from sites around 4,000 ybp, with varieties such as short nosed, loberro, and hairless forms (Haag 1948; Crockford 2005). Aztecs developed dog breeds specifically as food. Pre-Columbian forms from the Andes included dachshund-like, shepherd-like, hairless, bulldog, shortened snout, and long snout kinds. Dogs have been used by human groups in the Americas for hunting, food, transport, defense, company, and rituals, and Coast Salish First Nations in Western North America exploited dogs for manufacturing of blankets, with practices associated with their care and a marine food diet. In the huge area of the Amazon, with tremendous and ancient cultural diversity of human groups, dog acquisition reportedly began only toward the end of the nineteenth century; in some regions only in the twentieth century (Stahl 2012).

Cats

Cats (*Felis catus*) probably followed a commensal pathway to domestication. In early settlements in the Fertile Crescent, wild house mice (*Mus musculus domesticus*) were found among wild grain stores. Cats were likely attracted to these mice as a food source as well as to human-generated trash, leading them to approach human environments, although initially cats still had to hunt and scavenge, an ability they show today as feral cats (Driscoll et al. 2009).

Cats were transported to Cyprus probably from southwest Asia and buried with humans there by ~11,000 ybp, pointing to some kind of intimate relation already at the time (Vigne et al. 2012). A study on ancient DNA analysis of temporally and geographically widespread archaeological remains concluded that both the Middle Eastern and Egyptian populations of the wildcat *Felis silvestris lybica* contributed to the gene pool of the domestic cat at different historical times (Ottoni et al. 2017).

Domestic cats were deified in Egypt around 2,900 ybp as the goddess Bastet, and as such they were sacrificed, mummified, and buried. Although Egypt banned its export, the cat eventually reached Greece, the Roman Empire, the rest of

Fig 2.2. Examples of cat breeds (*Felis catus*). From left to right and top to bottom: Manx cat, Neva Masquerade, Siamese cat, Bengal cat, Sphynx cat, Norwegian Forest cat, British Shorthair, Abyssinian, British Longhair. Not to scale.

Europe, and the far East and Asia; the Americas possibly during Columbus' transatlantic trips; and Australia via European explorers in the 1600s (Driscoll et al. 2009).

For humans, the cat has been a pest-control agent, object of symbolic value, as well as a companion animal. Directed breeding of cats occurred later than that of most other domesticated animals, probably only after the Middle Ages (Ottoni et al. 2017). Morphologically distinct breeds of the domestic cat appeared more recently through breeding, especially in the past 75 years (Figure 2.2); around 80 different breeds were recognized in 2020 by the International Cat Association (www.tica.org /cat-breeds).

As opposed to dogs, which were mostly cross-bred from purebred lines, pedigreed cats come from street cats selected locally for specific traits. Genetic variation analyses of cat breeds have shown that cat breeds are less well defined than dog breeds (Menotti Raymond et al. 2008). It is challenging to identify cat breeds based on genetic markers, given the intense and short breeding history of these animals (Vella et al. 1999; Lipinski et al. 2008; Kurushima et al 2013). Genes have been identified that determine differences in coat color, fur length, texture, and other coat characteristics as well as mutations resulting in traits including tabby patterning, coloring, and long hair (Driscoll et al. 2009).

Misplaced biophilia? A pledge for cultural change

Biophilia refers to the natural inclination of human beings toward nature and the bond we feel for and inclination to interact with other living beings, a concept introduced and discussed by both psychologist Erich Fromm (1964) and evolutionary biologist E. O. Wilson (1984). The concept of biophilia is important for understanding some of the dynamics and consequences of our dealings with domestic animals.

Having domesticated animals has led to a misplaced biophilia; misplaced because we humans prefer an animal we can see and appreciate while ignoring many others suffering the consequences of the former. Outdoor cats have been shown worldwide to have the capacity to become a plague in cities and suburban areas, detrimental to small mammal and bird faunas (Marra and Santella 2016). Urban faunas can become thus impoverished, and considerable animal suffering occurs daily around the globe because of the biophilia of outdoor cat owners and lack of legislation to ban outdoor cats. A citizen scientist study tracking 925 pet cats from six countries quantified home and predation in outdoor cats (Kays et al. 2020). The study found that the per-animal ecological impact of cats is similar to that of wild carnivores, but the effect is amplified by the high density of cats in neighborhoods. Surely more studies are needed to examine this matter. Cats are globally reported to be tied to the extinction of 63 species, mostly in islands and their endemic faunas (Doherty et al. 2016).

As shown in behavioral economics studies, we feel sympathy only for what we see and experience directly. Socially engaged urban people buy cheap clothing in nicely decorated stores because although they have heard about it, they do not witness the conditions leading to the production of those clothes. Outdoor cat owners, or more realistically, children who could become outdoor cat owners later in life, could be educated in some relevant facts of biodiversity, small bird and shrew biology for example, to appreciate what is lost as a result of invasive species. The discipline of urban ecology could become important in this regard as it might involve scientists with a direct connection to people and their lives (Szulkin et al. 2020).

The same principle of exposure to information could lead to people making informed decisions on other matters. Zoological knowledge could be put to good use, for example, by communicating and helping to develop a positive inclination toward insects, as is widespread in Japan, which could lead to more acceptance of dietary diversification in many countries in which this is not common practice (Kawahara 2007). The same could be achieved concerning fish consumption, as some species are overfished out of ignorance or disregard for other species. These complex matters require local approaches and solutions.

Direct contact at the production stage with the animals that are eaten, with exposure to its gruesome aspects, as discussed and illustrated by Jamie Oliver in his book on Italian cuisine (Oliver 2005), would provide a more honest appreciation of the sources of food that could lead to informed decisions by consumers.

Fig 2.3. Examples of ferrets, *Mustela furo*. Polecat-colored ferret (*left*) and silver color ferret (*right*), not to scale.

Ferrets

The ferret (Figure 2.3) was probably domesticated from the European polecat, likely for hunting rats and rabbits (Blandford 1987; Sato et al. 2003). It is generally reported that the ferret was domesticated somewhere in the Mediterranean region, rightly contradicting historical accounts such as that of Strabo about 20 CE that mentioned an African origin (Thomson 1951; Clutton-Brock 1999).

Starting in the 1970s the ferret became a household pet, and this association has permitted behavioral studies of social cognition and comparisons with dogs. Ferrets are much like dogs in their ability to sustain eye contact with their owner and to follow human directional gestures (Hernádi et al. 2012). In the words of Clutton-Brock (1999, p. 184), though, "they always remain erratic and recalcitrant in their behavior." The ferret is used as a laboratory organism for several purposes, for example to study influenza and severe acute respiratory syndrome (SARS)-associated corona virus (Scipione Ball 2006). Ferrets as well as polecat-ferret crosses, called "Fitch" ferrets, have been used for fur production.

The global ferret trade has resulted in distinct genetic clusters, some with much lower genetic diversity than ferrets from Europe because of extreme founder events. The lack of hybridization with wild polecats or genetically diverse ferrets in some regions accentuates this pattern (Gustafson et al. 2018). In some regions in which ferrets were introduced, hybridization with native European polecats became common (Davison et al. 1999). The ferret has become an invasive species in some areas, and can be host to diseases such as tuberculosis (Byrom et al. 2015).

Goats

The study of haplotype groups and their distribution in goat breeds has suggested independent domestication "events" from the wild form *Capra aegagrus* (Luikart et al. 2001; Sultana et al. 2003; Joshi et al. 2004; Chen et al. 2005). Perhaps the original populations were from Sind, through Baluchistan to Afghanistan, as well as from the Kopet Dagh Mountains in Turkmenistan and Iran (Mannen et al. 2001). From several original populations, multiple dispersal events out of the Fertile Crescent region, and mixing, current goat diversity originated. Goat genetic diversity exhibits little phylogeographic structure compared with other domestic species, including sheep. Genomic data suggest that already in early goat domestication there

Fig 2.4. Examples of domestic goat, *Capra hircus*. From left to right and top to bottom: Toggenburger goat, brown goat, black-white goat, Rove, goat, long-haired goat, Angora goat, black goat, white goat. Not to scale.

was selection for pigmentation, reproduction, milking, and stature, which surely accompanied dietary change in domesticated goats (Daly et al. 2018).

Goats are very versatile in feeding habits and adaptable to harsh environments, features that contribute to the success of feral populations, some certainly qualifying as invasive species. The browsing of thorny scrubland by domestic or feral goats can result in land clearing following habitat changes caused by humans. Goats may have played a role in the desertification of the Sahara and the Middle East (Clutton-Brock 1999), as well as many other island and continental regions, including the arid regions of northwestern Venezuela (Sánchez-Villagra unpublished) and other regions of the Americas (Manzano and Návar 2000).

Goats can be a source of meat, milk, fat, and clothing, of bones and muscle tendons used for artifacts, as well as of dung for fuel and manure. Around 100 breeds of goats can generally be recognized (Porter 1996; Joshi et al. 2004), but exact characterization and separation of kinds is achieved by national and regional organizations (Figure 2.4). Numerous regional studies of genetic diversity have contributed much to our understanding of the history of populations and management strategies of these beautiful animals (Onzima et al. 2018; do Prado Paim et al. 2019).

Fig 2.5. Moufflon, *Ovis orientalis*, and examples of sheep, *Ovis aries*. From left to right and top to bottom: European mufflon, Crossbreed, Hissar, Ouessant, Suffolk, Arles Merino. Not to scale.

Sheep

The wild ancestor of the domestic sheep is probably *Ovis orientalis*, the Asiatic mouflon. *Ovis musimon*, the European mouflon, in some studies used as surrogate for the wild form (Sánchez-Villagra et al. 2017), is a feral domestic sheep (Clutton-Brock 1999; Chessa et al. 2009; Meadows et al. 2011).

Initially, sheep were most likely reared mainly for meat. Selection of domestic sheep with characteristics common to modern breeds occurred first in southwest Asia, later spreading successfully into the rest of Asia, Europe, and Africa. Wool production from sheep probably also arose in southwest Asia and then spread throughout Europe (Chessa et al. 2009).

The history of domestic sheep diversity and geographic history has been studied, for example, based on limited segments of the mtDNA that identified five haplogroups (Meadows et al. 2011) and endogenous retroviruses (Chessa et al. 2009). Ancient breeds, including Orkney sheep, Soay sheep, Icelandic sheep, and Cyprus Mouflon and Mediterranean Mouflon, have been identified as such based on these and other genetic studies. In contrast to more derived forms with respect to the wild ancestor, they are characterized by darker and coarser fleece, a molting coat, and the frequent presence of horns in females and males (Chessa et al. 2009). These ancient breeds may have been displaced, some to a semiferal state on islands without predators, by a second set of migrations of more derived breeds from southwest Asia (Chessa et al. 2009).

There is no comprehensive overview of the more than 500 breeds of sheep recognized by national and regional organizations around the world (Figure 2.5). Livestock producers search for the best combinations for diverse purposes such as meat quality, meat and wool combinations, wool for textiles, carpet wool, and milk production (Rasali et al. 2006).

Pigs

The wild boar may have followed a commensal route to domestication because it was able to readily consume human waste. However, as the wild boar was also hunted, the prey pathway to domestication may also be valid for pigs (Larson and Burger 2013; Marshall et al. 2014). Pig domestication may have involved a long phase of wild boar management in eastern Anatolia (Ervynck et al. 2001; Vigne et al. 2009). The zooarchaeological record suggests an extended era of commensalism and domestication of at least 3,000 years (Ervynck et al. 2001), as documented by a prolonged period of morphological change and culling profiles. Pigs may have been domesticated independently in East Asia and in the Middle East (Giuffra et al. 2000; Groenen et al. 2012).

In central China, early pigs were associated with both early sedentary gather-cultivators of wetlands and early villages of millet cultivators (Cucchi et al. 2011; Flad et al. 2007). In the northern Fertile Crescent, pigs were associated with early cultivating villages (Vigne et al. 2011). The diffusion of Chinese and Middle Eastern domestic pigs into southeast Asia and Europe involved dispersal and introgressions between introduced domestic pigs and local wild boars (Marshall et al. 2014; Larson and Burger 2013).

Wild boars still exist in abundance, as do feral pig populations (Wilson et al. 2017). Pigs are quite adaptable, and a variety of husbandry intensity and systems are associated with pigs: backyard pigs to stalled pigs, extensively herded pigs, and free-range pigs (Larson and Fuller 2014). The diversity of pig breeds is large, and this diversity is reflected in variation in head and body proportions and in the integument (Figure 2.6). The Angel Saddleback, the Bentheim Black Pied, the Berkshire, the Chato Murciano, the Duroc, the Kune Kune, the Large White, the Woolly, the Piétrain, the Swabian-Hall, and the Vietnamese Pot-bellied are among the many breeds of pigs (Porter 1993), an animal with a rich cultural history (Macho 2015).

Bovines: Cattle and relatives

Five species of bovines have been domesticated (chapter 1; Figure 2.7), all through the prey pathway. Bali cattle (*Bos javanicus domesticus*) are a source of meat in Indonesia and in other countries of southeast Asia. The wild form, the banteng (*Bos javanicus*), similar to the domestic form in size but differing in color, is currently found in dense tropical forests of southeast Asia. Crossbreeding of zebu cattle with Bali cattle occurs in some areas of southeast Asia (Mohamad et al. 2009).

The gayal or mithan (*Bos frontalis*) is the domestic form of the gaur (*Bos gaurus*) (Kamalakkannan et al. 2020); hybridization with other species of cattle may have occurred. The gayal has a limited geographic distribution in the hill-forests of China, northeast India, Bangladesh, Myanmar, and Bhutan.

The river and swamp varieties of water buffalo (*Bubalus bubalis*) are usually recognized. The zooarchaeological evidence suggests river buffalo were domesticated

Fig 2.6. Wild boar, *Sus scrofa*, and domestic pig, *Sus domesticus*. From left to right and top to bottom: Wild boar, domestic pig, black mini pig, domestic pig, Vietnamese Pot-bellied pig, Göttingen mini pig, Kounini pig, Mangalitsa, Oxford Sandy and Black. Not to scale.

in the Lower Indus Valley region and in western India by 4,500–4,000 ybp (Meadow and Patel 2003; Fuller 2006). The wild form of the swamp buffalo (*Bubalus arnee*) is distributed through tropical Asia, and it was likely domesticated between eastern India and peninsular southeast Asia before 3,000–2,500 ybp.

The domestic water buffalo is an integral component of traditional Asian rice cultivation; in China and southeast Asia it is used as draught animal. DNA analysis suggested that swamp buffalo domestication occurred earlier in Thailand than in China; thus it may have been introduced from southeast Asia (Yindee et al. 2010). However, when and where swamp buffalo was introduced into China remains unclear. A genotyping and geographic diversity distribution study of modern domestic water buffalo from 31 populations (15 river and 16 swamp buffalo) worldwide hypothesized that water buffalo migrated in two distinct directions post-domestication: river buffalos went west from the Indian subcontinent, whereas swamp buffaloes went east-southeast from northern Indochina (Colli et al. 2018b).

Although yak (*Bos grunniens*) exploitation is an important human activity on the Qinghai-Tibetan Plateau, yak bones have rarely been identified in ancient sites. Genome variation analysis based on 209 genes of domestic yak supported yak domestication dating to 7,300 ybp, most likely by nomadic people, with an estimated sixfold increase in yak population size by 3,600 ybp (Qiu et al. 2015). These two dates coincide with two early Qinghai-Tibetan Plateau human population expansions (Yang et al. 2008). Historical records and archaeological evidence suggest that yak pastoralist societies were established by ~4,000 ybp (Flad et al. 2007; Wiener et al.

Fig 2.7. Examples of wild and domestic bovids. From left to right and top to bottom: Banteng (*Bos javanicus*), gaur (*Bos gaurus*), domestic Asian waterbuffalo (*Bubalus bubalis*), indicine cattle (*Bos taurus*), yak (*Bos mutus*), domestic yak (*Bos grunniens*), aurochs skeleton (*Bos primigenius*), common cattle (*Bos taurus*). Not to scale.

2003). The yak was very prominent in Tibetan culture, becoming the mainstay of Tibetan pastoral society, with millions of domestic yaks today providing food, shelter, fuel, and transport in and around that region (Olsen 1990; Wiener 2003). Hybridization of yak with local cattle is practiced, producing hybrids of yak with "improved" European breeds (Wiener 2003). Yak dairy products used by ancient people could be identified based on analysis of lipids absorbed in pottery.

The wild ancestor of common cattle (*Bos taurus*) was the aurochs, now extinct but previously found in Europe, North Africa, and Asia. It is usually mentioned in writings that the last aurochs was killed in Poland in the year 1627, but to my knowledge the source of this statement is unknown. Domesticated cattle were probably introduced from western Eurasia alongside sheep and goats and subsequently interbred with Saharan wild populations (Gifford-Gonzalez and Hanotte 2011).

A report on variation of mitochondrial DNA sequences from 392 extant oxen from Europe, Africa, and the Middle East compared with data from four extinct British wild oxen showed ancient sequences in a tight cluster clearly distinct from modern cattle. Modern *Bos taurus* showed four haplotype clusters, one of which predominates in Europe and is present in the Middle East, while African diversity was a separate haplogroup that is rare elsewhere. These data are in agreement with a derived Near Eastern origin for European cattle (Troy et al. 2001). A study of 56 cattle remains from four Thai archaeological sites provided genetic evidence that *B. taurus* was domesticated in Thailand 3,550–1,700 ybp; a close relationship found among ancient Thai, Iranian, and Chinese taurines supports cattle from the Middle East first introduced into North China and subsequently into Thailand (Siripan et al. 2019). Analysis of SNP array data collected for more than 3,000 modern cattle samples from 180 worldwide populations confirmed a significant differentiation between African and Eurasian taurine cattle; modeling favored two "domestication events," suggesting subsequent hybridization with local aurochsen to explain the additional genetic variation detected in African cattle (Pitt et al. 2019).

Tooth wear patterns in 10,500-year-old remains of a bovine mandible from North China with a unique mitochondrial genome are consistent with domestication (Zhang et al. 2013a). To understand domestication in this part of the world, further archaeological evidence and studies of the influence of admixture involving other bovine species are needed (Larson and Burger 2013).

There are two alternative hypotheses concerning the Zebu cattle (*Bos indicus*): they were either independently domesticated in the Indus Valley between 9,000 and 7,000 ybp (Meadow and Patel 2003), or they may have originated from hybridization between wild zebu populations and taurine cattle that were transported eastward (Larson and Burger 2013).

Genome analyses of 67 ancient Middle Eastern *Bos taurus* samples including six aurochsen support genetic material introduction into domesticated forms by rapid introgression mediated by *Bos indicus* (zebu) bulls from the Indus Valley, as indicated by a Bronze Age shift (Pereira Verdugo et al. 2019). These males were well adapted to arid environments and contributed to herd survival in a process likely stimulated by a long-lasting drought ~4,200 years ago (Pereira Verdugo et al. 2019).

Old World camelids: Dromedary and Bactrian camels

The wild dromedary went extinct approximately 2,000 ybp. At the time of its extinction its range was probably limited to coastal mangrove habitats in the Arabian Peninsula. Nowadays, domestic dromedaries, the one-humped camel (*Camelus dromaderius*), live in the desert zones from North Africa, in the horn of Africa, throughout the Levant and the Arabian Peninsula. Feral populations of dromedaries are found in Australia, following their introduction in the nineteenth century (Orlando 2016). Dromedaries are used for riding and transportation of goods, but in East Africa they are used predominantly for their milk. Dromedaries are ideal cargo vessels for long journeys through the desert and have been essential to the development of overland trade (Orlando 2016). They show unique phenotypic adaptations to arid and extremely hot environments (Wu et al. 2014), including surviving almost one week without water, and enduring 30% water loss (Schmidt-Nielsen 1959). Furthermore, they can store approximately 35 kg of fat in their hump, helping to withstand limited environmental resources.

The domestication of the dromedary took place most likely around 3,000 ybp. Around that time, an increasing association with human settlement, a significant reduction in the size of bone remains, and unambiguous artistic representations have been recorded (Uerpmann and Uerpmann 2002; Vigne et al. 2005a).

Two major haplogroups (HA and HB) were found throughout the whole distribution range of the modern dromedary (Almathen et al. 2016). As no geographic region showed significantly more population genetic diversity, measured by microsatellite heterozygosity levels and allelic richness, no specific location for a dromedary domestication center was hypothesized. The population history probably involved movements of dromedaries along cross-continental trading routes (Almathen et al. 2016; Orlando 2016). Ancient DNA recovered from wild and domestic individuals was used to calibrate the mtDNA clock, and with that data population changes were dated. A massive demographic expansion some 600 ybp was detected, coinciding with the rise of the Ottoman Empire. There was probably an increasing use of dromedaries over long distances with the expanding empire (Almathen et al. 2016).

Bactrian camels (*Camelus bactrianus*) are thought to be a descendant of *C. ferus* (Peters and von den Driesch 1997), with the term "Bactrian" applied to two humped camels, as opposed to Arabian or dromedary camels with only one hump (Figure 2.8). Bactrian camels probably originated in the east in the Altai and surrounding regions (Flad et al. 2007), extending eventually from northern China to Asia Minor in the west, with the earliest evidence in Neolithic Inner Mongolia towards the west, Turkmenistan in the mid-fourth millennium BCE and Margiana and Bactria in the mid-third millennium BCE (Potts 2004). Although *C. bactrianus* remains are lacking in Iranian plateau sites, the species may have been present but used only for transport, as a draught animal and source of secondary products such as hair and milk, and not for meat (Potts 2004). The wild relatives of the domestic Bactrian camel still exist in small numbers in Mongolia and China (Burger 2016).

Bactrians camels can carry heavy loads of 220–270 kg over long distances and can be used for this purpose from age four, with 20–25 years of productive work (Epstein 1969). They tolerate extreme temperatures well but prefer winter, when they are preferentially used, working well at altitudes of 4,000 meters, while they pasture in the summer to build up fat reserves (Potts 2004).

Old World camels show high genetic variation (Fitak et al. 2020). Hybridization between Bactrian camels (*C. bactrianus*) and dromedaries (*C. dromedarius*) can produce hybrids of greater size and strength, with carrying capacity of up to 450 kg, double that of Bactrian camels, a practice possibly originating in the early first millennium BCE (Potts 2004). Hybridization also facilitates improved milk and wool yield or cool resistance (Burger 2016).

Fig 2.8. Examples of wild and domestic camelids. From left to right and top to bottom: Dromedary (*Camelus ferus*), Bactrian Camel (*Camelus bactrianus*), guanaco (*Lama guanicoe*), llama (*Lama glama*), vicuña (*Vicugna vicugna*), and alpaca (*Vicugna pacos*). Not to scale.

South American camelids: Llamas and alpacas

Llamas (*Lama glama*) and alpacas (*Vicugna pacos*) are the largest livestock domesticated in the Americas (Figure 2.8), and use of their dung may have greatly contributed to the expansion of agriculture in many areas of the Andes (Mengoni-Goñalons and Yacobaccio 2006). The high-altitude puna grasslands of the Andes are the primary habitat of the domestic llama and its wild ancestor the guanaco, *Lama guanicoe*, as well as the smaller, domestic alpaca and its wild ancestor the vicuña, *Vicugna vicugna* (Kadwell et al. 2001; Wheeler 2012; Marín et al. 2017). Guanacos range over wider areas and require water less frequently, and some populations migrate. In contrast, vicuñas live at higher elevations, do not migrate, and require water frequently. The four South American camelids can interbreed and produce fertile offspring; they all have the same chromosome number ($n = 74$). Both guanacos and vicuñas are social animals, both forming herds that establish strong dominance hierarchies—in these aspects they closely fit the suggested profile of proneness to domestication (chapter 1).

Llamas serve as a source of wool and meat, and as a means of transporting goods across the high Andes. Broader backs could be a morphological marker of domestication in this species as a result of breeding for better ability to carry cargo (Smith 1998), but this speculation has never been tested. No clear morphological markers, nor significant changes in size, have been identified as signs of domestication in South American camelids. Therefore, the only way to recognize their initial herding is to quantify changes in relative abundance of their remains and their age profiles in archaeological sites, as reported in other regions for other livestock species (Sykes 2014; chapter 1). Differences in skull shape between domesticated and wild forms of South American camelids have been documented by 3D geometric morphometrics (Balcarcel et al. 2021a). Perhaps there are features that can be used to differentiate wild from domestic forms in isolated skull parts from the archaeological record, with the petrosal bone one potential source of such features (Herre 1953).

Llamas and alpacas have been integral to the economy and culture of Andean communities for millennia (Cardozo 1975, Goepfert 2010; Wheeler 2012). Excavations of the rock shelter at Telarmachay in Peru showed domestication of the alpaca between 6,000 and 5,500 ybp and the llama between 5,500 and 5,000 ybp (Goepfert 2010). Archaeological evidence suggests that both llamas and alpacas were sacrificed; they have been found in tombs as offerings, and the llama is consumed at special events even today, including funerary meals (reviewed in Goepfert 2010). Herding of domesticated camelids was established over a broad area of the south-central Andes by 4,000 ybp, as documented from several archaeological sites based on abundance of remains and/or demographic profiles. Nowadays, people in the Andes tend to see llamas as working animals, and alpacas as source animals, for meat and fleece.

Reindeer

Reindeer (*Rangifer tarandus*) is known as caribou in North America (Figure 2.9). On account of long and close ties with human managing, the reindeer is listed as a domesticated animal, but the boundary with the wild is blurred, given how independent reindeer are and the fact that they do not need to be fed or housed by humans. Wild reindeer have a docile nature, willing to stay close to human settlements and even be milked. The study of reindeer herded by Sami pastoralists in northern Scandinavia has been the subject of anthropological studies that have significantly influenced our understanding of the complex and diverse nature of human-animal interactions (Ingold 1980). The cosmological vision of humans involved in reindeer domestication, one in which the animal undergoes a process of taming and is involved in a relation not of domination but of trust (Willerslev et al. 2015), has been central to understanding their transformation from hunting to pastoralism.

Fig 2.9. Reindeer herd, *Rangifer tarandus.*

Reindeer were domesticated by nomadic hunting societies, which may have resulted in reduced and less clear archaeological evidence compared with other prey pathway domestications. Of the two ecotypes of reindeer, the migratory tundra kind was the one domesticated, as opposed to the more sedentary forest dwelling ecotype (Røed et al. 2018). Reindeer herding is thought to have begun at least 3,000–2,000 ybp (Larson and Fuller 2014). Interactions with humans, perhaps involving management, were documented in rock art in Norway 6,700–6,200 ybp (Helskog 2011).

Antler trimming, an integral part of modern large-scale reindeer pastoralism in which prominent tines are removed to prevent aggressive animals from inflicting damage on other reindeer, humans, and property, may have also occurred in the oldest known type of hunter-gatherer reindeer herding in Siberia (Grøn 2011). Evidence of prehistoric antler trimming may therefore serve as a domestication indicator and also reveal which transportation methods were used locally. The possible use of domesticated reindeer for pulling sledges and for riding may have led to special anatomical characteristics in the feet and back (Upex and Dobney 2012). Carrying, load-pulling, and riding cause changes in reindeer bones (Salmi and Niinimäki 2016).

In spite of being quite similar in overall appearance, domesticated reindeer tend to be slightly smaller and show somewhat different coloration patterns compared with wild reindeer (Smith 2006). Domestic reindeer show a constant and characteristic instinct of herding, much weaker in wild reindeer, in which it is limited to migration (Smith 2006).

Studies on reindeer mtDNA (Røed et al. 2018) support a hypothesis of at least two separate and apparently independent reindeer domestication processes, in what is today eastern Russia and in areas of Norway, Sweden, and Finland, the latter as late as the Medieval period. Indigenous Sámi people started their reindeer husbandry by the Medieval period, using reindeer for traction, for carrying loads, and as a food source.

There are about 5 million reindeer on our planet; about half are domesticated today in Arctic pastures in Eurasia (UN environment program report from 2010).

Fig 2.10. Donkey (*Equus asinus*) and horse (*Equus caballus*) breeds. Selected examples of breeds from left to right and top to bottom: Donkey, Bay horse, Comtois, Belgian, Appaloosa, Falabella, Andalusian, Shetland Pony, Percheron. Not to scale.

Perissodactyla: horse and donkey

There are two odd-toed ungulates (Perissodactyla) that have been domesticated, the horse, *Equus caballus*, and the donkey, *Equus asinus* (Figure 2.10). Although attempted, zebras have never been domesticated; it is claimed this is because of their rather defensive-aggressive behavior toward humans (Diamond 2002).

Domestication of horses probably started more than 4,000 ybp in the Eurasian steppes and eventually revolutionized human mobility, economy, and warfare (Kelekna 2009). The ancestry of modern domestic horses and the onset and circumstances of their domestication are far from clear; there have probably been two domestication phases, one leading to today's modern domestic horses, the other leading to Przewalski's horses, which are actually secondarily wild, that is, feral (Gaunitz et al. 2018). The wild ancestor of modern domestic horses is most likely extinct or is represented by some relict population. Although the tarpan is reported as the wild form of the domestic horse (Zeuner 1963), this is most likely wrong (Clutton-Brock 1999). The genetic makeup of present-day horses has been reported to be a contribution from a lineage of wild horses that lived in the Artic until at least ~5,200 ybp and which have otherwise no descendants today (Schubert et al. 2014; Librado et al. 2015).

Humans use and have used horses for different tasks. Examples include riding horses for fast transport, trotting horses for drawing light carriages, and draft horses for transporting heavy loads.

Donkeys, one of the least studied large domestic mammals (Kimura et al. 2013), are probably the descendants of the African wild ass (*Equus africanus*), with domestication starting in northeastern Africa some 5,000 ybp. The timing and location of donkey domestication are difficult to determine because donkeys are uncommon in the archaeological record, and early domestication markers have not been established with certainty. Donkeys are used predominantly as transport animals. Domestication of the donkey transformed ancient transport systems in Africa and Asia and the organization of early cities and pastoral societies (Mitchell 2018). The donkey's advantage over cattle, which had spread throughout Africa a couple of thousand years earlier, are its various adaptations to dry conditions, which became more prevalent during that time.

The earliest form of food production in Africa involved animal domesticates and was developed in arid tropical grasslands (Marshall and Weissbrod 2011). The mobility of early herders shaped the development of social and economic systems. Genetic data suggest two different African wild asses were domesticated in North Africa, in the Sahara and in the Horn (Marshall and Weissbrod 2011).

Data from ten ass skeletons ~5,000 years old from an early pharaonic mortuary complex at Abydos, Middle Egypt, along with 53 modern donkey and African wild ass skeletons, showed similar proportions in Abydos metacarpals to those of wild ass, although individual measurements varied (Rossel et al. 2008). All Abydos skeletons exhibited a range of osteopathologies consistent with load carrying. The morphological similarities found indicate that donkeys did not undergo considerable phenotypic change during the early Dynastic period despite their use as beasts of burden (Rossel et al. 2008). That morphological change, difficult to establish in zooarchaeology and slow to appear, has been explained as resulting from high potential for interbreeding between founder populations, introgression with the wild form, and low levels of selection (Marshall and Weissbrod 2011).

The mule (*Equus* mule) is a common and ancient hybrid. It comes from breeding a male wild ass donkey (*Equus asinus*) and a female horse (mare *Equus caballus*). It is believed to have been initially bred in what is northern and northwestern Turkey today and in Egypt since before 3,000 BCE. Mules were used as pack, pull, and transport animals. The mule was considered even more valuable than the horse, as it had harder hooves, was easier to train, and could go for longer distances while needing less sleep and carrying considerably more weight. In addition to improved physical features compared with its parents, the mule has been suggested as an example of hybrid vigor also in terms of cognitive function, based on a visual discrimination learning task study (Proops et al. 2009).

Although occasionally fertile (Rong et al. 1985), mules are generally infertile because of the parental difference in number of chromosomes: 62 and 64 for the male and female parents, respectively (Henry et al. 1995). A mule ("Idaho Gem") has been cloned by nuclear transfer (Woods et al. 2003).

Still present in Turkey mostly in hilly and mountainous areas, mules have lost value in many areas because of their replacement by mechanical and electrical power, with Turkey's mule population numbers decreasing from 324,000 in 1977 to 51,500 in 2009 (Yilmaz and Wilson 2012).

Cavy (*Cavia porcellus*)

The ancestry of the domestic cavy, called also guinea pig or "cui," has been much debated (Spotorno et al. 2006). The most likely progenitor appears to be *Cavia tschudii*, from the high valleys of Peru, Bolivia, and northwestern Argentina (Donkin 1989). Phylogenetic analyses based on cytochrome *b* sequence data of a broad sampling of *Cavia* taxa to date provide support to morphologically defined and conflicting species, and evidence for populations of *C. t. tschudii* from the coastal region around Ica (Peru) representing the most probable origin of cavy domestication (Dunnum and Salazar-Bravo 2010), but see below.

Cavia was probably a much hunted wild animal. Its transition to a domesticated one has been hypothesized in sites as old as 4,500 ybp in highland valley settings in the south-central Andes. Based on their abundant representation among animal remains at sites near Bogotá in Colombia, and in southern Peru, it is likely that wild cavies were an important food source for early 12,000–7,500 ybp hunter-gatherer societies (Smith 1998). No specific morphological changes have been identified in cavy skeletons associated with domestication traceable in individual specimens and useful for zooarchaeology, a matter requiring further study.

The domestication of cavies in the central Andes may have been causally coupled with the domestication of quinoa, which would have provided a dependable food supply for them. Cavies are drawn to the warmth, protection, and food scraps of human habitation sites (Smith 1998). The earliest evidence of domestication or proto-domestication comes from the same sites in central Peru that preserve remains of domesticated camelids.

Many early reports from colonial times refer to cavies (Donkin 1989). At the beginning of the sixteenth century, cavies were kept, mostly for food, through much of the Andes, from the Sierra de Merida in Venezuela to central Chile. They were used in at least some areas in ceremonial events and in folk medicine. Cavies were introduced in pre-Columbian times to some Caribbean islands (LeFebvre and DeFrance 2014).

A report on the evolutionary history and distribution of cavies based on sequencing of 46 complete mitochondrial genomes (mitogenomes), including ancient pre-Columbian cavy specimens as well as samples of cavies in Europe and North America, and one modern sample from Puerto Rico, showed that all ancient mitogenomes were genetically distinct from the other populations studied, suggesting a different ancient *Cavia* species, tentatively identified by the authors as *C. anolaimae* (Lord et al. 2020a). The data in this study support an independent cavy domestication event in the eastern Colombian Highlands, as well as transport, starting around 600 CE, of Peruvian-derived *C. porcellus* to the Caribbean. Mitogenome sequence analysis also demonstrated a Peruvian origin of the earliest introductions of domestic cavies as exotic pets into Europe and North America and a modern reintroduction of cavies to Puerto Rico, where they were used for food.

Fig 2.11. Examples of breeds of domestic cavy, *Cavia porcellus*. From left to right and top to bottom: long-hair, hairless, Peruvian, Abyssinian, and two other breeds, not to scale.

Cavies are used mostly as food in some South American societies, and mostly as pets in some European societies. The American Cavy Breeders Association (www.acbaonline.com/breeds.html) recognizes 13 breeds of pet cavies, without including the hairless and livestock varieties. In addition, at least four breeds of cavies as livestock exist in South America, according to the Food and Agriculture Organization of the United Nations (Figure 2.11).

Cavies are the only non-primate laboratory animals that require a dietary source of vitamin C, and are then models for ascorbic acid metabolism (Burk et al. 2006).

European rabbit (*Oryctolagus cuniculus*)

Traditionally, it has been stated that rabbits were first held in captivity in monasteries during the Roman period, about 1,400 ybp, where they were used as a source of food. However, their domestication may have started much earlier, as hypothesized based on archaeological data (Irving-Pease et al. 2018). The original distribution of this species was on the Iberian Peninsula and southern France. There are cases of rabbits in Mediterranean islands 2,500 ybp (Quintana et al. 2016). Varro (116–27 BCE; in Balon 1995) wrote that Romans brought rabbits from Spain and bred them in special enclosures. Unborn rabbits excised from pregnant females were prepared as a special dish (Balon 1995). Modern domestic rabbits are much more closely related to wild rabbits from southern France than from anywhere else (Carneiro et al. 2011, 2014). Even if rabbits were bred in outdoor enclosures on the Iberian Peninsula and elsewhere, earlier populations of domesticated rabbits may have had little role if any in the genetic makeup of modern rabbits. Rabbits were transferred across Europe in medieval times and later all over the globe, where today many feral populations exist (e.g., rabbits as a major pest in Australia). To this day, humans have introduced rabbits in hundreds of islands and across continents for a variety of reasons, including sports, farming for meat or fur, as food source for other animals or bait for marine animals, to control vegetation, and as touristic attraction (Flux and Fullagar 1992).

There is considerable body size variation among different breeds of rabbits (generally about 1–6 kg), and many different varieties of coat color and length, as well as ear size and shape can be distinguished (Figure 2.12). A diversification of rabbits in

Fig 2.12. Wild rabbit, *Oryctolagus cuniculus,* and examples of breeds of the domestic rabbit. From left to right and top to bottom: wild rabbit, French Lop, New Zealand, Mini Lop, dwarf black, English Angora, Holland Lop, white rabbit, Mini Rex. Not to scale.

various sizes and colors aleady existed in the sixteenth century (Lebas et al. 1997). The current large phenotypic diversity of around 200 rabbit breeds and strains started to be developed at the end of the eighteenth century in Western Europe (Whitman 2004). Rabbits are used for many purposes, including as a source of meat, wool, fur, and therapeutic proteins, and as companion animals, as well as models of biomedical research being used to investigate a large number of genetic diseases they share with humans, such as hypertension, epilepsy, and osteoporosis (Shiomi 2009; Alves et al. 2015).

Pigeon (*Columba livia*)

The rock pigeon comprises many subspecies spread across North Africa and Eurasia (Gilbert and Shapiro 2014). Most information on its domestication comes from written accounts. The morphological distinction of the wild rock pigeon and its domesticated counterpart has not been properly investigated, and it is unlikely that zooarchaeological information will reveal critical information on early domestication phases. Thus, only DNA work may reveal which subspecies was the ancestor of domestics (or which combination of them) and clarify the tempo and mode of domestication. The oldest pictorial and written pigeon-keeping record is from ancient Mesopotamia more than 5,000 ybp (Serjeantson 2009); there is evidence for the pigeon in all the Old World centers of ancient civilization: the Middle East, Egypt,

Fig 2.13. The rock dove and examples of breeds of pigeon, *Columba livia*. From left to right and top to bottom: Homing pigeon, Fantail, Tippler, Old German Owl pigeon, rock dove, Jacobin, Tumbler, pigeon, German Modena. Not to scale.

India, and China (Larson and Fuller 2014). In a study of the genome of the rock pigeon and additional domestic and feral populations, Shapiro et al. (2013) found evidence for the origins of major breed groups in the Middle East and traced contributions from a racing breed to North American feral populations.

Pigeons are one of the world's most successful feral animals in terms of numbers. In Western Asia and North Africa, feral populations are probably as old as domesticates themselves (Gilbert and Shapiro 2014). Due to the common hybridization between domestic and free-living populations, truly wild rock pigeons might be on the verge of genotypic extinction. Some of the skeletal characters of feral pigeons resemble their domestic ancestors, for example in their disproportionately long tarsus, which may be disadvantageous under some conditions in the wild (Sol 2008). The morphology of feral pigeons is quite similar to that of racing breeds; some analyses suggest the two are also genetically similar (Stringham et al. 2012). Racing pigeons that do not return to their home and survive likely end up adding to the gene pool of the feral pigeon population.

Pigeons are very diverse in many organ systems, including the integument, skull proportions, and shape and body proportions (Darwin 1868; Price 2002b), more so than any other domesticated bird species (Gilbert and Shapiro 2014; Figure 2.13).

Chicken (*Gallus gallus*)

There is agreement that the red junglefowl (*Gallus gallus*) is the species from which chickens derive (Wang et al. 2020). Some degree of hybridization with other *Gallus* species occurred, including mainly the grey junglefowl, *Gallus sonneratii*, and also Sri Lankan junglefowl, *Gallus lafayetii* (Nishibori et al. 2005; Eriksson et al. 2008; Loog et al. 2017; Lawal et al. 2020). However, it is possible that admixture may be largely restricted to some local populations (Wang et al. 2020).

Chickens may have been domesticated by 4,500–4,000 ybp, given zooarchaeological remains beyond the distribution of wild red junglefowl found in the Indus Valley (Fuller 2006), but a more conservative estimate not older than 3,500 ybp is a more secure oldest date (Peters et al. 2016). Reports of chickens from older sites in China have been questioned. A thorough revision of bone remains and ancient DNA is needed (Peters et al. 2015). Several "domestication centers" of chickens were identified in south and southeast Asia (Tixier-Boichard et al. 2011). Perhaps the spread of rice contributed to the domestication of chickens and their dispersal.

Chickens are the most important source of protein for many societies all over the world (Crawford 1990; Al-Nasser et al. 2007; Bennett et al. 2018). The number of chicken breeds is estimated as hundreds, and varieties count in the thousands (Ekarius 2007; Figure 2.14), but there is no international commission or standard that regulates the naming and definition of breeds.

Regional poultry breeding societies publish standards on different exhibition bird traits depending on geographic origin (the American Poultry Association's Standard of Perfection) or on feather hardness, breed rarity, and size (the British system). Darwin (1868) traced back the antiquity of chicken breeds based on historical records, as did Aldrovandi around 1,600 CE (Lind 1963), and consideration of records of hybridization results in a network of relations among them (Núñez-León et al. 2019). Although determining origins and relationships of breeds is not an easy task, attempts based on genetics and morphology have found a "genetic distance" between breeds and red junglefowl using microsatellite markers (Kumar et al. 2015) and suggest that egg-type breeds (mainly European) would be the closest relative to red junglefowl (Niu et al. 2002; Moiseyeva et al. 2003). Genetic analyses and historical records do suggest establishment of current breeds by a "second wave" import of chickens to Europe from Asia in the nineteenth century; therefore, a breed framework nowadays would be the result of continuous hybridizations among European and Asian breeds (Rubin et al. 2010; Girland Flink et al. 2014). Because breed records are sometimes kept as proprietary information by commercial companies, and breeding experiments are uncommon, historical details on the relationships of breeds in the standards and other sources have been the only way to provide a framework of chicken relationships.

As part of the SYNBREED project, a wide variety of chicken population (wild type, commercial layers and broilers, indigenous village/local type, and fancy chicken breeds) samples collected worldwide were genotyped with a high-density SNP array to offer a publicly available SYNBREED chicken diversity panel (SCDP) (Malomane et al. 2019). A study analyzing the genetic diversity between and within SCDP populations showed reduced genetic diversity among fancy breeds and the highly

Fig 2.14. Red junglefowl (*Gallus gallus*) and examples of breeds of chicken. From left to right and top to bottom: red junglefowl, Brahma, Silkie, White Leghorn, Shamo, Hubbard, Polish, Araucana, Barbu d'Anvers. Not to scale.

selected commercial layer lines, while considerable genetic diversity was preserved within the wild and less selected African, South American, and some local Asian and European breeds (Malomane et al. 2019).

Remains of chickens in what is today Chile have been presented as evidence of pre-Columbian Trans-Pacific human dispersal (Storey et al. 2007), but this subject is reportedly controversial (Thomson et al. 2014) and deserves further study.

Guinea Fowl (*Numida meleagris*)

The helmeted guinea fowl *Numida meleagris* has a natural range and is found largely as a wild animal in a large part of sub-Saharan Africa (Figure 2.15). The guinea fowl was domesticated somewhere in the Maghreb and south of the Sahara and was perhaps first introduced from North Africa to Europe by the Greeks and the

Fig 2.15. Wild and domestic forms of four bird species. From left to right and top to bottom: Wild quail (*Coturnix japonica*), domestic Japanese quail, wild turkey (*Meleagris gallopavo*), domestic turkey, wild helmeted guinea fowl *(Numida meleagris)*, domestic helmeted guinea fowl, wild canary (*Serinus canaria*), domestic canary. Not to scale.

Carthaginians, and from West Africa by the Portuguese in the fifteenth century (Donkin 1989). Artistic and osteological evidence suggests it may have been domesticated less than 2,000 ybp in Sudan and Mali (Serjeantson 2009; Vignal et al. 2019). The guinea fowl is used primarily as a food resource, but in some places its capacity to give loud alarm calls alerting to danger, or its ability to hunt snakes or to consume parasites such as ticks is also part of the association with humans. It is farmed traditionally in Africa and is also bred more intensively in other countries, mainly Italy and France, and to a lesser extent in the Eastern European countries. In the former Soviet Union, the guinea fowl was bred for the production of eggs for food consumption.

A whole genome sequence study reported that domestic populations share a higher genetic similarity between them than they do with wild populations living in the same area (Vignal et al. 2019).

Domestic duck (*Anas platyrhynchos*) and geese (*Anser*)

The domestic duck is descended from the mallard, *Anas platyrhynchos*. Two different species of geese have been domesticated that can hybridize: the greylag goose (*Anser anser*) in Europe, northern Africa, and western Asia, and the swan goose (*Anser cygnoides*) in eastern Asia (Larson and Fuller 2014). The areas of origin of domestication may be approximately the same for duck and for geese. Written evidence suggests that duck and goose were present as domesticates in Central China after 500 BCE (Luff 2000). Populations of these species may have been attracted to rice paddies or grazed the stubble of harvested fields (Larson and Fuller 2014).

Whole genome sequencing and comparison of 78 individual ducks, including two wild and seven domesticated populations, revealed a complex history of domestication, with early selection for separate meat and egg lineages and strong positive selection of genes that affect brain and neuronal development and white plumage resulting from selection at the melanogenesis-associated transcription factor locus (Zhang et al. 2018). There is a wide range of colors and sizes in duck breeds, found in regional forms in North America, Europe, and across Asia (Figure 2.16).

Geese were the principal domestic bird of ancient Egypt, and geese appeared in ancient Greece surely by 700 BCE (Mannermaa 2014), perhaps earlier. India probably obtained the goose from Mesopotamia; then it spread to Burma, Thailand, and the Malaysian peninsula (Donkin 1989).

Based on mitochondrial genetic variation analyses of Russian domestic geese specimens excavated from 15 sites and dating to the ninth to eighteenth centuries have served to identify three genetic clades (Honka et al. 2018): the main domestic D-haplogroup, F-haplogroup, and taiga bean goose haplotypes. More generally, it is clear that, derived from the two ancestral geese species, around 181 breeds have been developed (Kozák 2019; Figure 2.17). Domestic goose, unlike its ancestors, is a nonmigratory bird. In contrast, concerning morphology, the goose shows less variation compared with other domestic species (Kozák 2019).

European goose domestication was studied via analyses of genomic data geese, confirming derivation from the swan goose *Anser cygnoides* (Heikkinen et al. 2020). The study examined the great degree of hybridization in geese and reported

Fig 2.16. Mallard, *Anas platyrhynchos*, and domestic duck breeds. From left to right and top to bottom: Mallard, crested duck, white domestic duck, duck, Indian runner, duck. Not to scale.

Fig 2.17. Wild and domestic geese. From left to right and top to bottom: greylag goose (*Anser anser*), swan goose (*Anser cygnoides*), and diverse domestic geese. Not to scale.

how many modern European breeds share more than 10% of ancestry with Chinese domestic geese, which are derived from the swan goose *Anser cygnoides.*

Turkey (*Meleagris gallopavo*)

The turkey (Fig. 2.15) is the only animal species first domesticated north of Panama, probably in west-central Mexico, and by the beginning of the sixteenth century it was important in almost all agricultural areas of Mexico and had spread to what is now the southwest of the United States (Donkin 1989). The domesticated turkey descends from a wild species with which it commonly interbreeds, the wild *M. gallopavo,* which was widespread in Mexico, Central America, and a region of the Antilles. The subspecies *M. gallopavo meriami* was reportedly domesticated by the Pueblo people in the southeast of the United States (Guillaume 2010).

In some areas of pre-Columbian North America, turkeys were used for food (meat and eggs), as well as for decoration and clothing (feathers) and musical instruments and tools (long bones) (Thornton 2016). After it was encountered by the Spaniards, the turkey spread within Europe in a few decades. In New England in the United States, they hybridized rapidly with local wild turkeys of the ancestral subspecies, resulting in larger animals with darker plumage. Afterward, the turkey became the Thanksgiving poultry in the United States, and a luxury poultry reserved for holy days in Europe (Guillaume 2010). It is the only American domestic animal subject to industrial breeding/farming comparable to that of chickens. There has been strong selection for "improvement" traits; the size of certain groups increased spectacularly, with adult reproductive males sometimes exceeding 40 kg today, and reproduction is done by artificial insemination (Stotts and Darrow 1955; Guillaume 2010).

In the archaeological site of El Mirador in Petén, Guatemala (300 BCE–100 CE), the earliest evidence of *M. gallopavo* found outside its natural geographic range is documented for Mesoamerican turkey rearing or domestication, as suggested by the presence of male, female, and subadult turkeys, and reduced flight morphology (Thornton et al. 2012).

The ocellated turkey (*Agriocharis ocellata* or *Meleagris ocellata*) was indigenous in the Mayan area and is considerably different in size and coloration. Although more resistant to domestication, it was among the turkeys kept in pens by the Aztecs, as described by the Spaniards. The ocellated turkey has been suggested by some researchers to be a completely separate species. Before the Spaniards arrived, both wild and ocellated turkeys were brought together into the Maya region by the extensive trade network.

Isotope analysis of animal remains from an important Maya urban center (Ceibal, Guatemala) across almost 2,000 years (1,000 BCE–950 CE), showed that domesticated and possibly even wild animals, including turkey, were raised in this region and deposited in the ceremonial core (Sharpe et al. 2018).

Japanese quail (*Coturnix coturnix*)

Japanese quails (*Coturnix japonica*; Fig. 2.15) have been domesticated in Japan since the first decade of the twentieth century. Its wild ancestor is a migratory bird wintering in southern China as well as on the coasts of Japan, China, and Korea, and breeding further north. The quail was first domesticated as a songbird some six centuries ago; later it was bred and raised in small volumes. The domestic stock nearly disappeared during the Second World War. The Japanese quail was imported to the West for the restocking of hunts, production of eggs, and as a gourmet meat source. It is also a laboratory animal in genetic research, offering the advantage of fast reproduction (Guillaume 2010).

Canary (*Serinus canaria*)

The wild form of the domesticated canary is endemic to the Canary Islands (Birkhead 2003). Artificial selection has focused not only on physical appearance, including color and ornaments, but also on richness of song. There are several canary breeds, consisting of phenotypically divergent populations, and varieties characterized by a special singing behavior (Fig. 2.15).

Classical genetic crosses and cross-fostering experiments showed that differences in song learning between canary breeds are inherited; genetic predisposition biases learning in a particular manner (Wright et al. 2004; Mundinger 2010).

CHAPTER 3

The genetics of domestication

Domesticated animals were central subjects in many of the first studies leading to fundamental discoveries in the field of genetics. Breeding experiments with albino and gray rats undertaken by the German medical doctor Hugo Crampe around 1880 were recorded and later analyzed in terms of Mendelian genetics by Doncaster (1906; Castle 1947). The seminal work of William Bateson on Mendelian inheritance in animals at the beginning of the twentieth century, conducted largely by many pioneering female scientists, used the chicken comb as the subject of study (Richmond 2001; Bateson 2002). In the 1910s and 1920s, many detailed studies of genetics were conducted on pigeons (e.g., Horlacher 1930). In those and more recent decades, the founders of genetics, including Thomas Hunt Morgan (1866–1945) and Sewall Wright (1889–1988), studied domesticated animals. In Jena at the beginning of the twentieth century, the successor of Ernst Haeckel devised breeding experiments to study the heredity of hair and skin features of dogs (Plate 1929; Gaspar 1930), and the remaining collections have been used to decipher the developmental genetic bases of dental traits of hairless dogs (Kupczik et al. 2017; Figure 3.1).

The growth of genetics and later genomics led to major advances (Andersson and Georges 2004), including those related to systematic animal breeding led by knowledge of population genetics (Hill 2014). The first mammal cloned from an adult cell was a sheep, "Dolly" (Campbell et al. 1996), with pigs, horses, and bulls also cloned later. The main legacy of this work was to advance the field of stem cell research (Weintraub 2016). Different breeds of dogs were among the first subjects of comprehensive studies of the genome (Lindblad-Toh et al. 2005), at a time when dogs had already been used successfully in gene therapy studies (reviewed in Ostrander and Wayne 2005). Currently, some of the most fundamental studies on the genetic basis of phenotypic divergence are being conducted on a variety of domesticated species, taking advantage of their high phenotypic variation, as described below. A main goal of this long history of studies has been to understand how genetic variation leads to phenotypic changes.

Studies on the relationship between genotype and developmental phenotype and physiology are far from simple, as it is expected that complex interactions and

Fig 3.1. Taxidermy specimens of hairless dogs studied during the early twentieth century in Jena, Germany, to decipher genetic mechanisms involved in specific traits. This pedigree historical collection of dogs has been reassessed, providing new insights into integumentary and dental traits and their genetic bases.

hierarchies of genes are at play (Pavlicev et al. 2008; Bradley et al. 2011). Several constructive developmental processes lead rather than follow morphological transformation associated with genetic change (chapter 5). Here I concentrate on other fundamental aspects of genetics. There is a vast literature on the subject, of which I present selected examples.

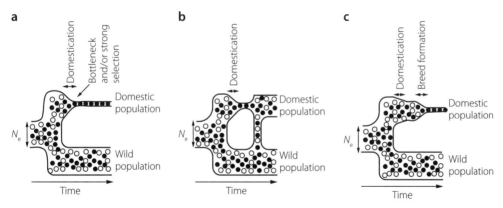

Fig 3.2. Changes in overall genetic diversity in domestic populations. Population size increases and decreases due to various events over time, such as bottlenecks, affecting genetic diversity. N_e is the effective population size, directly related to genetic diversity; the number of breeding individuals in an idealized population, different from the census population size. Black and white dots represent alleles at hypothetical locus in the genome. (a) Basic model involving a founder effect and/or strong selection. (b) The effect of gene flow from a wild to a domestic population following a bottleneck. We can see in this schematic graphic how genetic diversity can increase in the domestic population as a result of gene flow. (c) Model involving a mild domestication bottleneck and a strong bottleneck during breed formation.

Three fundamental mechanisms can occur in the domestication process (Frantz and Larson 2019; Figure 3.2): (1) founder effect, a kind of bottleneck in which only a small portion of individuals from the wild population are domesticated (Larson and Burger 2013); (2) intensive artificial selection for specific traits, as in the selection for tameness, which is hypothesized as fundamental for domestication; (3) reproductive isolation between domestic and wild populations, resulting in trait differentiation, as in becoming tame (Marshall et al. 2014). Hybridization can occur, though, which can lead to the acquisition of new traits, as described below.

Hybridization

Hybridization has traditionally been expected to act against speciation by homogenizing the genomes of the hybridizing lineages (Dobzhansky 1951). In contrast, recent evidence has shown that hybridization, common in rapidly speciating clades, can be important in adaptive radiations, as in the case of cichlids in African lakes (Salzburger 2018).

Hybridization in the wild may lead to new species directly through hybrid speciation or indirectly through adaptive introgression (Abbott et al. 2013). It can be mediated by humans using selective breeding of domesticated or domesticated-wild animals. The importance of hybridization in the origin of new species has been recognized, underscoring this mechanism of innovation in genetic and morphological diversity and presenting taxonomic challenges due to the blurring of boundaries among species (Kindler et al. 2017; Salzburger 2018).

Hybridization can be an abundant source of genetic variation and can drive behavioral or ecological adaptations on time scales far shorter than those required for variation to arise through mutation alone (Abbott et al. 2013). The animals that result from hybridization can have beneficial combinations of genetic makeup that have already been positively "tested" by natural selection, in contrast to mutations (Sætre 2013). Hybridization has been shown to lead to hybrid phenotypes that lie outside the range of parental phenotypes (transgressive phenotypes), which can "enlarge the working surface of natural selection" (Stelkens and Seehausen 2009). This may allow adaptation to novel environments and increased fitness (Grant et al. 2003; Tompkins et al. 2006).

Hybridization is a fundamental component of domestication often observed between the domesticated and the wild form, and also common in the creation of new breeds (Lord et al. 2016). It can result in apparently saltatory changes in morphology, perhaps altering the timing of the developmental program. In dogs, hybridization can result in changes in behavior as it relates to predation, courtship, territory, social hierarchy, and pack formation, the latter tending to fragment or even disappear (Coppinger et al. 1987; Lord et al. 2013). Introgression can serve to explain specific and visible features of domestics. For example, it has been shown that the yellow skin phenotype in domestic chickens, unlike the red junglefowl, is the result of introgression of a grey junglefowl chromosomal fragment into the former (Eriksson et al. 2008). Introgression explains some dark patterns in the pigmentation of domestic pigeons, as they result from breeding with a different species, the African speckled pigeon, *Columba guinea* (Vickrey et al. 2018). There are two genomic regions introgressed from red siskins (*Spinus cucullata*) into canaries that are required for red coloration of the latter (Lopes et al. 2016b).

There is a hybrid phenotype of canaries that show carotenoid-based sexual dichromatism, a difference in coloration between males and females (Gazda et al. 2020). These mosaic canaries are the offspring of the sexually dichromatic red siskin and monochromatic canaries. This phenotype is associated with the gene that encodes an enzyme, such that the dichromatism of *mosaic* canaries results from differential carotenoid degradation in the integument. Hybridization leads then to this alternative mechanism to the more usual sex-specific variation in physiological functions such as pigment uptake or transport (Gazda et al. 2020).

In the past and by necessity driven by costs and technology, studies of population genetics of domestication and domesticated populations concentrated only on mitochondrial genes. Mitochondrial DNA is a single, non-recombining, maternally transmitted DNA. The nuclear genome, in contrast, provides more resolution—its molecular sequencing has revealed ubiquitous gene flow among the wild forms and diverse domesticated populations of the same species and among just the latter (Larson and Fuller 2014). This advantage of the nuclear genome has been shown for horses (Schubert et al. 2014; Der Sarkissian et al. 2015), pigs (Frantz et al. 2015), and cattle (Park et al. 2015).

An old network of dog breeds

Buffon (1707–1788) was an influential pre-evolutionary zoologist who proposed a theory of transmutation of domesticated animals in response to the changed human environment. He observed that when a dog breed is transported to an area with different climate, it changes over time into a different type of dog. Lamarck (1744–1829), an intellectual successor of Buffon, took this idea further, extending it from domesticated animals to species in the wild. Buffon's multivolume *Histoire naturelle, générale et particulière* includes one of the first and most ambitious compendia of varieties of many domesticated mammals. In a portion of this work he published with L. J. M. Daubenton, Buffon (1755) presented a network of genealogical relationships among breeds of dogs. Although now outdated in many details, as in the interpretation of the position of many breeds, this work represents one of the oldest if not the oldest graphic representation of horizontal connection of genealogical relations (Morrison 2012; Figure 3.3). Buffon used illustrations of networks for visualization of genealogy, with solid lines representing vertical descent (parent to offspring) and dashed lines for horizontal cross-breeding.

Buffon recognized 30 "fixed varieties" and 17 "variable races," grouped into four main geographic and functional classes, predating the approach currently used by the Fédération Cynologique Internationale (World Canine Organization) to classify the approximately 350 breeds of dogs. Buffon recognized about 10 "progenitor breeds" for each class, originally developed to fill different

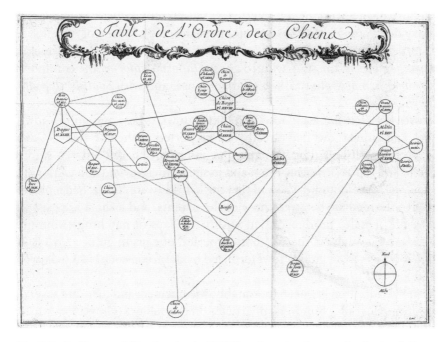

Fig 3.3. Buffon and Daubenton's (1755) network of genealogical relationships among breeds of dogs. This is perhaps the oldest graphic representation of horizontal connection of genealogical relations.

roles (e.g., hunting, herding, retrieving), with today's breeds deriving from them. Buffon considered the wolf the ancestor of modern dogs or at least very closely related, although his attempts to cross-breed dogs with wolves or dogs with foxes did not succeed. William Smellie's English translation of Buffon's work reported a successful crossing of a female dog and a male wolf.

An example of more recent but still historically significant frontrunners of phylogenetic networks in dogs is the work of Max Emil Friedrich von Stephanitz (1921) on the German Shepherd, for which he presented a network of related breeds.

Avian and mammalian hybrids among domesticates

Crossings between domestic and wild forms are very common. As a result, "pure" wild forms of domesticated species are either extinct or rare. An example is the cross that occurred between domestic (*Anas platyrhynchos*) and wild Muscovy duck (*Cairina moschata*) before the number and range of the latter were reduced. Crosses across some domesticated species are also common. Hybrids of the Muscovy and the common domestic duck presumably date from the earliest contact between Europe and the Americas, the offspring being sterile. Systematic breeding for meat of these species has been reportedly practiced, particularly in France, resulting in the *canard mulet* or *canard mulard*, raised for its liver and meat since the second half of the nineteenth century (Donkin 1989). This crossing has also been a widespread practice in Taiwan for centuries (Clayton 1984a). Muscovy duck crosses with several other species, the progeny of which are likewise sterile (Donkin 1989): *Sarkidiornis melanotos,* the knob-billed duck, *Plectropterus gambensis,* the spur-winged goose, *Tadorna tadorna,* the common shelduck, *Alopochen aegyptiaca,* the Egyptian goose, and with *Spatula clypeata,* the northern shoveler.

Crosses between domestic fowl and guinea fowl, turkey and guinea fowl, and peacock and guinea fowl had already been reported in the late 1800s (Cornevin 1895). Hybrids between domestic fowl and guinea fowl have been well studied, in particular in France (Petitjean 1969) and in the former Soviet Union (Mongin and Plozeau 1984). These studies showed that the hybrids were sterile, had a phenotype generally intermediate between those of the parents, and a short life expectancy, and the few females produced died within 48 days of hatching (Mongin and Plozeau 1984). Whereas the incubation period of eggs in guinea fowl is 26–27 days, that of eggs from guinea fowl fertilized by domestic cock is 23.6 days (Mongin and Plozeau 1984).

Avian hybridization has been thoroughly documented, and there are several syntheses of the (literally) hundreds of articles on the subject (Gray 1958; McCarthy 2006). They present and discuss results for each cross, addressing viability and fertility of hybrids, their appearance, and whether contacts in nature occur.

The mule is the classic interspecific mammalian hybrid, the result of crossing a donkey male and a horse female. Hinny is the cross between a female donkey and a male horse. The mule tends to have longer ears, whereas the hinny has a thicker mane and stronger legs. The differences between the mule and the hinny illustrate how the expression of a gene is determined by its origin rather than by its DNA sequence—genomic imprinting. This phenomenon has been traditionally discounted because of the dominant influence exerted by the Mendelian rules of inheritance (Hunter 2007). The occasional occurrence of fertile mules is recorded in diverse reports (e.g., Rong et al. 1985).

Crosses between zebra and both donkey and horse, including ponies, are known. Among bovids, there are crosses between a domestic cow/bull and a yak. Both the American and the European bison cross with domestic cows, and there is also a hybrid between a bison and a yak.

Domestication and species delimitation

Extreme variation can occur as a consequence of intensive selective breeding. For example, modern broiler chickens are clearly different from their ancestors in their skeletal morphology, pathology, bone geochemistry, and genetics as a result of human-directed changes in breeding. Diet and farming practices have led to at least a doubling in body size from the late medieval period to the present (Bennett et al. 2018). Should a new species *Gallus "broilensis"* be proposed for this generic kind of chicken? Certainly not, as no new species is erected when genetic technology is applied to engineer important changes in any organism either, including works involving CRISPR manipulations, in which genetic material from one species is inserted into another, equivalent to what can occur in hybridization (Petersen 2017). There is much useful debate on species definition (Zachos 2016), but even if a final word on this subject is never possible, most people would agree that genetic manipulation alone does not lead to the creation of new species.

In the early nineteenth century, the Dutch naturalist Caenraad Temnick suggested that what he recognized as the six main domesticated chicken types should have distinct species names. Linnaeus named six fancy breeds of domestic pigeon as separate species—we now know they are all descendants of the rock dove, all domestic forms of *Columba livia*. These taxonomic views changed with the acceptance and understanding of evolution, variation, and its causes (van Grouw 2018).

Assortative mating can be one step toward speciation, since in this way populations become isolated and start to differentiate. Johnston and Johnson (1989) reported on two sorts of nonrandom mating that influenced reproductive output in a population of feral pigeons from eastern Kansas: size-based assortative mating, in which individuals of like sizes paired, and plumage-based disassortative mating, in which individuals of unlike plumages paired. Size-based mating was interpreted as based on perception of size or a size-correlated variable, such as social dominance rank, by both sexes, while plumage-based pairing was based on perception of unlike plumage patterns, probably by females. The origin of nonrandom mating based on male plumage pattern is not a feature characteristic of the wild form and must have evolved *de novo*: plumage variation of wild rock doves, the ancestral form, was probably restricted (Johnston and Johnson 1989).

Potential ethological isolating mechanisms and assortative mating

Since speciation is often a gradual process, there is a continuum along which it is impossible to objectively determine when there is a new species (Feder et al. 2012); there are no universally accepted criteria to achieve such objectivity.

Mate preferences are affected by learning at a young age, the process called sexual imprinting, in many cases using a parent as the model. The extent to which sexual imprinting can influence mate choice, and with that speciation, has raised much interest since the classic studies of Konrad Lorenz (1935) of ducks and other birds (Bateson 1978). How imprinting influences evolution of exaggerated parental traits is one of the many aspects for which domesticated species offer potential model systems of investigation, as in a study of the influence on mate choice of beak color in zebra finches (ten Cate et al. 2006).

Potential ethological isolating mechanisms and assortative mating have been investigated for decades in chickens (Lill and Wood-Gush 1965), domestic pigeons (Sambraus and Sander 1980), and Japanese quail (Bateson 1978). Lill and Wood-Gush (1965) studied Brown and White Leghorn chickens and recorded different kinds of preferences and discrimination by both females and males. This included assortative mating within a single line, with preferences related to appearance and not to male courtship behavior.

The behavioral experiment conducted by Tiemann and Rehkämper (2012) on chickens is particularly informative. They investigated the sexual behavior of the White Crested Polish breed toward Red Leghorns and Lohmann Selected Leghorns by conducting mate choice experiments and measuring fertilization and hatching rates of mixed-breeding groups. They found breed-specific preferences. Furthermore, hatching success of purebred offspring was twice as high as that of hybrids. The authors interpreted these results as suggesting that the mechanism of sexual selection is still present in domestic chicken breeds, and that this can lead to reproductive isolation and thus speciation. What makes the experiment particularly relevant is that the chickens were all raised in similar conditions and shared the same environmental experiences, including social contacts, so that the choice of mate was not learned.

This chicken example is quite relevant to understanding mechanisms of isolation that can result from domestication and how some breeds may behave differently from others. The reported isolation of the specific population of White Crested Polish chickens may not apply to other breeds, and not all breeds would behave like the White Crested Polish breeds toward the red junglefowl. This lack of universality extends beyond chicken breeds to domesticated animals in general: reproductive isolation may occur in some cases and not in others, and this isolation in most cases may be one of degree (probability of mating or of viable and fertile offspring).

Mutations, developmental genetics, and domestication

Mutations are one of the raw materials for evolution, and their effects have been intensively studied in domesticated animals. Mutations can arise as an effect of oxidative mutagens or from ionizing radiation exposure resulting in DNA damage, or perhaps more frequently due to errors during DNA replication that are not repaired

Fig 3.4. Head crests in pigeons. Variation is shown ranging from small and simple, as a peak of feathers, or elaborate as a hood enveloping the head: (a) Indian fantail, (b) Old German owl, (c) Old Dutch capuchin, (d) Jacobin.

(Bromham 2009; Scally 2016). Only mutations in germline cells can be passed on to the next generation, and some can become ultimately fixed in a species.

A thoroughly studied mutation leading to morphological change in a domesticated bird is related to head crests in pigeons. In the many breeds with crests, neck feathers grow toward the top of the head instead of down the neck, a phenomenon referred to as "growth polarity." There are many kinds of crests (Figure 3.4), and genetic experiments have shown that the head crest segregates as a simple Mendelian recessive trait (Sell 1994). A comparative genomics study identified a single mutation in *EphB2* (a gene encoding Ephrin receptor B2) located in the catalytic pocket of the intracellular kinase domain that acts on the embryo and causes the growth polarity leading to the crest in pigeons, a phenotype manifesting only later in juvenile pigeons (Shapiro et al. 2013). The study also showed that the genomic region containing the mutation is conserved in all crested pigeons, indicating that the crest evolved just once and spread throughout pigeon breeds.

A subsequent study identified another *EphB2* mutation in the same kinase domain, significantly associated with the crested morph phenotype of ringneck doves (Vickrey et al. 2015). Both *EphB2* crest mutations in pigeons and doves replace conserved residues on the surface of the kinase domain, very close to the site of adenosine triphosphate (ATP) binding. These mutations are thus predicted to negatively affect ATP-binding function of the kinase, an enzyme that regulates the biological activity of other proteins downstream of this pathway by phosphorylation of specific amino acids. Together, these results demonstrate remarkable evolutionary convergence in domesticated pigeons and doves.

The crest phenotype in chickens has also been extensively studied, a case of ectopic expression with well-studied genetic basis (Wang et al. 2012).

Among other examples, two other studies on foot feathers (Fig. 2.13) are notable, both involving two loci. There is striking convergence in the molecular basis of the transformation of scaled to feathered skin on the feet of pigeons (Domyan et al. 2016) and of chickens (Li et al. 2020). In both cases a reduction of expression in the *PITX1* gene occurs, along with ectopic (misplaced, in an extraneous location) expression of *TBX5* in hind limb buds. This results in the hind limb acquiring an identity or mode of development that is forelimb-like.

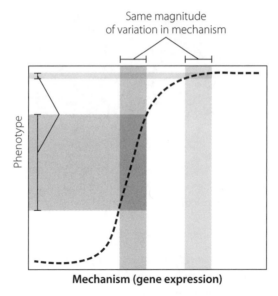

Fig 3.5. The nonlinear genotype–phenotype map. The amount of a particular developmental process, including cell survival, proliferation, and molecular signaling leads to
a mean phenotype. This relationship is nonlinear. The lighter gray vertical bar represents
"wild-type" gene expression, and the darker vertical bar the "mutant" gene expression. Different amounts of phenotypic variation can be generated with the same
amount of variation in the generative mechanism. The light gray horizontal bar (top,
thin) represents the "wild-type" shape variation, the canalized region where variance
is buffered. The dark grey horizontal bar (thick) represents the "mutant" shape variation, the area where canalization is lost. In the wild type, canalization results in a wide
range of variation in expression translating to a narrow range of variation in phenotypes (thin line on top).

Given that the same genes with roles in development are often expressed at different places and times during development, mutations resulting in changes in the
protein sequence are expected to have detrimental effects on many organs. On the
other hand, a change in the temporal or spatial expression of a gene may affect only
a subset of organs and lead to viable morphological variation. It is thus important
to understand the mutational robustness of organisms, as this can determine individual, or even species, survival (Félix and Wagner 2008). Experimental studies in
mice show how development can explain robustness in the face of different genotypes (Figure 3.5). By manipulating the gene dosage of the signaling molecule Fgf8,
a critical regulator of vertebrate development, Green et al. (2017) found a nonlinear
relationship to phenotypic variation: different amounts of phenotypic variation can
be generated with the same amount of genetic variation. This is an example of the way
genetic variation relates to specific phenotypic variation, the genotype–phenotype
map or GPM.

Indeed, developmental evolutionary biology has found the GPM to be highly
nonlinear and complex (Milocco and Salazar-Ciudad 2020). In quantitative genetics, linear approximations of the GPM are typically used to predict the response to
selection in a population, using the multivariate breeder's equation (Lande and Ar-

nold 1983). Modeling has shown that these predictions that assume linearity often fail when considering a realistic depiction of developmental processes and the complex, nonlinear GPM (Milocco and Salazar-Ciudad 2020).

Studies on the genetics of diverse traits of domesticated species have served to address the long-standing question regarding the relative importance of coding and noncoding mutations for explaining phenotypic variation as well as disease. Hundreds of phenotype-genotype pairs have been studied, in most cases with the specific underlying mutation(s) known (Andersson 2016). Some of these genes are associated with multiple phenotypes. Coding mutations, which are easier to detect, usually result in more striking phenotypic changes, and therefore their effects are easier to interpret. Nevertheless, about 50% of the mutations known for domesticated mammals (Andersson 2016) are noncoding, which strongly suggests that these are significantly more abundant than coding mutations underlying the phenotypic diversity of domestic animals. Examples of regulatory changes controlling phenotypes in birds include those related to foot feathering in pigeons (Boer et al. 2019), red coloration in canaries (Lopes et al. 2016b), and yellow skin in chickens (Eriksson et al. 2008).

Structural genetic changes, in which fragments of DNA are rearranged within one chromosome or transferred between two or more chromosomes, can lead to altered regulation of one or several genes. This rearrangment and subsequent alteration can produce phenotypic changes. Among the numerous phenotypic traits explained by structural genetic changes, these are some examples:

- Dark brown color in chickens (Gunnarsson et al. 2011), caused by deletion of a regulatory element.
- Greying with age in horses, caused by duplication of a regulatory element (Pielberg et al. 2002).
- Rose-comb in chickens, caused by the translocation of a gene to another position, where it is influenced by another constellation of regulatory elements (Imsland et al. 2012).
- Pea-comb in chickens, caused by a sequence duplication that changes the regulation of *SOX5* expression, affecting the differentiation of cells essential for comb and wattle development (Wright et al. 2009).
- Lack of horns in goats, caused by a mutation consisting of the deletion of a DNA element that affects the transcription of at least two genes: *PISRT1* and *FOXL2* (Pailhoux et al. 2001).
- Dwarfism in rabbits, caused by a large effect, loss-of-function (LOF) mutation in the *HMGA2* gene combined with polygenic selection (Carneiro et al. 2017).

Several genetic background studies of many phenotypes of domesticated forms have revealed an accumulation of multiple consecutive causal mutations in the same gene, with five examples documented by Andersson's group, as follows: white-spotting in dogs (Baranowska et al. 2014), dominant white color in pigs (Rubin et al. 2010), black spotting in pigs (Kijas et al. 2001), smoky color in chickens (Kerje et al. 2004), and the *Rose-comb2* allele in chickens (Imsland et al. 2012). These multiple mutations have probably accumulated in a process over thousands of years for some species, given the antiquity of their domestication.

Locomotion in horses

Documentation of locomotion in domestic animals has existed since the late nineteenth century, after the pioneering work of Eadweard Muybridge, who used multiple cameras to capture motion in stop-motion photographs, most famously perhaps in his photos of horse locomotion taken between 1877 and 1878 (Brookman et al. 2010). Since then, functional anatomical and experimental studies of locomotion in domestic animals abound, including horses (Killbourne and Hoffman 2013). The genetic bases of the different kinds of gaits have been deciphered (Andersson 2016).

At different speeds, horses perform different gaits: walk, trot, canter, gallop (Figure 3.6). Galloping is the natural gait at high speed of horses. Trot is a two-beat diagonal gait, pace is a two-beat lateral gait.

A high incidence of the "Gait keeper" mutation occurs in horses used for harness racing (trot or pace). This condition promotes pace and inhibits the transition from trot/pace to gallop, allowing horses to trot and pace at very high speed, a positive effect on racing performance (Andersson et al. 2012). The "Gait keeper" mutation has been useful as it gives horses the ability to offer a smooth ride—this was particularly important when they were the only means of long-distance transport in some parts of the world, so horses would be trotting at high speed as they pulled small carriages. Ancient sculptures show horses with alternate gaits, suggesting that the "Gait keeper" mutation arose more than 2,000 ybp at the very least (Andersson 2016).

Icelandic horses can perform two additional gaits, "töit" and the faster "pace," the latter as fast as trot, but slower than gallop. Töit is a four-beat ambling gait in which the horse always has one foot on the ground. Half the population of Icelandic horses is classified as four-gaited (walk, trot, gallop, and töit) and the other half is five-gaited, with pace as the original gait. It is also known that the ability to pace has a high heritability (Albertsdóttir et al. 2011).

Fig 3.6. Locomotion in horses: examples of kinds of gait. (a) normal walk; (b) medium trot; (c) gliding phases, (d) flying pass, and (e) slow or half pass of running trot; (f) middle gallop; (g) running gallop.

Genetic mechanisms, architecture, and diversity behind domesticated animal traits

The "genetic architecture" of a trait refers to the number of genes or loci involved, the relative effect strength, and the genome location. Many major genes identified in chickens, dogs, cattle, sheep, horses, and pigs act in a monogenic fashion: they explain almost all the variation of the traits they regulate (reviewed in Wright 2015). For example, the melanocortin 1 receptor (*MC1R*) gene product expressed primarily in melanocytes and melanoma cells (García-Borrón et al. 2005) and involved in black/red pigment switching, plays an important role in feather, skin, and coat coloring. In wild and domestic populations of pigs from Europe and Asia, *MC1R* variation is determined by 13 alleles (Fang et al. 2009) and has been used to detect wild-domestic hybrids (Koutsogiannouli et al. 2010).

Seemingly unrelated traits can appear together in the process of domestication. The genetic mechanisms that result in such coupling can be of different kinds (Agnvall et al. 2017; Figure 3.7): pleiotropy, linkage, epistasis, and epigenetics. In pleiotropy, a single gene or mutation affects the phenotype of two traits simultaneously. In linkage, two genes are inherited as a unit and linked due to close physical localization. In epistasis, the effects of a gene on a trait depend on another gene. In epigenetics, a gene is modified (for example, through DNA-methylation) so that the expression of this gene is affected—and in some cases others are also affected, a form of pleiotropy. Many domestication traits are polygenic. Recombination, linkage, and epistasis can contribute to the genetic bases of many traits of domesticated animals.

Numerous examples of the genetic mechanisms described above exist. In chickens, genes associated with melanin synthesis (recessive white and albino alleles of the *C* locus), exert a regulatory influence on diverse physiological traits, as reflected in body weight and other features (Pardue et al. 1985). Another example is the congenital deafness of cats being correlated with blue eyes and absence of pigmentation in the skin. This set of features is associated with a dominant white gene that suppresses the deposition of melanin in the skin and the iris (Noden and La Hunta 1985). Deafness in cats and dogs (e.g., dalmatians) is associated with pigmentation, whereby the modes and causes of deafness in domestic animals are diverse (Strain 2015).

A prominent case of pleiotropy is that associated with tameness (Agnvall et al. 2018), as this is reportedly tied to other phenotypes, including those of the "domestication syndrome," a set of traits first recognized by Darwin but now known not to be universal among the first forms that resulted from domestication (Sánchez-Villagra et al. 2016; Lord et al. 2020b). An experiment on domesticating silver foxes reportedly showed how selection for tameness resulted in several other traits, including changes in pigmentation and floppy ears (Trut 1999). According to Wilkins (2019), another group of foxes was created that expressed aggressive behavior, which could be associated with fear, and these foxes also showed changes in pigmentation. Furthermore, the occurrence of some of these traits in the farmed foxes in which the experiment was conducted indicates that the case for syndromes has been overstated (Lord et al. 2020b), or rather, that the "experiment" with foxes had already started with their farming and captivity, predating the study by Belyaev and colleagues (Statham et al. 2011), described in chapter 8.

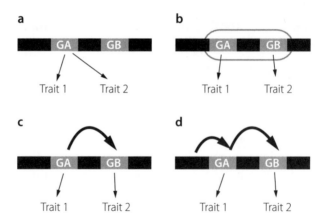

Fig 3.7. Schematic representation of a chromosomal portion with two separate genes, GA and GB. Seemingly unrelated traits can appear together in the process of domestication, as a result of different genetic mechanisms: (a) Pleiotropy. (b) Linkage. (c) Epistasis. (d) Epigenetics.

An example of linkage is that reported for a small region on chicken chromosome 1 which affects traits ranging from growth to reproduction and fear-related behavior (Schütz et al. 2004).

When a gene modifies the effect of one or several other distantly situated genes there is epistasis. For example, early growth in chickens (prior to 46 days of age) is affected by several gene pairs in an epistatic manner. This was estimated by a quantitative trait loci or QTL study involving intercrosses between red junglefowl and White Leghorn chickens (Carlborg et al. 2003). In this case, selection for a mutation in one gene caused effects on phenotypes controlled by another gene.

In domestication, the fact that in many cases strains have been selected over many generations for different states or values of a phenotypic trait makes them particularly suitable for quantitative trait loci (QTL) mapping. These strains can be crossed to create various combinations of their genotypes. After two generations of interbreeding, researchers search in F2 individuals for genetic markers found in the parental strains, testing for statistically significant associations with a quantitative trait. Once scientists identify these quantitative loci, they can then try to identify the genes in these regions that are influencing the trait in question (Zimmer and Emlen 2013).

Epigenetic modifications occurring in response to environmental changes affect gene expression without affecting the DNA sequence and can be drivers of phenotypic changes associated with domestication (Jensen 2014). The occurrence and relative importance of these heritable modifications is currently under study (Wright 2015), as in the epigenetic mechanisms associated with differences in the brain of domesticated chickens compared with the red junglefowl (Nätt et al. 2012) and in tamed forms of the latter (Bélteky et al. 2018).

Using QTL mapping approaches to compare wild and domesticated animal species, researchers have found multiple domestication traits clustered together; however, many identified mutations do not have pleiotropic effects. Instead, gene clusters are believed to exist with multiple interacting loci influencing expression, known as "linkage clusters" (Wright 2015).

The domestication syndrome and correlation of traits

Despite the different pathways leading to domestication in different time periods and parts of the world, similar phenotypic alterations in diverse structures and organs have been reported in mammalian species belonging to distantly related groups, a pattern known as the domestication syndrome (Darwin 1868). This syndrome includes changes in size and body proportions, pigmentation, reproduction, and social behavior. Other morphological changes have also been associated with domestication in mammals, including a reduction in brain size, changes in hair (wavy or curly), shorter and coiled tails, and floppy ears (e.g., Dobney and Larson 2006). To avoid the implication of illness associated with the term "syndrome," some authors (e.g., Leach 2003) prefer the term "domesticated phenotype." I follow the standard use here and refer to it as syndrome. The term "domestication syndrome" has been widely used to describe a similar phenomenon in crops and other cultivated plants (Hammer 1984).

A potential explanation for this pattern is that humans have selected repeatedly, independently in each species, for the same features. In this view, similar mutations may appear in the different species, with humans selecting for animals expressing phenotypically such mutations. Dmitry Belyaev (1979) considered this option unlikely, and thus set out to test it in the well-known silver fox experiment described in chapter 8 (Trut 1999). Belyaev hypothesized that just selection for tameness—a reduced fear of humans—is the central underlying factor driving the acquisition of these traits.

In fact, a pattern of correlated changes was reported for the silver fox, a landmark study (Trut 1999). The perceived features of dogs have also been largely influential in perception of the syndrome (Lord et al. 2020b). In reality, the pattern of these canids is not universal across mammals. Furthermore, segregation of morphological (Hansen Wheat et al. 2020) and behavioral (Hansen Wheat et al. 2019) traits has been reported in dogs and is otherwise common in mammals (Andersson 2016; Sánchez-Villagra et al. 2016).

Notwithstanding these important caveats, there are features sufficiently generic among domesticated mammals (changes in pigmentation, shortening of the rostrum, brain reduction) as to require an explanation. The neural crest hypothesis addresses this (Wilkins et al. 2014; chapter 4). For example, tamed brown rats in contrast to aggressive ones in experiments initiated by Belyaev (Albert et al. 2009; Singh et al. 2017), also showed shorter and wider muzzles and a higher frequency of white spots. Several hormonal changes may also be universal. In particular, endocrine-controlled behaviors that decrease the animal's reactivity to humans and facilitate its life in an anthropogenic environment (Zeder 2015).

There can be correlation of traits associated with domestication besides what is addressed in the "syndrome". A classic example was provided by Darwin (1868) in his description of the "lop-eared" rabbit, in which artificial selection for long external ears induced variation in the external auditory meatus (Fig. 3.8). A study of trait covariance in natural woodrat populations demonstrated a

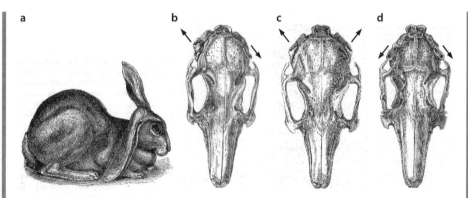

Fig 3.8. Domesticated rabbit skull and the lop-eared condition. Variation in the skull, in dorsal view, of the domesticated rabbit, *Oryctolagus cuniculus*, in relation to the lop-eared phenotype; (a) the "half-lop" rabbit; (b) skull of "half-lop"; (c) skull of the wild type; (d) skull of the "full-lop." Arrows show the orientation of the opening of the auditory meatus.

remarkably similar covariance pattern as in the "lop-eared" rabbit. In the woodrat case, this covariance was associated with a climatic longitudinal gradient (Cordero and Berns 2016).

The domestication syndrome has not been examined closely in birds. Studies of the domestication of the red junglefowl (Agnvall et al. 2015; Katajamaa et al. 2018) included the study of phenotypical changes following selection for reduced fear of humans, in contrast to the condition of the opposite line with selection for high fear of humans. Researchers recorded, for example, reduced brain size after eight generations (Agnvall et al. 2018). White plumage is one of the first changes visible when wild ducks are kept in captivity (Bottema 1989). As in mammals, domestication in birds may lead to pigmentation changes, perhaps the only universal feature of a syndrome.

Population size and geography

Besides selection, another factor believed to greatly influence the appearance of mutations and genome evolution are changes in the "effective population size" (N_e), which decreases as domestic populations become isolated from their wild relatives. The importance of N_e was underscored in a study that compared whole genome data from eight domesticated species (goat, sheep, chicken, cattle, cat, dog, pig, and horse) with data from their wild/ancient relatives, with results strongly suggesting that higher N_e could enhance positive selection and facilitate accumulation of functional mutations (Chen et al. 2018). The link between N_e and selection is universal and results from population genetic theory (Charlesworth 2009).

Genetic diversity or variability, which is advantageous in order to respond to environmental challenges, is expected to decrease in domesticated species (in the absence of introgression from wild species) with geographic distance from the domestication epicenter. Goats are considered among the first livestock domesticated species, starting about 10,000 years ago and dispersing afterward with migrating humans throughout the world. A study on genetic variability of goat populations worldwide that evaluated gene flow, population structure, and migration events showed ancient geographic diversity partitioning between continents, with recent introduction of highly productive breeds leading to mixing and introgression with local goat gene pools (Colli et al. 2018a). Furthermore, this study indicated that gene flow reduction occurred as a result of management practices leading to geographic and reproductive isolation.

Ancient domestication phases have given rise in most domesticated forms to an increased proportion of deleterious variants because of population bottlenecks and artificial selection. It is notable that this pattern is the case even though the census population sizes of domesticated forms are huge, and it has been shown by a quantification of the prevalence of deleterious variation in dog, pig, rabbit, chicken, and silkworm compared with their wild ancestors, using pooled whole-genome resequencing data (Makino et al. 2018). Domesticated forms show, in contrast to wild ones, reduced genetic variation, proportionately more single nucleotide polymorphisms in highly conserved elements, and a tendency to harbor a higher proportion of changes classified as damaging (Makino et al. 2018). European domestic pigs do not follow this pattern, probably as they continued to exchange genes with wild boar populations after domestication (Frantz et al. 2015). This is thus yet another example of the lack of universality among domesticated animals.

Low effective population size can have an impact on the genetic architecture of traits, as shown in a study of DNA of 80 domestic dog breeds and some feral individuals (Boyko et al. 2010). Strong selection to produce or maintain the specific appearance of a breed results in a footprint of selection in regions of the genome important for controlling those special traits. The morphological diversity in the dog appears to have a simple genetic basis dominated by few genes of major effect, in contrast to what would be the case in wild populations of other species.

Effects of trait selection on genes

Much work on genetics of domesticated animals has focused on identifying loci that respond to strong selection for features to fulfill the demand for food production. For example, consumer demand for lean meat in pigs has led since the 1940s to a drastic increase in pig muscle growth with a corresponding decrease in fat deposition (Andersson 2016). This selection for lean meat affected two major loci involved in body composition and metabolism in pigs. A causal mutation in the *IGF2* gene was identified affecting muscle growth that has undergone a selective sweep in meat-producing pigs and resulted in a considerable high mutation frequency in these pigs worldwide (Andersson 2016). The *RN* gene affects glycogen content in skeletal muscle via increased glucose uptake; sequencing and genetic analysis in pigs as well as experiments with transgenic mice revealed causal mutations in a muscle-specific kinase (encoded by *PRKAG3*) that resulted in excess glycogen content in skeletal muscle, known as the RN phenotype (Andersson 2016). In humans a rare mutation is found in the same amino acid residue (R225) associated with the same (pig RN) phenotype, high glycogen content in skeletal muscle, as well as decreased triglyceride storage, increased glucose uptake in exercised muscle, and muscle fatigue resistance (Costford et al. 2007; Crawford et al 2010). The PRKAG3 protein is a drug target for treatment of type 2 diabetes.

Yellow-skin domestic chickens, preferred by consumers of U.S., Mexico, and China markets, owe their phenotype to a recessive allele responsible for deposition of yellow carotenoids in the skin. The possible cause of the yellow skin phenotype was identified by linkage analysis and mapping as tissue-specific regulatory mutations that alter expression of *BCDO2* (beta-carotene dioxygenase 2, which cleaves colorful carotenoids into colorless ones resulting in white skin) only in skin and not in other tissues; a skin-specific reduction in *BCO2* expression results in yellow skin (Eriksson et al. 2008).

All domesticated vertebrates reproduce sexually, and each species has two sexes, in which both the alleles beneficial for males and those beneficial for females are separately maintained under positive selection. This may explain in part why some populations already have the pool of alleles/genes available needed to respond to selection for domestication, as exemplified by rabbits (Carneiro et al. 2014). In the wild, both sexes of rabbits have a high probability of expressing traits that are suboptimal for their own sex (feminine males and masculine females). The combined effects of a benefit in one sex and a cost in the other will prevent alleles/genes from going to fixation or being lost (Rice 1996; Perry and Rowe 2015). Many other mechanisms, including neutral processes, can maintain genetic variation in populations.

Commonality of pattern does not necessarily imply commonality of mechanisms behind those patterns. One study did show that a chromosomal region associated with increased tameness in domesticated silver foxes (Trut 1999) is orthologous to a domestication-related region in dogs, implying similar genetic backgrounds in these two species (Kukekova et al. 2011). In contrast, gene expression patterns in the frontal cortex of the brain between domesticated and wild specimens across a range of species are mostly different (Albert et al. 2012).

Domestication can involve changes in diet, requiring physiological adaptations to metabolize new kinds of foods, and some of these changes can be documented in the genome. Dogs, in contrast to wolves, have a starch-rich diet. Axelsson et al. (2013) conducted whole-genome resequencing of dogs and wolves, recognizing 3.8 million genetic variants that were then used to identify 36 genomic regions that could represent targets for selection during dog domestication. They identified candidate mutations for dogs and selection in some genes, reporting on ten genes they hypothesized played key roles in starch digestion and fat metabolism. Whereas in humans amylase activity occurs in the saliva, dogs express amylase only in the pancreas. Genomic work on house sparrows (*Passer domesticus*) has identified strong selection for part of the amylase gene family, analogous to the report for dogs (Ravinet et al. 2018).

Some of the genetic underpinnings of the various behavioral characteristics associated with domestication involve mutations and changes that have analogues in humans, such as the Williams-Beuren syndrome (WBS), a congenital disorder characterized by hypersocial behavior. Hypersociability is a core element of domestication, and one that distinguishes dogs from wolves. Von Holdt et al. (2017) found that structural variants of genes present in human patients with WBS also contrib-

Fig 3.9. Combinations of mutations in three different genes (*RSPO2, FGF5*, and *KRT71*) result in different kinds of hair in dogs (Cadieu et al. 2009). Some examples: (a) short, Basset Hound; (b) long with furnishings, Bearded Collie; (c) wire, Australian Terrier; (d) long, Golden Retriever; (e) wiry and curly, Airedale Terrier; (f) long with furnishings, Border Collie.

ute to extreme sociability in dogs. This variant was detected in a region on chromosome 6 previously found to be under positive selection in domestic dogs.

Geneticists scan the DNA of domesticated animals and look for locations where "selective sweeps" have occurred, meaning areas of the genome with imprinting due to natural or artificial selection. These areas are characterized by homogeneous sections that include the beneficial mutation, surrounded by greater DNA diversity outside the region. An example is the discovery of a mutation in chromosome 18 that leads to short legs in dogs (Parker et al. 2009). The mutation led to the expression of a retrogene encoding fibroblast growth factor 4 (*fgf4*), strongly associated with chondrodysplasia, a short-legged phenotype characteristic of more than 20 dog breeds, including Basset Hound, Dachshund, and Corgi.

The combinatorial effect of a few genes can result in varied phenotypes. This is the case of the *RSPO2, FGF5*, and *KRT71* genes, which encode R-spondin–2, fibroblast growth factor–5, and keratin-71, respectively, as they relate to coat color and type in domestic dog breeds (Figure 3.9). Variation in coat growth pattern, length, and curl characterize very different-looking dogs, resulting from mutations in those three genes and their different combinations (Cadieu et al. 2009).

Are common domestication phenotypes associated with specific genes across species?

The question has been posed whether there are genetic signatures of domestication that are common among species. These genes in common would be related to either the initial phases of domestication or to more intense interactions, such as selection for "improvement" traits (Alberto et al. 2018). Actually, there seem to be different genetic changes to achieve similar phenotypic endpoints. For example, polledness—the condition of lacking horns—is driven by different genes in goat and sheep (Kijas et al. 2012, 2013). Alberto et al. (2018) conducted a genomic study of haplotype differentiation in wild Asiatic mouflon and Bezoar ibex versus the sheep and goat domestics from local, traditional, and improved breeds, respectively, and concluded that in spite of the common targets of selection, the mechanisms involved are not the same. This should not be surprising given the many genetic interactions behind traits and the different phylogenetic history of each species.

Another aspect to consider by breeders is heritability of a trait. Here, it is relevant that a high heritability does not necessarily imply a monogenic inheritance. An example of this is human height, which has a high heritability but is affected by hundreds of genes, each with a small effect (Andersson 2016).

Extreme muscularity in domesticated forms: the myostatin gene mutation

One of the most striking conditions regarding muscle in domesticated species is the (wrongly called) double-muscling in several breeds of cattle (Figure 3.10). Selection for muscularity in beef breeds has led to this condition, related to a mutation in the myostation gene. Myostatin is a growth differentiation factor-8 (GDF-8) that plays an important inhibitory role during muscle development (Lin et al. 2002): the myostatin gene product regulates for decrease in skeletal muscle growth. Muscle and adipose tissue develop from the same mesenchymal stem cells (Thomson et al. 1998), and the myostatin gene knockout in mice may cause a switch between myogenesis and adipogenesis. Experiments with mice showed that a targeted deletion of the myostatin gene led to development of extreme muscularity (McPherron et al. 1997).

The selection for muscularity has picked up several different alleles in cattle, increasing the frequency of at least five independent loss-of-function mutations in the myostatin gene (Grobet et al. 1998). A mutation in the myostatin gene is also known in some sheep (Crispo et al. 2015), dogs, and humans (Mosher et al. 2007). In contrast, this condition has not been recorded in pigs, despite a similarly strong selection pressure for more muscle in some breeds. Perhaps the prenatal over-growth associated with these mutations is incompatible with the reproductive strategy of the pig, which produces 5–20 piglets in each litter (Andersson 2010).

Fig 3.10. A Belgian Blue bull showing the "double-muscling" phenotype.

Fig 3.11. "*Canis mexicana*." "A domestic dog with peculiar humps and apparent muscle hypertrophy," as depicted in Hernández's (1651) *Rerum medicarum Novae Hispaniae thesaurus, seu, Plantarum animalium mineralium Mexicanorum historia.* This depiction has been dismissed as a caricature, but it may illustrate a phenotype associated with mutations in the myostatin gene.

The Mexican "humped" dog (Hernández 1651; Figure 3.11) has been dismissed as a caricature (Ueck 1961). There is, however, the possibility that such form represents phenotypes associated with mutations in the myostatin gene, which leads to abnormally heavy muscling in homozygous Whippet dogs ("bully whippets"), among extant dog breeds (Segura et al. 2022).

Dental variation and the case of hairless dogs

A relation between teeth and integumental structures is deeply embedded in vertebrate phylogeny, because of shared developmental mechanisms. There are common signaling pathways, as in the case of ectodysplasin. Ectodysplasin is a protein encoded by the *EDA* gene involved in cell-cell signaling during the development of ectodermal organs, including teeth and parts of the integument (Harris et al. 2008).

The Mexican and Peruvian hairless dog breeds, as well as the Chinese crested breed, are characterized by lack of coat or having sparse hairs, as well as missing teeth or anomalous permanent teeth (Drögemüller et al. 2008). Hairless dogs lack some teeth or have a reduced number of cusps of them, or both (Figure 3.12).

Kupczik et al. (2017) validated the general relevance of the mechanisms of cusp development postulated in previous studies based on mice in a study of hairless dogs and their dental modifications. The *EDA/FOX13* pathway affects the formation of tooth cusps and of tooth crown size in a dose-dependent manner.

Using ancient DNA extracted from a historical museum sample of pedigreed hairless dog skulls, they identified this *FOX13* variant associated with the peculiar dental phenotype of hairless dogs (Kupczik et al. 2017).

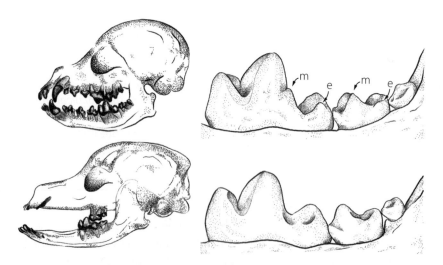

Fig 3.12. Dental phenotype in individuals of (*top*) dog with normal coat and (*bottom*) hairless dog. Drawing of 3D renderings of skulls and teeth in lateral view. In the hairless type there are neither premolars nor second generation canines; only the molars and the deciduous incisors and canines are present. On the right, the mandibular right first, second, and third molars in lingual view of (*top*) coated and (*bottom*) hairless dog are shown. Note the clearly developed metaconid (*m*), and entoconid (*e*) cusps in the coated dog, lacking in the hairless dog. Mesial (anterior) is to the left.

Reportedly the same mutation at the forkhead box transcription factor family FOX13 causes hairlessness in three dogs of distinct origins—the Chinese crested, the Xoloitzcuintli (Mexican hairless), and the Peruvian hairless (Drögemüller et al. 2008). They may have all inherited the mutation from a common ancestor, as suggested by a shared haplotype (Parker et al. 2017).

Summary: Variation in the genetics of domestication

Hybridization and the continuum of variation within populations lead to difficulties when trying to establish absolute, universal, and clear boundaries among wild versus domestic species counterparts. Introgression is responsible for the features of some domesticated species or of populations of them.

Studies in genetics of domestication have uncovered genes or groups of genes related to traits such as those related to food production and ornamentation. Genetic background studies have revealed an accumulation of multiple consecutive causal mutations in the same gene for color phenotypes in dogs, pigs, and chickens.

Extreme variation can occur due to intensive selective breeding. The modern broiler chicken is a good example, differing from its ancestors in skeletal morphology, pathology, bone geochemistry, and size. The genetic and epigenetic bases of these changes are mostly known.

Domesticated populations show reduced genetic variation as well as increased deleterious variants when compared with wild forms. This is a consequence of management practices leading to geographic and reproductive isolation, population bottlenecks, and artificial selection, evident even in some huge modern populations of dogs, pigs, rabbits, and chickens. European domestic pigs, however, are an exception, probably due to post-domestication genetic exchange with wild boar populations, showing the lack of universality among domesticated animals.

There are diverse mechanisms of isolation that together with assortative mating can play a role in the biology of domestic and feral populations of domesticated species. Diverse mechanisms of isolation can result from domestication and operate differently among breeds of a species. Reproductive isolation may occur in some cases and not in others, and may involve a probability of mating or of viable and fertile offspring.

CHAPTER 4

Evolutionary development

Phenotypic differences arise not only from genetic changes; tissue interactions and the physical and endocrine environment of the developing structures are also important. The phenotype can result only after the accommodation of genetic and environmental perturbations by the regulatory interactions occurring during ontogeny (Alberch 1982a, b; Minelli 2003; Salazar-Ciudad and Jernvall 2013; Fig. 4.1). These ideas are at the core of the field of evolutionary developmental biology, or "evo-devo."

An evo-devo approach to domestication must account for the distinctly nonrandom phenotypic variation across domesticated forms. The role of development in evolution is usually conceptualized in terms of constraints, which contradict neo-Darwinian views on morphological evolution that concentrate on natural selection and a gene-centered view of evolution. It is certainly an improvement over the classical view centered on adaptation and population genetics, as it contradicts the idea of random universal variation without developmental biases; however, it still has natural selection as a main mechanism, with development delivering the raw material for selection to act upon.

An alternative and sounder view compares developmental mechanisms by the morphological variation they produce and the way by which their functioning can change as a result of genetic variation (Salazar-Ciudad 2006). Pattern transformations are produced under different initial and environmental conditions or DNA changes that affect how multiple genes in a network interact. The developmental system theory is a way to conceptualize this view of evolution (Oyama et al. 2003).

Some species such as dogs and pigs have become more morphologically diverse than others, such as horses or cats. What is behind these patterns? This subject concerns evolvability: the potential of a species for evolutionary change. It is related to the degree of morphological variation exhibited by that species (Sniegowski and Murphy 2006). By variation I mean the observed differences in the phenotype, whereas variability refers to the tendency to generate those differences (Wagner et al. 1997).

In what follows I discuss concepts of comparative ontogenetic studies and of evo-devo as they relate to domesticated mammals and birds. I start with a summary of ontogenetic data. The interconnected topics address a multitude of aspects that are

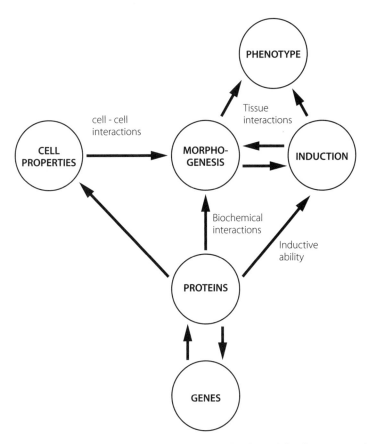

Fig 4.1. Changes in development result in changes in the adult phenotype via interactions and emergent aspects of the developmental process.

relevant in developmental evolution. In this and the following chapter I discuss morphogenesis, patterning, and growth. Morphogenesis refers to changes in the shape or location of a cell or tissues, involving in many cases foldings and specializations of groups of cells. With patterning one means the establishment of groups of cells in relation to each other and to surrounding tissues. Growth is the increase in size or number of cells (Noden and de Lahunta 1985).

An evo-devo approach to study of domesticated animals

Developmental biology investigates how a single cell becomes a completely formed organism, and evolutionary developmental biology, evo-devo, investigates how that process has changed over the course of evolution. Evo-devo concentrates on how the capacity to evolve, evolvability, is different among groups of organisms (Hendrikse et al. 2007). Evo-devo aims at understanding the origins of animal forms by integrating information from disparate fields such as phylogenetics, genomics, morphometrics, paleontology, cell biology, and developmental biology (Figure 4.2; Minelli 2003; Mallarino and Abzhanov

2012). Evo-devo initially addressed commonalities among major groups (Carrol 2005). It has expanded to explain the differences of a smaller scale—for example, the beaks of different species of birds. The study of domestication would represent a relatively new frontier in evo-devo—differences within species (Marchini and Rolian 2018; Núñez-León et al. 2021).

One could argue that within-species variation has already been extensively studied in model organisms. This is not the case. For example, laboratories using the chicken model rarely document the breed of chicken they use or consider the potential influence of such choice in their results.

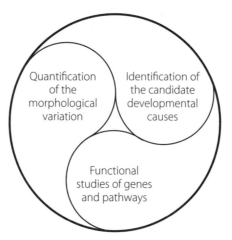

Fig 4.2. The evo-devo research program, as conceptualized by Mallarino and Abzhanov (2012), integrating three complementary approaches to characterize the patterns and processes of evolutionary change in morphology. The structure includes three parts. (1) Quantification of morphological variation, which can be accomplished with morphometrics or some simpler documentation of particular structures, including 3D imaging. (2) Identification of candidate mechanisms based on methods that document traits as they emerge in real time, quantitative trait locus (QTL) mapping, microarray and RNAseq screens, and studies of embryos, if appropriate with immunochemistry and/or in situ hybridization. (3) Investigation of causative relationships using methods of experimental embryology and molecular biology, including physical manipulation, transgenics, and functional reagents.

Domesticated animals in comparative embryology

How do different species compare with each other in patterns of differentiation and growth? Both prenatal and postnatal, and prehatching and posthatching are time windows of development during which change can occur.

Studies of domesticates require a comparative basis. In the late nineteenth century (Keibel 1897) and beginning of the twentieth century, many "normal tables of development" of vertebrate species, including domesticated mammals, were published

Table 4.1. Examples of primary works on development and "staging" of different species of domesticated mammals

Species	n	# stages	Time window studied	Illustrations	Reference
Bos taurus		3		drawings	Haeckel 1891
Bos taurus	2	2	ca. 6th week & ca. 50 d	photographs	Michl 1920
Bos taurus	11	10	? d–36 d	drawings/ photographs	Kröllin 1924
Bos taurus		10	32 d–fetus stages	drawings	Kröllin 1925
Bos taurus	25	10	68 d–birth (>274 d)	photographs	Lyne & Heideman 1959
Bos taurus	48 & 2 newborn		7th–20th week, 7–9 months	photographs	Naaktgeboren 1960
Canis lupus		3		drawings	Haeckel 1891
Canis lupus	70		ca. 14–20 d	drawings	Bonnet 1895
Canis lupus			early stages	drawings	Bonnet 1901
Canis lupus			egg–fetus	drawings	Bischoff 1845
Canis lupus			cleavage–60 d	drawings/ photographs	Evans 1993
Capra hircus	19	4	blastula stage–late embryos (SSL 11 mm)	drawings	Tsukaguchi 1912
Equus caballus		1	19 d	drawings	Hausmann 1840
Equus caballus		5	4–8 week	drawings	Ewart 1897
Equus caballus			1–21 d	drawings	Ewart 1917
Equus caballus	1	1	21 d	drawings	Robinson & Gibson 1916
Equus caballus		12	from 4th week on (SSL 6–25 mm)	photographs	Krölling 1942
Equus caballus			1–22 d	photographs	Betteridge et al. 1982
Equus caballus	34	12	17–40 d	photographs	Acker et al. 2001
Felis catus	>126	22	2.5–66 d	photographs	Knospe 2002
Mesocricetus auratus		16	7.5–15 d	photographs	Boyer 1953
Mus musculus		28	fertilization–24 d (post partum)s	drawings/ photographs	Theiler 1972
Oryctolagus cuniculus		?	egg–early fetus (ca. 10 d)	drawings	Bischoff 1842
Oryctolagus cuniculus		3		drawings	Haeckel 1891
Oryctolagus cuniculus			7.5 d–late embryo	drawings	Keibel 1905
Oryctolagus cuniculus		7	13 d–newborn (after 21 d)	drawings	Henneberg 1908
Oryctolagus cuniculus	129	11	9.5–19.5 d	photographs	Beaudoin et al. 2003

(*continued*)

Table 4.1. (continued)

Species	n	# stages	Time window studied	Illustrations	Reference
Ovis aries	45	ca. 23	2–34 d	drawings/ photographs	Bryden et al. 1972a
Ovis aries		20	14–34 d	drawings/ photographs	Bryden et al. 1972b
Ovis aries		16			Bryden et al. 1973
Ovis aries		3	24, 32 & 50 d	drawings/ photographs	Fernandes et al. 2017
Rattus norvegicus		19	13 d (before birth)– 20 d (after birth)	drawings	Henneberg 1908
Sus scrofa		3		drawings	Haeckel 1891
Sus scrofa domesticus	94	29	14–22 d (and older)	drawings (14 d–late embryo)	Keibel 1897
Sus scrofa domesticus		10	SSL 21 mm–newborn (>220 mm)	drawings	Henneberg 1908
Sus scrofa	74	13	2–7 d	drawings/ photographs	Heuser & Streeter 1929

(Werneburg et al. 2016). Many descriptive works on development and "staging" of different species of domesticated mammals (Table 4.1) and birds (Table 4.2) exist in the classical and more recent literature (Sterba 1995). This documentation can be a rich source for comparative studies. Furthermore, the embryos of species used in experimental embryology were assigned to stages once created for the reference species. Examples include the Hamburger and Hamilton (1951) stage system for the chicken, and staging for the mouse (Theiler 1989; Wanek et al. 1989; Downs and Davies 1993). They are used daily in laboratories worldwide as reference. Characterizing the timing and sequence of events in embryological development can also be relevant for biomedical studies of domestic animals, as in the investigation of cleft palate versus normal palate development in dogs (Freiberger et al. 2020).

Werneburg (2009) introduced a protocol to compare vertebrate embryos and presented a set of more than 100 soft-tissue characters that serve to "atomize" the body into clearly defined homologous structures. This "standard event system" can be expanded and adapted to any species, and the resulting information can be used in studies of heterochrony (Werneburg and Sánchez-Villagra 2011; Werneburg et al. 2016). One can compare different organ systems, levels of organization, and time windows of development—for example: organogenesis, skeletogenesis, early craniofacial embryogenesis, and the development of integumental structures or other organs. For bony fishes, a hierarchical analysis of ontogenetic time across clades was used to propose a method to circumvent the idiosyncratic and typo-

Table 4.2. Examples of primary works on development and "staging" of different species of domesticated birds

Species	# individuals	# stages	Time window studied	Illustrations?	Reference
Anas platyrhynchos	420	31	32 h–24 d	photographs	Koeck 1958
Anas platyrhynchos	260	12	1–25d	photographs (of 12 stages)	Caldwell & Snart 1974
Columba livia	190	43 (identical to Hamburger & Hamilton stages)	3 h–17 d	photographs, stages 1–43	Olea & Sandoval 2011
Coturnix coturnix	ca. 400	46 (stages 4–28 identical to Hamburger & Hamilton stages)	18 h–16.5 d (hatching)	photographs, stages 4–45	Ainsworth et al. 2010
Coturnix coturnix		16	1–16 d	photographs	Padgett & Ivey 1960
Gallus gallus		ca. 22	gamete–4 d	drawings/photographs	Patten 1951
Gallus gallus		10?	egg–early embryo?	drawings	His 1868
Gallus gallus		3		drawings	Haeckel 1891a
Gallus gallus		3		drawings	Haeckel 1891b
Gallus gallus		35	9 h–10 d	drawings	Keibel & Abraham 1900
Gallus gallus			egg–21 d	drawings/photographs	Lillie 1952
Gallus gallus	>296	46	<6 h–21 d	photographs	Hamburger & Hamilton 1951
Gallus gallus	250		1–21 d	drawings/photographs	Künzel et al. 1962
Gallus gallus	40	11	2–19 d	photographs	Fonesca et al. 2013
Gallus gallus	>200	19	5.5–21 d	none	Rempel & Eastlick 1957
Lonchura striata var. *domestica*	>600	46	<1.5–17 d	photographs	Yamasaki & Tonosaki 1988
Meleagris gallopavo	>4,000	7?	13–19 d	photographs (snood, 13–19 d)	Mun & Kosin 1960
Melopsittacus undulatus	36		1–12 d	drawings (histology)	Abraham 1901
Numida meleagris	5,000		1–27 d	none	Ancel et al. 1995
Numida meleagris	175	28	1–28 d	photographs, stages 1–28	Araújo et al. 2019
Taeniopygia guttata	1,063	46	0 h–14 d	photographs	Murray et al. 2013

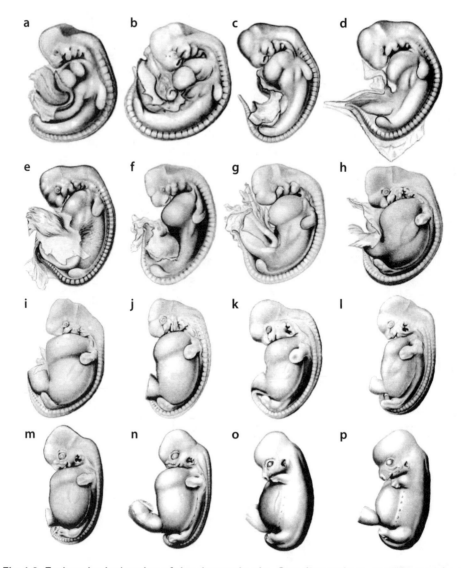

Fig 4.3. Embryological series of the domestic pig, *Sus domesticus*, as extracted from one of the classic "normal tables" of F. Keibel (1897). Embryos not to scale.

logical standard staging system (Lecointre et al. 2020), useful in laboratories devoted to the study of just one species, but not for comparative and evolutionary studies. Non-invasive imaging techniques can be used to document the morphology of embryos and their internal organs (Metscher 2009). If preservation allows, museum or veterinary school collections could be sources of such materials (Figure 4.3).

Model organisms

There are cellular mechanisms and signaling pathways throughout the life span of an organism that are similar across species (Trainor 2016), yet morphological diversity among species does arise. Searching for commonalities in development has been a main agenda of developmental genetics studies. This work has largely concentrated on model organisms, because in these a sophisticated analytical toolkit can be used (Sommer 2009). None of these model organisms belong to the classic clades studied by evolutionary biologists as examples of adaptive radiation, such as cichlids or the *Anolis* lizards. The choice of model species has been dictated largely by practical criteria, with a preference for organisms with short life cycles and abundant progeny, and that are easy to breed in the laboratory under inexpensive conditions. Among domesticated mammals and birds, rats, mice, and chickens are classic models for biomedical research (Burt 2007). The choice is dictated by practical considerations and the question at hand. For example, chick embryos are used for studing early human development instead of mice, because mouse gastrulation is derived within amniotes. As important as model organisms are, hypotheses about evolution based solely on model organisms are limited if we wish to understand anatomical variation within a "body plan," in particular within a species or between closely related ones (Milinkovitch and Tzika 2007; Werneburg et al. 2013).

The timing of change in embryonic development

When in development is variation generated? Reportedly, similarity among embryos is greatest at the "phylotypic stage" (Sander 1983; Figure 4.4a), and from that stage increasingly different shapes develop that characterize the different species. Given differences in early (e.g., at gastrulation) and later (late embryos and, ultimately, adulthood) development, this pattern of change has been dubbed the developmental hourglass, or egg-timer (Duboule 1994). The phylotypic stage has been interpreted as the time window less prone to change than others, less evolvable, and thus the ideas of the "privileged embryo" (Wolpert 1994), pharyngula (Ballard 1981; Kimmel et al. 1995), or zootype (Slack et al. 1993) were proposed. Transcriptomic studies have been presented as confirming this view (Domazet-Lošo and Tautz 2010; Irie and Kuratani 2011). The "phylotypic stage" is at approximately 18 days of gestation in the dog, 24 days in the cow, and 48–60 hours of incubation in the chick embryo (Noden and Lahunta 1985, p. 9). In the chicken, the stage has been identified at HH16 in the standard Hamburger-Hamilton staging system (Wang et al. 2013b), corresponding to 51–56 hours of age (Figure 4.4b).

A quantitative study of organogenesis in developmental timing across vertebrates as a whole and within mammals, respectively, produced unexpected results (Bininda-Emonds et al. 2003): variation among species was highest in the middle of the developmental sequence. Another study of developmental timing in 51 placental mammal species of 74 features revealed an intermediate period

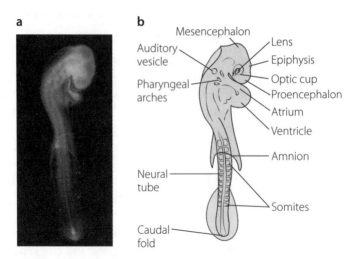

Fig 4.4. In the chicken, the phylotypic stage has been identified at HH16 in the standard Hamburger-Hamilton staging system (Wang et al. 2013), corresponding to 51–56 hours of age. Here illustrated by an individual of the Shamo breed.

of decreased variation, but here the pattern resembles an irregular hourglass (Cordero et al. 2020). More comparative studies of ontogenetic trajectories among species are needed (e.g., Young et al. 2014). These might lead to a rejection of universality of developmental patterns, with each clade of organisms showing unique trajectories of amount of variation. It may be the case that some middle stages of development are less variable. It is paradoxical that the middle stages of development, supposedly resistant to selection, are those in which pattern formation takes place; it is therefore expected that changes in adult phenotype would be easily influenced by changes in these middle stages (Bininda-Emonds et al. 2003).

A comprehensive study of embryonic morphospace in faces involving 12 amniote species found a period of reduced shape variance and convergent growth trajectories–reminiscent of a phylotypic period (Young et al. 2014). After this phase of development, phenotypic diversity sharply increases. This study included mouse, gerbil, rat, hamster, chicken, quail, and duck among domestic forms. Likewise, the quantification of the timing and mode of lability of craniofacial development during early face formation in chick, quail, and duck, using three-dimensional morphometrics, shows the commonalities and the point of divergence of ontogenetic trajectories in facial shape variation (Smith et al. 2015). Chick and quail embryos are more similar to each other than to the more phylogenetically distant duck in the enlargement of their frontonasal prominences.

When similarity or divergence is compared in development, it is paramount to take into account the taxonomic and population sampling conducted, the parameters measured, and the methodology used. Discussions on the phylotypic period suffer from the fact that the studies conducted to test this idealization of a pattern, although excellent in detail, have not been fully comparable.

The neural crest

Long before the discoveries of developmental genetics of the last decades, influential biologists such as Huxley (1932), Goldschmidt (1940) and de Beer (1954) had predicted that genes could regulate the time and rate of development of skeletal structures. In fact, such genes have been discovered, and many of them concern the neural crest (Schneider 2018a, b).

Neural crest cells are proliferative stem cells that migrate and differentiate into structures such as bone and cartilage derivatives, neurons, glia, pericytes, melanocytes, and integumentary structures (Figure 4.5). The neural crest is the source of structures of great importance in behavior, including parts of endocrine glands (pituitary, salivary, lachrymal, thymus, thyroid) as well as chromaffin cells in the medulla of the adrenal gland, which produces epinephrine, norepinephrine, and dopamine. Neurons in a brain region associated with learning and reward are also derivatives of neural crest cells (Vickaryous and Hall 2006). The critical fact is that neural crest cells give rise to neuroendocrine cells, and these cells produce hormones that mediate many behaviors.

Neural crest-derived mesenchyme orchestrates spatiotemporal programs for chondrogenesis, affecting cartilage size and shape across embryonic stages. An example is its fundamental role in patterning information for the face and jaws (Eames and Schneider 2008; Figure 4.5). The neural crest is also important in osteogenesis, for the regulation of molecular and cellular events known to play a role at different steps (Hall et al. 2014). Furthermore, bone resorption is mediated by neural crest mesenchyme. Studies of quail and duck embryos (Ealba et al. 2015) demonstrated that modulation of bone resorption can either lengthen or shorten the jaw skeleton. Differences in skull shape and in pigmentation are determined by differences in the amount, place, and timing of development of cells and in the remodeling of structures once produced (Schneider 2018b).

Regulatory changes to the neural crest (Martik and Bronner 2017) can be a major source of phenotypic transformations in the skeleton, the integument, physiology, and behavior (Schneider 2005). Variation in the migration, differentiation, survival, and/or division of neural crest cells (Kissel et al. 1981) can lead to adult variation. Abnormal development of the neural crest leads to "neurocristopathies," syndromes that span multiple organ systems.

Although we know how fundamental this embryonic cell population is in the generation of vertebrate structures, including the skeleton, there is little information on diversity in neural crest development across species. Our understanding of neural crest development is based largely on the study of a generic "chicken" (Stern 2018); this knowledge has never been used to address developmentally how breeds of chickens become different and how they differ from the wild form. A study of red junglefowl versus chicken adult skull shape (Stange et al. 2018) found significantly more variation in cranial neural crest–dependent features than in mesoderm-derived traits. Comparing the underlying molecular processes in fowl versus chicken morphogenesis could shed light on the mechanisms that have led to the divergent morphology resulting from domestication, and to the diversification of breeds (Núñez-León et al. 2019). However, this would not be an easy task, as many aspects

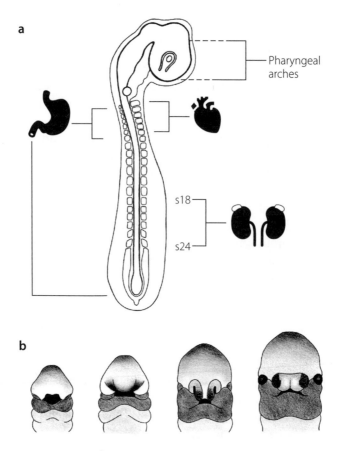

Fig 4.5. The neural crest. (a) Cells of the cranial neural crest migrate into the pharyngeal arches and the face, being involved in the formation of bones, cartilage, and cranial nerves. In the chicken embryo, the vagal (somites 1–7) and the sacral (posterior to somite 28) neural crest form the parasympathetic nervous system. The cardiac neural crest (somites 1–3) contributes to the division between the aorta and the pulmonary artery. The trunk neural crest (somites 6 to tail) contributes to the melanocytes (pigment cells) and makes sympathetic neurons. Somites 18–24 form the medulla of the adrenal gland. (b) Development of the face in a mammal, with neural crest contributions. The frontonasal prominence contributes to the primary palate, the forehead, the nose, and the area between the nose and the lip. The lateral prominences give rise to the sides of the nose. The maxillomandibular prominences generate the sides of the middle and lower regions of the face, much of the upper jaw, and the lower jaw.

of neural crest development can vary (Sánchez-Villagra et al. 2016, p. 4): "(i) timing of emigration, (ii) overall size of the progenitor pool, (iii) allocation and/or regional distribution of sub-populations (e.g., midbrain versus hindbrain), (iv) specification of lineages (e.g., cell types and derivatives), (v) growth parameters (i.e., proliferation rates and timing of differentiation), (vi) signaling interactions with adjacent and non-neural crest-derived tissues, and (vii) regulatory changes affecting spatial and temporal domains, and levels of gene expression."

Some aspects of neural crest development have been studied across species. For instance, previous work has shown how expression of certain transcription factors might differ between bird species, resulting in divergent phenotypes in craniofacial bones (Mallarino et al. 2011; Schneider 2018a). Studies between quail and duck also reveal that differences in the numbers of neural crest progenitors relate to species-specific variation in jaw length (Fish et al. 2014).

Neural crest and the "domestication syndrome"

Most of the features reported as part of the "domestication syndrome"—traits associated with the first phase of domestication and reported mostly for mammals (chapter 3)—are direct derivatives of neural crest cells or causally linked to their development (Wilkins et al. 2014). The selection for tameness, important in a first phase of domestication, could drive phenotypic changes in many organ systems, as shown in a study of silver fox domestication (chapter 8). The neural crest hypothesis predicts that tameness-associated traits result from the correlated expression of changes driven by that population of progenitor cells in development (Wilkins et al. 2014). In summary, it was proposed that domesticated traits result principally from mild cranial neural crest deficits during embryonic development.

It has also been suggested that domestication traits result from juvenilization—extending the phase before sexual maturity—as determined by altered concentrations of thyroid hormones (Crockford 2002). The latter play key roles in regulating the rates of growth and maturation. This hypothesis predicts significantly extended periods of thyroid hormone production in domesticated compared with wild forms—as has been shown in a comparison of bonobos to chimpanzees (Behringer et al. 2014). A role for thyroid hormone signaling in neural crest cells migration was shown in the amphibian model *Xenopus* for early embryogenesis effects. Inhibition of thyroid hormone receptor action, either by knockdown of receptor gene expression or by drug inhibition of receptor activity, led to severe malformations related to neurocristopathies (Bronchain et al. 2017). Thus, Wilkins (2017a, p. 438) suggested that it is possible to link the neural crest and thyroid hormones hypotheses.

Genomic studies of the domestication of dogs (Pendleton 2018), cats (Montague et al. 2014), horses (Schubert et al. 2014; Librado et al. 2017), and dromedaries and Bactrian camels (Fitak et al. 2020)—also see work on rabbits (Carneiro et al. 2014)—reported evidence for positive selection on genes involved in neural crest "development programs" (Wilkins 2017a). These studies addressed the question: is the neural crest involved in the domestication syndrome? The answer seems to be yes, as modifications in neural crest development genes associated with features of the syndrome are recorded to have changed when comparing wild versus domesticated forms, or some approximation of this (e.g., wolves and dogs, Pendleton 2018). The work on dromedaries and Bactrian camels additionally reported positive selection for thyroid hormone–based signaling (Fitak et al. 2020). There are limitations to the power of this genomics approach when it comes to testing the neural crest hypothesis (Johnsson et al. 2021; but see Wilkinson et al. 2021).

A more fundamental aspect to consider when evaluating the role of the neural crest in the generation of morphological patterns associated with domestication is whether this role is any different or remarkable when compared with evolution outside domestication. Any process of morphological diversification is likely to involve the neural crest, given the ubiquity of this population of embryonic cells in the origin of many morphological structures of vertebrates.

For all the power of genomic tools, it is the actual tracing of neural crest cells in embryos that is key to comprehending how tempo and mode of migration of cells and their interactions can generate differences in adults. Very little such data are currently available to build a comparative basis for relating neural crest developmental patterns and morphological diversification. In only a handful of vertebrate species has the neural crest been investigated (Hall 2009). In mammals, developmental parameters of the neural crest are known in the marsupial *Monodelphis domestica* (Smith 2001b; Vaglia and Smith 2003) and the mouse (Barriga et al. 2015). Clades of tetrapod animals vary in how neural crest development is linked to adult morphology. Work on ranid frogs has shown that the timing of appearance of streams of neural crest and their divisions can vary significantly among closely related species (Mitgutsch et al. 2008), differences that in other clades would characterize major groups. It is unclear whether and how these developmental differences among species result in morphological differences in adults. A specialized pattern of neural crest cell migration has been reported for Silkie chickens (Reedy et al. 1998, 476): "In Silkie embryos, melanoblasts are specified late and only invade the dorsolateral path after they have been specified. Unlike quail and White leghorn melanoblasts, however, Silkie melanoblasts also migrate ventrally, but again only after they are specified".

Neural crest, faces, and *Runx2*

Neural crest mesenchyme mediates the timing of the transition from proliferation to differentiation and modulates the expression levels of osteogenic transcription factors, thus establishing species-specific size and shape in much of the skull (Wilkins 2017b). One of the osteogenic transcription factors related to the neural crest is *Runx2* (Schneider 2018a). The species-specific size and shape that result during osteogenesis involve (among other variables) *Runx2*, which affects the temporal and spatial expression of members and targets of the bone morphogenetic proteins (BMP) pathway (Merrill et al. 2008). BMPs are involved in bone formation both embryonically and postnatally (Schneider 2018a, b).

Runx2 has been well studied in relation to facial length in dogs. Fondon and Garner (2004) reported that the extent of muzzle length is a function of changes in the *Runx2* gene sequence: the greater the number of small, repeated DNA sequences in the coding region of this gene, the longer the skull. This correlation seems to hold for the Carnivora (including dogs, cats, ferrets, and all their closest relatives) as a whole (Sears et al. 2007). *Runx2* contains a tandem repeat of glutamine (Q) and alanine (A), and the QA ratio influences the regulation of bone development. The Siberian husky has 20 glutamine- and 8 alanine-coding nucleotide sequences within the repetitive glutamine and alanine domain of *Runx2*. Its QA ratio is 2.50, correlated with a longer face. The domestic cat (*Felis catus*) has 21 glutamine- and 10 alanine-coding nucleotide sequences in the same *Runx2* domain, with a QA ratio of 2.10 correlated with a shorter face (Usui and Tokita 2018). What holds true for the tandem repeat region of *Runx2* in carnivorans does not apply to other mammals (Pointer et al. 2012), but there are likely other functional domains within the gene (besides the QA region) that may play important regulatory roles.

Comparative morphological work is in some cases productively informed by our knowledge of developmental genetics. The phenotypes of human cleidocranial-dysplasia patients are associated with *Runx2* mutations, but these do not significantly co-occur in afrotherian mammals, the clade including elephants, hyraxes, and manatees among others (Asher and Lehmann 2008). This kind of study can reveal clade-specific features of the relation between phenotype and genotype, and here domesticated animals can be of great use as subjects of study.

Experimental embryology of domesticates

Experimental embryology and molecular biology can be used to investigate causal relationships in development. Here the use of chimeras is common. Surgical transplants that combine embryonic components from distinct animal species are used to follow the movements and fates of cells, and to understand the inductive properties of tissues (Harrison 1935; Le Douarin and McLaren 1984; Le Douarin 2005; Abramyan and Richman 2018). This approach has provided major insights into developmental evolution (Harrison 1935). Experiments include grafting or extirpation of tissues through microsurgery, implantation of reagent-soaked beads, labeling of cells for lineage analysis, injection of biochemicals, and retroviral infection or electroporation to manipulate gene expression (Schneider 2018a).

Bird chimeric systems have been used for decades, as avian embryos are easily accessible *in ovo* for all kinds of experimental manipulations. The resulting "chimeras" in many cases involve domesticated species. Classically, quail and chick have been used, as these are closely related birds with similar morphology and growth rates (Hamburger and Hamilton 1951; Nakane and Tsudzuki 1999). Domestic duck embryos have also been the subject of experimental embryology studies for almost a century (e.g., Waddington 1930, 1932; Tucker and Lumsden 2004).

Tissue transplantation across species has in some cases been done between distantly related groups, as in a study of the capacity of ectomesenchyme to induce tooth formation in a bird (Mitsiadis et al. 2003). This capacity would be remarkable given that birds, descendant from dinosaurs, lost their teeth some 70–80 million years ago in the Cretaceous. Neural tube cells were transplanted from mice to chick embryos to replace the neural crest cell populations in the latter. The chimeras showed tooth formation, demonstrating that the avian oral epithelium can induce a nonavian developmental program. This kind of induction experiment had some precursors. A study looked at epithelial-directed mesenchyme differentiation in vitro, using mice odontoblasts mediated by quail epithelia (Cummings et al. 1981). In another experiment, lizard tooth mesenchyme was grafted to a chicken (Lemus et al. 1986; Lemus 1995).

Fig 4.6. Franklin Dove's "fabulous unicorn." As a one-day-old calf, this adult male had undergone a surgical transplantation of its two horn buds from their lateral position to a single, central location on the skull.

Perhaps the most peculiar experiment in experimental embryology involving a domesticated animal was that conducted by Franklin Dove, a biologist at the University of Maine, in the 1930s (Blumberg 2009). He demonstrated the autonomous growth of horn bud in cattle by creating a unicorn-looking bull, revealing the importance of connective tissue and stem cells in early development (Figure 4.6). Dove surgically transplanted the two horn buds of a one-day-old male calf from their lateral position to a single, central location on the skull. The transplanted, two horns anlagen developed into one large horn solidly attached to the skull and located between and somewhat above the eyes (Dove 1935).

Apparently this transplantation peculiarity has not been performed exclusively in cattle. There are historical reports of a Nepalese "unicorn" sheep, in which horn buds were transplanted to form a single, middle horn (Hall 2015), and Duerst (1926) reported and illustrated a cow head with a third, middle horn, presumably from an 1892 publication by a veterinary doctor in Lyon, France.

Developmental plasticity—the case of cat and rabbit coats

Environmental factors that affect the development of the embryo and postnatal/posthatching growth include temperature, pressure and gravity, light, nutrition, and the presence or absence of predators and of conspecifics (Gilbert and Epel 2009; Feiner et al. 2018). When examining variation, the question arises

Fig 4.7. Himalayan rabbit and Siamese cat. The lower temperature in the distal parts of the body affects enzyme activity associated with coloration, resulting in the black and white patterns characteristic of these animals.

whether it results from plasticity. For example, is variation based on differences in what animals have been eating (Yom-Tov et al. 2003; Menegaz et al. 2009), or is it something hard-wired in the organism? Variation may be generated by developmental plasticity, the ability of one genotype to produce more than one phenotype in response to different environmental stimuli (Payne and Wagner 2019). To address this question, it is relevant to take into account the genotype–environment interactions. An example concerns cattle breeds. Reportedly, Limousine cattle show a greater flight (escape) distance than Jersey cattle when reared in their usual environments (extensive and intensive, respectively). But when reared in similar, extensive conditions, this difference disappears (Fisher et al. 2001; Mignon-Grasteau et al. 2005).

An example of developmental plasticity is the temperature-dependent coloration in Siamese cats and Himalayan rabbits, caused by differential case enzyme activity (Gilbert and Epel 2009; Fig. 4.7). Temperature can affect enzyme folding and thus alter the shape of an enzyme's active site and its interactions with other proteins crucial for enzyme activity (Feiner et al. 2018). This is the case of the enzyme tyrosinase, which catalyzes the production of melanin, the skin's dark pigment. Some tyrosinase mutations block melanin production, resulting in albinism, the lack of dark pigment throughout the body. In the case of Siamese cats and Himalayan rabbits, tyrosinase folds properly at the relatively cold temperatures of the extremities but does not fold properly—and thus becomes inactive—at the warmer temperatures of the rest of the body. This means that melanin pigment is produced only in the extremities and in other cold body parts, such as the ears, face, and end of the tail. If one were to shave the back of a Himalayan rabbit and keep it cold, it could be speculated that hair in that area would grow black. While Burmese cats, like Siamese cats, also show tryonsinase gene mutations, these are located elsewhere. This difference results in other threshholds for gene activity, and thus Burmese breeds have darker body color (Schmid-Küntzel et al. 2005). Analogous conditions are found in humans, for whom only the hair at the extremities is pigmented (Berson et al. 2000; Gilbert and Epel 2009). Given its coloration (all white except the ears, rhinarium, and in some individuals the distal portion of the legs), the ancient White Park breed of cattle (Ekarius 2008) may represent another example.

Developmental plasticity—the case of Slijper's two-legged goat

A well-known case of developmental plasticity is that of the two-legged goat born with a congenital defect of the front legs that learned to walk and run by using its hind legs alone. This case was reported in 1942 by the Dutch morphologist E. J. Slijper (1907–1968), who documented the skeletal and muscular changes of the animal after its death. The ability of "Slijper's goat" to hop about on its hind legs was supported by changes in the skeleton and musculature with respect to standard goats: larger hind limbs, a more curved spine, and larger neck (Figure 4.8). The spines of the thoracolumbar column of the bipedal goat have a different orientation from those of the normal goat. Slijper (1942) noted also how the compressed thorax and the elongate ischium of this peculiar goat resembled some features of kangaroos, related to their bipedality.

The two-legged goat case illustrates how a viable adult form can occur in spite of a developmental perturbation. It is an example of how a change in the phenotype—in this case the lack of front legs—can lead to new functional interactions and morphological complexity driven by correlated changes that did not require natural selection or numerous genetic changes. A single genome contains the potential to generate multiple morphological outcomes (Maturana-Romesin and Mpodozis 2000; West-Eberhard 2005). Analogous cases of quadrupedal animals managing to develop the capacity to walk on just the hind legs are known, such as that of the dog named Faith, which at some point gained notoriety by being featured on a popular U.S. TV show and can be seen on some open-access video platforms.

Developmental plasticity in tooth growth

Phenotypic plasticity related to changes during life as influenced by the environment has been recorded in numerous studies on tooth growth and dental morphology of domesticated animals. The different tissues constituting a mammalian tooth serve as records of an individual's life history and living conditions (Klevezal 1996; Figure 4.9). It may seem strange to discuss teeth in a chapter devoted mostly to embryos, but development does not stop at birth or hatching, and many of its features, including phenotypic plasticity, characterize all stages of life.

Comparative studies of goats on islands have shown different aspects of molar architecture, in particular tooth height, responding to different environments (Madden 2014). To understand mechanisms, controlled feeding experiments have

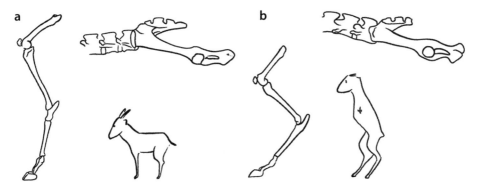

Fig 4.8. Skeletal transformations in Slijper's two-legged goat. (a) bones of the hind limb and pelvis of a normal goat and (b) same of goat born with congenitally absent forelimbs. The differences (e.g., dorsoventral flattening and elongation of the ischium in the bipedal goat) are related to the changes in posture. Not illustrated here is the fact that the spines of the thoracolumbar column of the bipedal goat have a different orientation from those of the normal goat.

been key, as performed on goats, cavies, chinchillas, and rabbits (Ackermans 2020). Studies have shown how root growth compensates for molar wear in adult goats (Ackermans 2019) and how incisor growth is affected by different diet in rabbits (Müller et al. 2014; Meredith et al. 2015), in cavies (Müller et al. 2015), and in chinchillas (Wolf and Kamphues 1996). Early works also demonstrated such plasticity using more invasive methods, sadly more typical of the times, such as experimental shortening and prevention of eruption in rabbits (Ness and Brown 1956). Other approaches use imaging techniques to measure microwear and dental surface texture as they relate to dietary abrasiveness, as has been done with rabbits (Schulz et al. 2013).

In an experiment with goats, the effect of changes in diet on cementum microstructure and growth was tested (Lieberman 1993). Animals were fed diets of different hardness (softened or normal pellets) and with different mineral (calcium and phosphorus), protein, and vitamin content. Cementum growth was affected, as reflected in the different orientation of the Sharpey's fibers, which realign to counter tensile forces that press teeth in the alveoli during occlusion. Changes in diet mineralization affected the growth of the collagen matrix, resulting in darker cementum bands.

In an experiment on cavies fed on diets with low, medium, and high silica content, the effects of these treatments were measured in dental and skull features using micro CT scans (Martin et al. 2019). The high silica diet, based on bamboo, brought maxillary incisors and molars close to the minimum required for functionality. Bamboo-fed animals had longer cheek tooth rows and larger occlusal surfaces, another example of an abrasive diet triggering expansive growth of the tooth base.

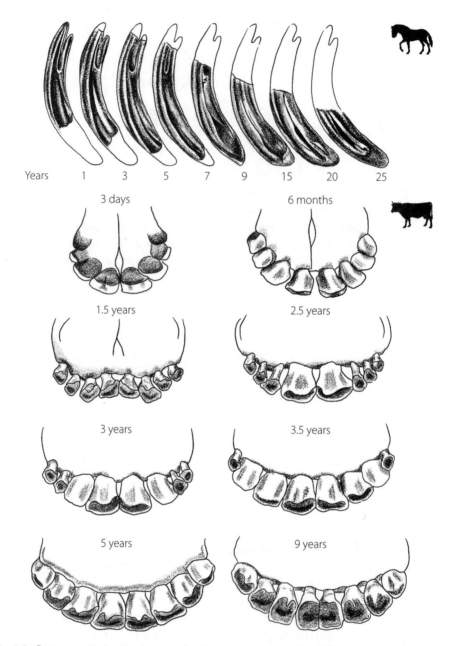

Years   1   3   5   7   9   15   20   25

3 days   6 months

1.5 years   2.5 years

3 years   3.5 years

5 years   9 years

Fig 4.9. Ontogenetic tooth changes in horses and cows. (*top*) Ontogenetic changes in the incisor of a horse, in antero-posterior and longitudinal view. This series of horses illustrates the gradual diminution of the pulp cavity and wearing. (*bottom*) Ontogenetic changes in the incisors of a cow, in occlusal view.

Epigenetic changes in chickens: movement and temperature variation during embryonic life

As stated by the Swiss anatomist Wilhelm His (1831–1904), "to think that heredity will build organic beings without mechanical means is a piece of unscientific mysticism" (His 1888, cited in Nowlan et al 2010). At the beginning of the twentieth century, considerations of mechanical forces were critical in the understanding of development, with the field of "Entwicklungsmechanik" including research on domesticated animals (Roux 1894). At some point in the twentieth century, advances in genetics and more recently in molecular biology displaced the relevance of mechanical forces in development, as this was considered outdated. The discovery of how cellular events can be guided by positional information based on molecular signals has been very important, and also the understanding that complex networks of regulatory genes orchestrate the development of integrated systems in the embryo (Nowell et al. 2010). However, to obtain the whole picture of how structures develop, mechanical considerations are equally important. Examples provided by experiments conducted largely on chickens show how forces determine the proper development of joints and several bony structures. Thus, molecular and "mechanical" approaches should be complementary and not mutually exclusive.

Our understanding of mechanical signals in embryonic development in vivo is limited by technical constraints related to measuring the mechanics of developing tissues and organs. A comprehensive recent review of available techniques to measure cellular or tissue mechanics in vivo (through laser ablation, tissue dissection and relaxation, force inference, micropipette aspiration, cantilevers and atomic force microscopy, magnetic tweezers, microrheology, droplet-based sensors, and FRET-based tension sensors) suggests that promising techniques will contribute to insights into the mechanics within living embryonic tissues and organs (Campás 2016). While some of these require contact with the sample, imaging-based techniques do not, and these techniques have often been used for in vivo and in situ characterizations. Molecular biology is used in FRET sensor systems engineering so they can be expressed in cells of developing embryos, in which tension is then measured by imaging (Campás 2016).

Pioneering works on chickens have addressed how skeletal muscle contractions affect joint development (Drachman and Sokoloff 1966; Murray and Drachman 1969) and how bones and muscles in general are affected by movement (Hall and Herring 1990) during embryonic life. These studies used neuromuscular blocking agents to paralyze chick embryos in ovo. Further studies have identified critical periods during which the developing skeleton becomes influenced by the impact of movement. Developmental periods have been identified that more significantly modify specific osteochondral components (Pitsillides 2006). These studies have provided insights that have informed our understanding of the function and evolution of some innovations in the digits of birds (Botelho et al. 2014, 2015b).

The importance of incubation temperature on embryonic development is well known. A study in which chicken embryos were incubated for three days at temperatures ranging from 37.5°C to 38.5°C showed that higher temperatures lead to an increase in embryonic movement (Hammond et al. 2007). These embryos grow to significantly heavier weight and exhibit significantly longer leg bones (tibia and tarsus) than the control ones raised at 37.5°C. Another change in embryos incubated at higher temperature is the increase in number of leg myonuclei, starting from embryonic day 12, which may result in greater leg muscle growth later in development. Comparisons of chicken and mice studies on joint and bone formation indicated that, whereas extrinsic mechanical forces from movements of the mother or littermates impact skeletogenesis in mammals, the bird embryo is reliant on intrinsic movement for mechanical stimulation (Nowlan et al. 2010). Ex-utero movements in mice may be important for joint development.

Different incubation microclimate conditions can affect the development of a bird's behavior and physiology of growth. In a study of chickens (Bertin et al. 2018), a control group of chicks was incubated in an optimal thermal environment (37.8°C) and an experimental group of chicks exposed in ovo to suboptimal temperatures (27.2°C for 1 hour twice a day). The suboptimal temperature treatment delayed hatching and decreased growth rate and also produced behavioral and physiological changes (e.g., higher neophobic responses than controls in novel food and novel environment tests).

Light conditions can affect hatching time, as shown in a study of White Leghorn lines incubated in continuous light and in unlighted conditions (Bohren and Siegel 1975). The light treated eggs averaged five hours earlier hatching time than the unlighted treatment. This effect worked in the two lines of chickens studied, selected for fast and slow hatching time. The fast hatching line averaged 48 hours shorter incubation time than the slow hatching line (Bohren and Siegel 1975).

Historically, the examination of chicken embryos has been important in understanding the developmental contribution to behavior and the nervous system, with the original contributions of Zing-Yang Kuo (1898–1970) as one of the pioneers and most original of the experimentalists (Kuo 1932; Gottlieb 1972; Qian et al. 2020).

Fluctuating asymmetry

Fluctuating asymmetry refers to the differences in meristic traits between the left and the right side of the body. These morphological deviations form bilateral symmetry (Van Valen 1962) are often studied to assess developmental stability, meaning the ability to precisely express genetically determined information despite environmental disturbance. The genetic variability connected to developmental stability is usually measured as heterozygosity of allozyme loci (Zachos et al. 2007). The study of fluctuating asymmetry has had a troubled history, and it is debatable what exactly we can learn from it (Houle 1998). However, the prospect of connecting phenotypic traits at the population level with important processes is appealing and worth exploring.

A study of limb skeletal elements in CD1 mice (a population of outbred laboratory mice) showed that fluctuating asymmetry variances decrease with gestational age from day 14 to birth, at day 20.5 (Hallgrímsson et al. 2003). Specific mechanisms may ensure stability during embryonic development.

A higher level of fluctuating asymmetry was found to be associated with growth rate and with stress in chickens. A comparison of the level of fluctuating asymmetry among wild junglefowl from India, two fast-growing breeds (ScanBrid, Ross 208), and a slow-growing breed (La Belle Rouge), showed a positive relationship between growth rate and asymmetry (Møller et al. 1995). Furthermore, fluctuating asymmetry could be tied to the experienced levels of stress. In a study of fast-growing chickens kept at three densities (20, 24, and 28 chickens per square meter), a positive association between fluctuating asymmetry and density was found (Moller et al. 1995).

Cranial right-left asymmetry has been recorded in some populations of domesticated forms (e.g., rabbit: Parés-Casanova and Medina 2019), but the significance of this pattern beyond indicating lateralized muscular function (mastication) remains unclear. A comprehensive study of orbital angle morphology, a feature that has been studied in zooarchaeological research, has symmetry that is highest in wolves and lowest in a sample of archaeological dogs (Janssens et al 2016).

Interactions of developing tissues: the case of the brain and skull

Domesticated animals have been the subject of important studies concerning inductive interactions in development. Schowing (1968) documented his experiments assessing the inductive relationship between the developing brain and the overlying calvarium in chickens. New molecular approaches could build upon this experimental work to test ideas concerning this interaction (Teng et al. 2019). During development, the brain serves as a framework on which the skull grows, both structures influencing each other's development through mechanical interactions, molecular signals, and later somatic growth (Marcucio et al. 2011).

Knowledge gained from the study of domesticated animals, such as the chicken, must in some cases be complemented with the study of other species that by virtue of preserving some plesiomorphic or ancestral traits can better resolve the identity of structures and relations (Smith-Paredes et al. 2018). An evolutionary and paleontological perspective is valuable for understanding the brain-skull interaction, and, in doing so, for addressing homology and nomenclatural issues of skull bones across vertebrates (Koyabu et al. 2012), including of course domesticated species (Figure 4.10). The bones of the braincase meet at sutures, and the identity of those bones and sutures are nontrivial to establish, as different criteria can be used. If aiming at homologous structures with the same names, the historical continuity of structures needs to be tested to allow meaningful comparisons across taxa. This is the case of the prominent "coronal" suture of vertebrate animals, which separates the front and back of the skull. There is a decoupling of the germ layer origins of the mesenchyme that forms the bones from the inductive signaling that establishes bone centers. Changes in the proportions and shape of the skull and the fusions and losses of bones during evolution make the comparisons challenging. It seems clear that what has been traditionally named the coronal suture is not the same structure across vertebrates (Teng et al. 2019).

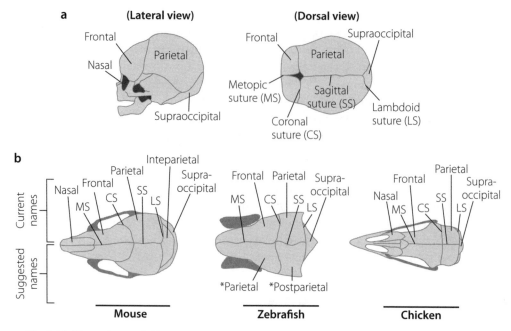

Fig 4.10. Homology of the skull roof bones in (a) humans and (b) three domesticated species. These hypotheses of bone identity for skull roof bones by Camilla S. Teng and colleagues were based on the integration of development genetic, embryological, comparative anatomical, and paleontological data.

Digit number in cavies, dogs, and chickens: Developmental biases and malformations

The maximum number of digits in an autopod (hand or foot) of any living land vertebrate has been five since the Devonian period (Galis et al. 2001), and reductions are common, as in the single middle digit of horses. Early horse embryos possess five digit condensations: what will become the main central digit of adult horses is surrounded on each side by the fusion of two digits (Kavanagh et al. 2020), as recorded in the fossil record for horse evolution from pentadactyl ancestors. This persistent pentadactyl initiation shared by the horse with other modern taxa with digit reduction demonstrates a certain conservatism or stability at the initiation of digit formation.

Developmental work on mammals has shown that the mechanisms for gaining or losing digits are diverse. Either early limb patterning or later post-patterning stages of chondrogenesis can result in digit loss (Cooper et al. 2014). Digit loss in pigs is determined by downregulation of *Ptch1* expression during early limb patterning. This downregulation does not involve an increase in cell death, which in contrast is recorded in the horse, in which extensive cell death sculpts the tissue around the remaining toes (Cooper et al. 2014). These mechanisms refer to fundamental properties of these species, not the result of domestication, but they inform us nevertheless on mechanisms and potential pathways of developmental change among breeds of the same species.

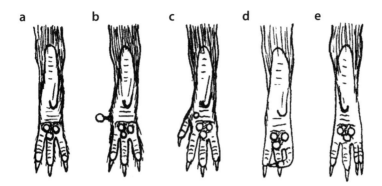

Fig 4.11. The hind feet of cavies, *Cavia porcellus*. (a) normal 3-toed condition of the left foot is contrasted with several variations, including (b) imperfect development of an extra little toe, (c) a foot with well-developed little toe and a corresponding plantar tubercle, (d) abnormal condition of a digit, and (e) duplication of a digit on the right foot.

Many efforts in the first decades of the twentieth century were devoted to understanding the genetic bases of digit numbers. These studies were conducted in the early days of Mendelian genetics and in many cases involved domesticated forms. A series of papers concerned cavies, based on breeding experiments (e.g., Castle 1906; Wright 1934; Figure 4.11). Variation in the number of digits could not be accounted on the basis of a simple recessive or simple dominant allele. Sewall Wright (1934, 509) stated that "non-genetic factors play such an important role that normals and polydactyls may be of the same genetic constitution." Later work showed that polydactylous morphology can have a genetic basis of an additive polygenic system or of a discrete Mendelian gene (Lande 1978; Park et al. 2008).

Comparing digit variation in dogs (Fig. 4.12) shows that mutations result in polydactyly in large breeds but not in small ones (Alberch 1985). Wolves and most dog breeds possess five digits on the front limbs and only four on the hind limbs. However, some breeds have a vestigial first toe, "the double-clawed" condition (Evans and Lahunta 2012), or "hind-limb-specific preaxial polydactyly" (Park et al. 2008). Large breeds, such as Mastiff, Newfoundlands, and St. Bernards have a tendency to develop an additional toe (Darwin 1868). In the Great Pyrenees, another large dog, the presence of an extra digit is a standard for the breed and, through selection, has become established in the population. It has been claimed that in these breeds the extra toe could give greater support, may be of use in swimming, and may improve walking through deep snow (Prentiss 1906). In other breeds, the extra toe has been seen as an undesirable character, and breeders have consistently selected against it. But the character keeps reappearing. It is common veterinary practice to surgically remove the vestigial first toe or dewclaw from dogs at the puppy stage. It is argued that the dewclaw, which lacks the standard bony connection to the rest of the autopod like other toes, could get ripped off in many different ways, causing a lot of damage.

This vestigial toe is usually reduced or absent in small breeds (Kadletz 1932), such as poodles and Pekingese. Alberch (1985) argued that extra digits are less likely to be generated during development in smaller than in larger breeds. He assumed

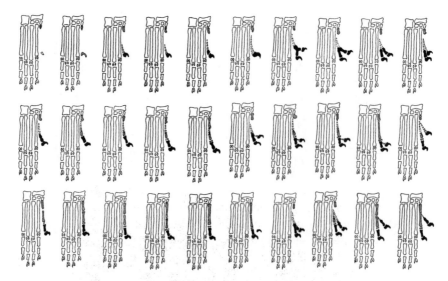

Fig 4.12. Skeletal variation in the hind feet of dogs.

that large dogs have, on average, relatively larger embryos and consequently larger limb buds, containing more cells. Work on limb development in salamanders (Alberch and Gale 1983) had shown that size (or number of cells) in the embryonic limb bud can influence the number of digits specified during pattern formation. There might be a minimum number of cells necessary before a given skeletal element is specified. Breeders selecting for dogs of large size, as in the case of St. Bernards, indirectly select for larger limb buds, and thus promote the atavistic appearance of an extra toe. To my knowledge, nobody has ever tested whether the embryos—or better, the digital plate of the limb bud—of large dogs are in fact larger than those of smaller dogs, but this seems likely (Wehrend 2013).

The kind of preaxial extra toe just discussed is one of many kinds of polydactyly, the occurrence of supernumerary toes or forelimb digits. Preaxial polydactyly refers to additional digits situated anterior to the median axis of the limb, that is, the first-toe or tibial side of the foot or the thumb or radial aspect of the hand. The additions of digits posterior to the median axis—the fibular aspect of the foot and the ulnar aspect of the hand—are cases of postaxial polydactyly. The rare addition of a surplus digit in neither a preaxial nor a postaxial position is called central polydactyly (Lange and Müller 2017). Polydactyl conditions involving both the fore and hind limbs occur, as in the Norwegian Lundehund, a specialized breed of dog (Metzger et al. 2016).

Polydactyly is associated with different kinds of mutations and mechanisms. One of them is the variation in the number of repeats in the coding regions of the *Alx-4* (aristaless-like 4) gene, quantitatively associated with the dewclaw condition (Fondon and Garner 2004). The study that first proposed this association reported that all four Great Pyrenees with bilateral polydactyly examined were homozygous for the *Alx-4* (51 nucleotide repeat) allele, whereas none of the other 88 breeds examined for this locus, lacking polydactyly, had a deletion in *Alx-4*. A single Great Pyrenees that lacked the extra dewclaw was homozygous for the normal, full-length allele shared by all other nonpolydactylous breeds (Fondon and Garner 2004). What particularly supports the causal association of this extra dewclaw polydactyly are observations in *Alx-4* knockout mice (Qu et al. 1998). This modification eliminates the ability of *Alx-4* to bind with some molecules and alter gene expression in limb bud mesenchyme. The discovery of the *Alx-4*-polydactyly relationship was highlighted as one showing that length variation in tandemly repeated sequences of genes involved in development are a major source of genetic variation that leads to morphological variation in vertebrates (Fondon and Garner 2004).

The oldest well-known documentation of polydactyly in a domesticated animal I know of is that of chicken, with a three-legged chick with five toes (instead of the regular four) on its third leg by Aldovandri in 1600.

Work in recent decades on limb morphogenesis using the chicken model has provided an understanding of the regions of the limb bud and the molecules involved in the patterning of the limb. This has allowed tying genetic changes to developmental mechanisms and to adult morphology, including polydactyly.

An important discovery in the wing bud of the chick was a key signaling center that is responsible for the anteroposterior patterning of the digits: the zone of polarizing activity, or ZPA (Saunders and Gasseling 1968). This discovery was made after blocks of posterior tissue were transplanted to the anterior rim of the limb bud, which led to additional digits formed in a mirror image to the normal digits.

In another classic grafting experiment, it was shown that by implementing diverse transplantations to different anterior positions, the effect of the ZPA transplant was dose-dependent. Varying the amounts of inserted tissue resulted in varying numbers of mirrored toes in the anterior position (Tickle et al. 1975). The Sonic hedgehog (*Shh*) gene product was later identified as the responsible morphogen in the ZPA (Riddle et al. 1993). The *Shh* pathway has been shown to play a role in the development of many organs, and its study has benefited from the study of experiments (Fig. 4.13) and of chicken mutants. For example, the chicken mutant oligo-zeugodactyly has a large deletion within a regulatory DNA sequence that shuts off *Shh* expression in the ZPA of the posterior limb (Maas et al. 2011). This is a deviation from the norm: an evolutionarily conserved, non-coding ~800-base-pair sequence. Much can also be learned by studying variations among breeds.

Silkie breed chickens, like Great Pyrenees dogs, are preaxially polydactylous—it possesses additional digits in the region of the hind toe. There is much variation among individuals, some having four toes like most other chickens, some having just an extra bone in the hind toe, some possessing a fifth toe branching off from somewhere along the length of the hind toe. Some individuals have even six or seven toes (van Grouw 2018). A mutation results in the addition of toes from digit 1, producing a mirror image of the foot. The Silkie reportedly develops ectopic ("misplaced") anterior limb *Shh* expression (Dunn et al. 2011; Maas et al. 2011) while having a single base pair change within the regulatory region of the zone of polarizing activity, related to *Shh* expression.

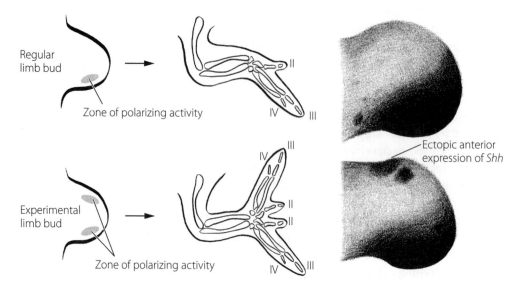

Fig 4.13. Experimental generation of polydactyly in chicken and mouse. Left two panels illustrate digit doubling in a chick wing following the anterior insertion of an additional zone of polarizing activity (ZPa). The right panel shows (*top*) the wild type forelimb bud of a mouse embryo showing the regular *Shh* expression (dark stain) in the posterior ZPa region; and (*bottom*) the ectopic anterior expression of *Shh* in a transgenic embryo expressing the Hemingway mutation.

The Silkie chicken

The Silkie is one of the most peculiar chicken breeds and a great subject for developmental biologists for reasons other than its common polydactyly. Its feathers lack a flat web as well as melanocytes, and their abnormal barbules and the absence of barbicels result in their silky appearance, typical also of related breeds. One of the peculiarities of Silkies is the extensive hyperpigmentation of dermal and connective tissues, making this breed a model of melanoctye precursor and neural crest cell migration and proliferation in the developing embryo (Dorshorst et al. 2010). The skin is blackish, as is the surface of the bones and viscera (Arisawa et al. 2006).

Mutations in the regulatory sequence associated with the ZPA were also recorded in other domestic animals and in humans. Different mutations are associated with different conditions, such as the triphalangeal thumb syndrome, in which there is one additional phalangeal bone in the thumb (Lange and Müller 2017). By comparing some genes of the ZPA regulatory network associated with preaxial polydactyly in the hind limb among dog breeds with mice and humans, commonalities are found (Park et al. 2008).

The ocurrence of extra digits in cats has been studied, and polydactyl cats are recorded for many breeds varying in shape, size, color, and behavioral patterns. The writer Ernest Hemingway (1899–1961) had polydactylous cats in his home in Florida, USA, and this led to the name "Hemingway mutation" to refer to a mutation affecting the regulatory region of digit formation. A biased and discontinuous distribution of supernumerary digits and toes has been documented in a population of cats with the Hemingway mutation. As these phenotypic effects cannot be explained by a point mutation alone, a model, termed the Hemingway Model, has been proposed (Lange et al. 2014). There is a random bistability of individual cells in the limb area affected by the mutation, and different, discrete character states are generated via developmental threshold effects. Changes in developmental parameters thus give rise to different morphological patterns and phenotypic novelty.

Domestication and malformations

Some commonly accepted standards of normality among domestic animals are actually malformations (De Boom 1965), for example, the lack of horns of some cattle such as the Aberdeen Angus, Red Poll, and Poll Hereford; the brachycephaly of some dog breeds (chapter 7), and the complete absence of caudal vertebrae in Manx cats (presumably from the Isle of Man, UK) with the rumpy phenotype (Buckingham et al. 2013).

There are numerous malformations in chickens produced by known inherited mutations (Romanoff and Romanoff 1972). Stocks of different kinds of chickens developed and maintained by research institutions and agricultural science programs preserve populations with malformations that are subjects of research (Delany 2004).

Malformations present at birth that result from errors arising during development are called congenital defects. The expression of deleterious genes, especially those that are recessive, is more common in highly inbred lines. Malformations can be inherited or can be the result of an in utero infection. Induced birth defects have been much studied in domesticated forms, in particular in relation to teratogens. An example was the discovery of the deleterious effects of hyper- and hypovitaminosis A in pig and rodent embryos, which induces anomalies during the closure of the neural tube, resulting in brain, eye, and heart defects (Noden and Lahunta 1985).

Differences among domestic species in placental transport function mean that the results of some drug tests cannot be directly translated to humans. For example, thalidomide is not teratogenic in rodents and it is so only in high doses in many other species, whereas it is highly teratogenic even at low doses in humans, in which it induces limb defects. In contrast, acetylsalicylic acid is a potent teratogen in rodents and in dogs but not in humans (Noden and Lahunta 1985).

Tissues are most sensitive to disruptions by exogenous factors at times in which morphogenesis and inductive interactions occur. This critical period is usually during the first weeks of development (Noden and Lahunta 1985). An example is the abnormal development of neural plate tissues during gastrulation following the influence of alcohol, as studied in mice. Postnatally, this is characterized by reduced head size, narrow forehead and a small nose among several other features (Sulik and Johnston 1983).

Studies of anatomy associated with malformation, teratology, have a long history in domesticated mammals (Nieberle and Cohrs 1967; Foley et al. 1979). Wright and Wagner (1934) studied more than 300 cavy mutants that exhibited a variety of head malformations. These could be classified along a series of increasingly severe conditions. The less affected condition exhibited a slightly shortened lower jaw and loss of the incisors. Increasing degrees of malformations included progressive reduction of head structures, including a lack of nostrils, the formation of a univentricular telencephalon, and associated cyclopia. These malformations are not exclusive to cavies but instead are widespread among vertebrates, as shown in surveys and catalogs of them (Nieberle and Cohrs 1967). These common patterns are seen as indicative of shared developmental biases and common occupation of particular portions of the potential morphospace (Alberch 1989).

Strong deviations from the wild forms in some breeds and some features resemble human malformations (Diogo et al. 2017), and in fact in many cases they share similar genetic and developmental bases.

The protein-coding gene *BRAF*, part of a signaling pathway related to memory and learning, can lead to a broad range of craniofacial defects in humans (Cesarini et al 2009). *BRAF* has been under positive selection in cats and horses (Theofanopoulou et al. 2017). Through interactions with other genes, *BRAF* inactivation can lead to malformations resulting from disrupted neural crest development (Newbern et al. 2008). *BRAF* is implicated in Noonan, Leopard, and cardiofaciocutaneous syndromes, with symptoms including a prominent forehead, bitemporal narrowing, and short stature (Keyte and Hutson 2012; Sarkozy et al. 2009). Another example is the glutamate receptor *GRIK3*, which showed positive selection signals in genomic studies of dog and cattle, and which interacts with other receptors associated with positive selection signals in the horse (*GRID1*) and cat (*GRIA1/2*). These receptors have been related through associations with other genes to craniofacial anomalies (Sahoo et al 2011).

Chondrodysplasia is a pathological condition of humans but a characteristic of some healthy breeds of dogs and cattle, also reported for rabbits (Brown and Pearce 1945) and sheep (Thompson et al. 2005). This disorder in skeletal formation results in a short-legged phenotype. It results from disturbed cartilage and endochondral bone growth and consists of disproportionately short, thickened, and curved long bones (Almlöf 1961; Jezyk 1985; Kealy et al. 2011). The terms achondroplasia and condrodrystrophy, among others, have been used in the veterinary and medical literature to describe forms of this condition.

Chondrodysplasia defines at least 19 dog breeds, including Basset Hound, Dachshund, and Corgi (Almlöf 1961; Parker et al. 2009). This condition is associated with expression of a recently acquired retrogene encoding fibroblast growth factor 4 (fgf4). This is an example of how a small percentage of mutations can encode functional proteins in a way that leads to a morphological novelty, and with this, how a single evolutionary event can have a role in biasing and directing phenotypic diversity.

Chondrodysplasia occurs in individuals of the Dexter breed of cattle. Breeding two chondrodysplasia-affected Dexters together can lead to premature abortion of the fetus, a stillborn calf with short limbs, a protruding lower jaw, swollen tongue, compressed nose, and bulging head (De Boom 1965).

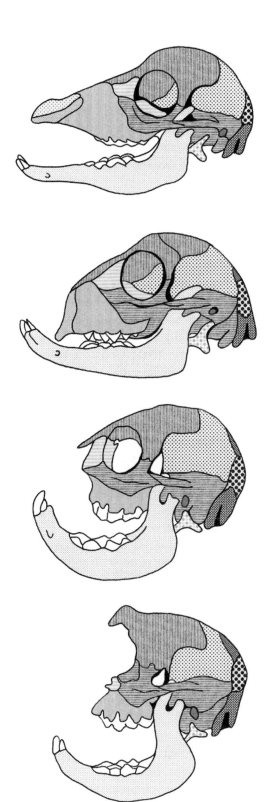

Fig 4.14. Lateral view of skulls of newborn lambs. From top to bottom: normal condition, mildly defective, "cebocephalic," and cyclopic, as summarized in the work of Rui Diogo and colleagues. The different shades and tones illustrate homologous skull bones.

Likewise, the patterns of connections of parts (e.g., muscles) in networks and the accommodations of different kinds of tissues to the new topography of the skeleton in the abnormal forms is similar across taxa. These patterns of connection have been quantified in several studies concerned with anatomical networks of musculo-skeletal structures (Diogo 2017). An example is the case of newborn lambs with various malformations, ranging from mild defects to cyclopia (Diogo et al. 2019; Fig. 4.14). Deviations from the norm in the muscles are less marked than osteological abnormalities. When an osteological structure is missing, the muscle that attaches to it then attaches to the "nearest topological neighbor" of the missing one, a pattern also found in cyclopic humans.

Evolutionary veterinary medicine

Evolutionary medicine addresses the explanations for various medical conditions, matters concerning adaptation to a changing environment, host defenses, virulence, and evolutionary mechanisms underlying disease conditions. A leading idea is that human biology evolved in a Paleolithic environment, in marked contrast to modern lifestyles (LeGrand and Brown 2002). Domestic animals are at the core of our modern lives and subject to similar problems themselves (Böhmer and Böhmer 2017). Cats and dogs that are overweight because of lack of exercise or bad diet, or both, are examples of the mismatch between the biology of these animals and their pet life. More importantly, economically useful physical traits resulting from selective breeding in many species ultimately interfere with their functioning under normal environmental conditions. Examples of domestic animals at the edge of their physiological capacities include metabolic diseases of dairy cows, obstructed labor at birth (dystocia) in double-muscled cattle (Fiems 2012), musculoskeletal injuries of racehorses, and bone and tendon deficiency disorders in domestic poultry (Toscano 2018). Many veterinary problems arise from artificial selection for traits that humans perceive as aesthetic. Examples in dogs abound, including parturition and breathing difficulties in brachycephalic breeds, intervertebral disk disease in Dachshunds, and corneal irritation from excessive skin folds in certain breeds.

Specialists treating the same disorders in different species can learn from each other. There are many overlaps between human and other-animal pathology. An example is stress incontinence in women and sphincter mechanism incompetence in the female dog (Janssens and Peeters 1997). Another example is capture myopathy, when animals held by well-intentioned vets die of a sudden surge of adrenaline, as can happen when wild animals are caught by predators. The analogue situation in humans may occur when patients are restrained in hospitals, and possibly also when infants experience shock while lying on their stomachs (Natterson-Horowitz and Bowers 2013). The Zoobiquity Conferences (Natterson-Horowitz and Bowers 2013) were created to bring together veterinary medicine, clinical human medicine, and evolutionary biology to foment exchange and research.

Animals themselves engage in activities relevant to their health and can be explored from an evolutionary perspective. This is the case with self-medication behavior and the preventive or therapeutic use of substances or materials to manipulate behavior or alter the body's response to parasites or pathogens (Roode et al. 2013). In some areas of the world people have obtained ethnobotanical knowledge through animal observation, as in the consumption of some anti-parasitic plants by goats in Uganda (Gradé et al. 2009). In general, application and management of local ethnoveterinary medicine knowledge could be an important and economical strategy (Wanzala et al. 2005).

Summary: Variation in evolutionary development of domestication

Development should not be understood as a matter of gene expression only, which is under environmental and epigenetic control. Although the study of gene frequencies and mutations has provided great insights into the origin of numerous traits that result from selective breeding, they alone cannot explain the morphological diversification accompanying the domestication process. Instead, the network of interactions, integrations, and regulations that characterize developmental genetics and the environment should be considered.

Genetic programs unfold in ontogeny, and their alterations during domestication lead to morphological novelty. An evo-devo research program that integrates classical and comparative as well as experimental approaches is quite suitable for studying the origin of the stunning morphological diversity brought about by domestication. Numerous comparative and quantitative approaches are available to document the evolution of the phenotype in parallel with experimental approaches. Phenotypic divergence in development can occur at different time windows.

Genomic work has shown correlations of genes with derivatives of the neural crest. The developmental mechanisms involved that result in phenotype changes remain a black box. Given the many interactions in the cell niche of neural crest descendant lineages, would it not be naive to expect a one-to-one relationship between neural crest cells and the traits that are eventually constructed by neural crest derivative tissues?

Interactions between mechanical forces and cells and tissues involved in the processes of cartilage and bone development can result in phenotypic changes. An organism's phenotype can respond to changes in its environment, as in the case of the two-legged goat that could walk, an example of developmental plasticity. Other examples are the adjustment of tooth growth to wear, and the impact on fur coloration by enzymes affected by temperature.

Little is known on quantitative comparative embryology of breeds of domestic animals. What we know does not reveal a common or neat pattern of change, for example, a phylotypic period in the middle of a "perfect" hourglass shape of phenotypic change across developmental time.

CHAPTER 5

Ontogenetic change

Diverse developmental mechanisms generate phenotypic divergence. Repatterning, modularity, and changes of ontogenetic trajectories are aspects of developmental change I treat in this chapter. Investigations of these subjects benefit from methods aiming to quantify both morphometric and discrete ontogenetic data.

Developmental repatterning

Developmental repatterning leading to evolutionary change (Fig. 5.1) can occur via changes in position or in topology, amount, kind, and timing (Arthur 2011). This repatterning does not concern a total or global change of the organism, "the entire phenotype," nor a particular change operating in all time windows. Developmental repatterning may have occurred in evolution for some regions of the body and not for others, as discussed under the subject of modularity below.

Heterotopy means "different place" and refers to changes in the spatial/structural aspects of the ancestral developmental program, particularly in the location of developmental change. An example is the presence of feathers on the legs of some chicken and some pigeon breeds. In a morphometric sense, heterotopy means that a descendent form differs from the ancestor in its trajectory of ontogenetic shape change in that the ratios of change are not conserved.

Heterometry refers to changes in quantity, size, or concentration at different levels. Discoveries of this kind of repatterning include the supernumerary number of toes in some dog breeds (chapter 4), horns in goats or sheep, and changes in brain size (chapter 7). There is heterometry in the early development of the rose comb of chickens, one of many kinds of combs (Figure 5.2), and other kinds of repatterning of chicken combs are known (as described below). Another example concerning chickens is the patterns of neural crest migration in Silkie chicken as they relate to pigmentation. The number of neural crest cells migrating is more extensive in this breed than in others and presumably more than in the red junglefowl, resulting in almost ubiquitous black color (melanocyte activity) in every tissue in the body except the feathers (Reedy et al. 1998).

Heterotypy involves a change in kind. The significant morphological changes known in goldfishes (chapter 9) could offer a system in which to study repatterning

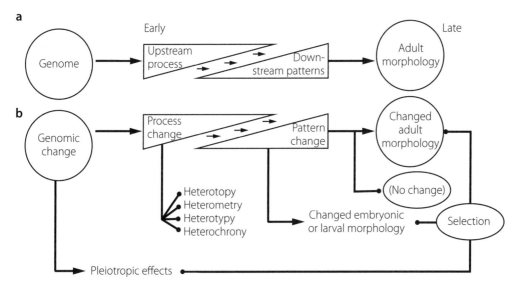

Fig 5.1. Developmental repatterning in evolution schematic diagram. (a) The ancestral state, and (b) the descendant, showing developmental repatterning relative to (a). Developmental changes can be divided into patterns: downstream changes in developmental events, resulting from processes, the upstream changes in developmental mechanism. Early developmental events tend to be classified as processes, whereas late events tend to be seen as patterns. The changes in developmental processes may themselves be heterochronic or of a different nature. Developmental repatterning may be subject to selection. The chain of events leading from genomic change, via altered developmental processes, to a pattern, may or may not lead to changes in adult morphology. Pleiotropic effects may occur, as signified at the bottom of (b), as genomic change may alter developmental processes in more than one system.

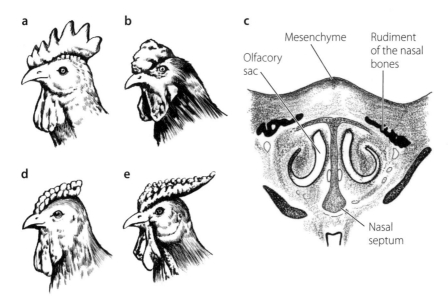

Fig 5.2. Chicken comb types. Different kinds of combs in chickens: (a) single, (b) walnut, (d) pea, and (e) buttercup; (c) cross histological section of a portion of the embryonic head, at the level of the lachrymal duct, illustrating the rudiment of the rose comb at early stages of development in an 8-day chick embryo.

of this type (Ota and Abe 2016). Some of the different adult structures of the comb of chicken breeds, for example, result from fundamental changes recorded in the anlage or precursor of such structures (Fig. 5.2). The comb is an integumental outgrowth, common and variable in shape throughout modern birds. It is a relatively simple structure in red junglefowl and in some breeds, whereas in certain breeds the comb is instead pea, rose, walnut, or v-shaped (Núñez-León et al. 2019). The peacomb shows three main longitudinal rows of tissue, in contrast to the single comb, which comprises one row of tissue on the dorsal side of the head. Morphological differences related to each kind of comb can be easily distinguished at around 16 days of development among fowl and chicken embryos, and at 8 days the rudiments of some kinds of comb are already different.

Heterochrony involves changes in developmental timing, a subject much studied in domesticated and wild animals. Heterochrony can be defined in terms of shifts in onset, cessation, or rate of growth (Alberch et al. 1979; Klingenberg 1998). It can be studied by comparing ontogenetic trajectories of size and shape among species. A traditional distance analysis approach can capture most morphological variation (for example, in skulls) in a simple and intuitive way. In this distance or growth trajectory approach, relationships among measurements are examined, resulting in bivariate or multivariate coefficients of allometry (Weston 2003), which are amenable to analysis in a phylogenetic framework (Giannini 2014; Wilson 2018). However, to characterize morphological complexity, the geometric morphometric approach is better suited than traditional distance analyses (Mitteroecker et al. 2004). Thus, the classic approach to study heterochrony involves the analysis of size and shape parameters, which is operationally related to the study of allometry by involving morphometrics (Smith 2001a; Lawling and Polly 2010).

The modern analysis of phenotypic trajectories is based on geometric morphometrics, from which trajectories through multidimensional shape space can be generated and compared. Assuming that group-specific trajectories are fairly linear, their directions can be compared statistically.

An alternative approach to growth heterochrony is sequence heterochrony (Smith 2001a). It deals with changes in the timing of discrete events and it avoids size and absolute time as the measures of comparison. One can incorporate developmental events into a matrix such that the timing of events can be compared with one another to form "event pairs" (Smith 2001a). Besides a simple mapping of event-pairs, methods involving reconstruction of evolutionary changes in the sequences include event-pair-cracking (Jeffery et al. 2002), "parsimony-based genetic inference" or PGi (Harrison and Larsson 2008), and examining the entire sequence and performing comparisons based on squared-change optimization and independent contrasts analysis (Germain and Laurin 2009). In addition, several statistical methods exist to study developmental sequences besides those with a phylogenetic emphasis, as described above (Maxwell and Harrison 2009). These include, for example, nonparametric methods to test for the overall divergence of ranked data (Nunn and Smith 1998), evolutionary lability or amount of change in sequences (Poe and Wake 2004), and the relative timing of development between serially homologous structures (Bininda-Emonds et al. 2003). Sequence heterochrony is the proper approach to study events that occur early in development when size and shape are not appropriate reference subjects. These events include cell and tissue specification and differentiation, and formation of organs and skeletal elements.

Heterochrony, in particular pedomorphosis, has been suggested to underlie behavioral and morphological changes during domestication; for example, dogs have been classically studied from this perspective (Wayne 1986; Morey 1994; Price 1999). The skulls of dogs are treated below; limbs have also been studied in this context (Sommer 1931; Lumer 1940; Dechambre 1949; Wayne 1986). In both dogs and pigs, it has been shown that despite similarities between the domestic form and juveniles of the wild form, there is no simple global neotenic growth pattern. Instead, localized neomorphic features and heterochrony may expain growth differences, as discussed below.

Modularity

Explanations of the origins of the morphological diversity generated by domestication require an understanding of how parts of the organism (modules) compare in terms of variation, and how this modularity is generated in development. By modules I mean assemblages of parts that are tightly integrated internally by relatively many, strong interactions but relatively independent of one another (Klingenberg 2005, p. 224). Modularity expresses the extent to which a system can be partitioned into distinct components (Payne and Wagner 2019). Many studies of modularity quantitatively explore questions relating to facilitation and constraint in morphological evolution (Goswami et al. 2014).

Modularity is a subject usually treated in relation to ontogeny. Because of modularity (Wagner et al. 2007), some anatomical regions may undergo developmental repatterning in evolution, while others do not. Vertebrate development is modular and stepwise: individual bones and structures form according to a defined and coordinated schedule of events (Goswami et al. 2009). The patterns of trait interactions during development can be modular, and this modularity is a fundamental aspect of the generation of diversity in development.

Most current work on modularity is being carried out on wild adult animals (Porto et al. 2009; Felice and Goswami 2018). The aim is to understand and quantify, under some assumptions, how the direction in which variation is generated is biased at both micro- and macroevolutionary levels (Wilson 2013). But the correspondence between adult anatomy and developmental modules is poorly understood (Minelli 2017).

There are few reports of modularity during development despite the significance of the subject. In terms of skeletogenesis, studies are largely restricted to mammals. A study of developing skulls of a placental versus a marsupial mammal found significant changes in cranial integration through ontogeny, with a mixture of increased, decreased, and stable levels of integration across distinct cranial regions (Goswami et al. 2012), yielding a "palimpsest" pattern (Hallgrímsson et al. 2009). A study comparing the prenatal and early postnatal cranial ontogeny in 10 species of therian

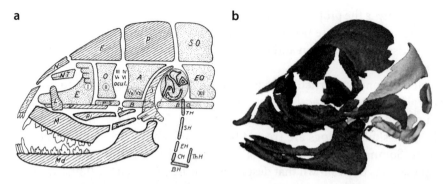

Fig 5.3. Mammalian skull. (a) Schematic representation of a mammalian skull. Dermal bones are hatched, endochondral bones dotted. Parts of the visceral skeleton are also depicted. Labels: I–XII cranial nerves. A alisphenoid, B basisphenoid, BO basioccipital, E ethmoid, EO exoccipital, F frontal, L lacrimal, M maxilla, Md mandible, N nasal, O orbitosphenoid, P parietal, Pl palatine, PS presphenoid, Pt pterygoid, S squamosal, SO supraoccipital, T ectotympanic, V vomer, Z jugal. Between the S and EO is the petrosal bone, the three ear ossicles and the ectotympanic bone. (b) Dermal (darker) and endochondral (lighter) bones in the cranium of a prenatal cow, *Bos taurus*; 3D reconstructed from computer tomography generated images.

mammals found that marsupials show higher integration of the oral apparatus in early postnatal ontogeny than placentals, which is likely associated with the extended period of continuous suckling in marsupials, during which most of their development occurs (Goswami et al. 2016). This exemplifies how integration and modularity in development evolve, how these can be coupled with life history features, and how they can affect morphological outcomes and patterns. Furthermore, developmental features of marsupials have been hypothesized to have had an effect on the morphological diversification of the group.

Skeletal development includes earlier processes than the onset of ossification that could also be studied in terms of relative timing and its evolution: migration of cells (e.g., neural crest cells or somite derivatives), epithelial-mesenchymal interactions, condensation formation, and differentiation (Hall and Miyake 2004; Kavanagh 2020). Studies of these could be informative about the generation of adult patterns of phenotypic variation.

Discrete developmental events can be investigated in the context of modularity. An example is the appearance of ossification centers, thus serving to document skull formation or skeletogenesis in general. These centers can be observed through noninvasive imaging, using high-resolution X-ray microtomography.

In mammals, including many domesticated forms, the modes of ossification endochondral and dermal unite bones into integrated two modules of heterochronic changes (Figure 5.3); however, the supraoccipital among skull-roof bones diverges from the constrained pattern (Koyabu et al. 2014). The onset of ossification in this dermal bone is correlated with brain size. Thus, cranial heterochrony in mammals occurred within a conserved modular organization and was influenced by encephalization (Koyabu et al. 2014).

Modularity in the skull of mammals and birds

Many studies on patterns of modularity concern the skull and use inter-trait correlations extracted from geometric morphometric data in large groups of organisms (Klingenberg and Marugán-Lobón 2013). A common pattern of modularity in therian mammals has been found, with variation in the magnitude of integration among species (e.g., Goswami 2006; Porto et al. 2009). Six distinct modules were identified for marsupials and placentals in a geometric morphometric study of 106 mammal species: anterior oral-nasal, cranial base, cranial vault, orbit, molar, and zygomatic-pterygoid, but only three modules display significant integration in all species (Goswami 2006). An aspect of integration and modularity that remains to be studied beyond skulls is its relation to the surrounding soft tissues, such as the brain and muscles (Richtsmeier et al. 2006; Esteve-Altava et al. 2015).

A comprehensive study implementing a geometric morphometric method that more thoroughly captures shape reported that bird skulls consist of seven modules (Felice and Goswami 2018). According to quantifications of rate and disparity in their evolution, these modules are different from each other, and the differences relate in part to their developmental origin. Regions derived from the anterior mandibular-stream cranial neural crest or from multiple embryonic cell populations have evolved more quickly and exhibit more morphological diversity. There are disparate evolutionary rates among cranial regions for individual clades (Tokita et al. 2016). For example, parrots exhibit high evolutionary rates for traits throughout the skull, whereas the closely related falcons and caracaras exhibit high rates only for traits of the rostrum (Felice and Goswami 2018). There may not be a universal association among integration, modularity, and diversification—it may follow different directions in different clades. In any event, it is important to examine the matter in the proper phylogenetic context before reaching any conclusion (Navalón et al. 2020).

Comparatively little is known about the lability of modular patterning and magnitudes of integration in smaller groups within mammals and birds, particularly in groups that evolved on relatively short time scales and under selective breeding regimes, as in domestication. Only recently have we started to explore whether modularity differs between wild and domesticated forms, and, if so, how this may be related to the morphological changes that occur during domestication in its different kinds, including artificial selection (Wilson et al. 2021).

The face and the neurocranium are two fundamental modules of the skulls of mammals. Some breeds of several species exhibit orofacial disproportions, with mismatching upper and lower jaws and maloccluded teeth (chapter 7). This disproportion could be due to lower integration strength between the modules of the skull. This hypothesis has been tested for dogs in comparison to wolves. Two independent studies found that the patterns of covariation in dog and wolf skulls are similar. It was concluded that the higher skull shape diversity in dogs is not explained by less-integrated skull modules (Curth et al. 2017). The first study (Drake and Klingenberg 2010) quantified skull shape using geometric morphometrics of 677 adult dogs of 106 domestic breeds and of large samples of Carnivora, including wolves, coyotes, and golden jackals. The second study (Curth et al. 2017) used

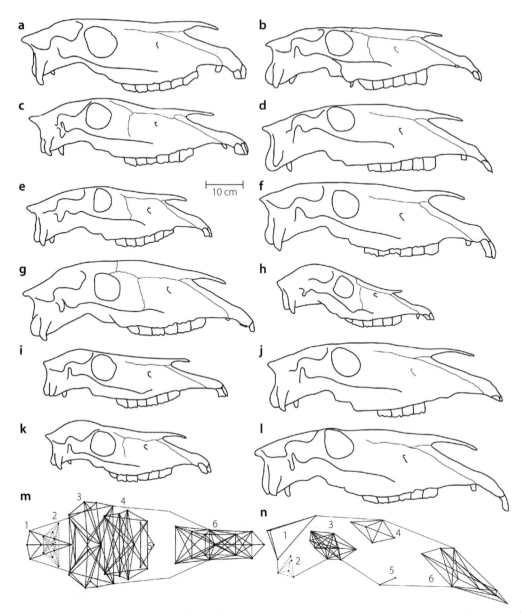

Fig 5.4. Examples of skull shape in equids and modularity in the horse skull. Lateral views of (a) Przewalski's horse, (c) zebra, (e) donkey and different domesticated horse breeds, (b) English Thoroughbred, (d) Hannoverian, (f) Arab (light horse), (g) Kladrubian, (h) Falabella pony, (i) Icelandic pony, (j) Belgian (draft), (k) Shetland pony, (l) Shire (draft). (m–n) Modules in the skull of horses: cranial vault (1), cranial base (2), zygomatic-pterygoid (3), orbit (4), molar (5), and anterior oral-nasal (6).

geometric morphometric quantification of the cranium (subdivided into rostrum and braincase), the mandible (subdivided into ramus and corpus), and the upper and lower tooth rows using CT images of 196 dog and wolf skulls. The analyses explored integration strength and covariation (using partial least squares analysis) of modules in relation to the patterns of disparity or shape diversity found.

Modularity in adult skulls was also investigated in horses, including Przewalski's horses, donkeys, and zebras (Heck et al. 2018b). Maximum likelihood model-based tests showed that the horse cranium is composed of six modules, as reported for placental mammals in general (Goswami 2006; Figure 5.4). The magnitude of integration in the skull of the domesticated horse was lower than in wild equids across all six cranial modules. Furthermore, lower values of integration

---

Has domestication produced morphological innovation or novelty?

The term innovation has been used to denote the process by which a new feature has evolved, and the term novelty is used to designate that new feature. It is often assumed that innovations involve the origin of new developmental modules and that tinkering and cooption of preexisting regulatory gene networks are involved in this process (Santos et al. 2014). According to West-Eberhard (2003, p. 198), a novelty is a "phenotypic trait that is new in composition or context of expression relative to established ancestral traits." Novelties are also defined as resulting from transitions between peaks in the fitness landscape (i.e., changes in the evolutionary trajectory) coupled with circumvention of "developmental constraints," which then allows the generation of variation in ontogeny that previously could not be produced (Hallgrimsson et al. 2012). Santos et al. (2014, p. 2) refer to innovations as "lineage-restricted traits linked to qualitatively new functions."

In the strictest sense of the term, novelties are new structures in evolution, such as feathers, or limbs: a major, new, qualitative transformation. Seen this way, for all the remarkable diversity of domesticated animals, no true novelty has ever evolved in any domesticated species. Domestication is largely about changes in—or even intense selection for—existing structures; thus feathers change in shape or size or color, but no new structures are generated. In fact, novelty should not be considered a property or feature as a whole, as it can be decomposed into multiple traits, traceable to separate homologues (Minelli 2017). Novelties comprise not only the emergence of major body plans, or additional structures without homologues, but also changes in the variational patterns of existing structures (Müller 2010).

If we accept this more inclusive definition of "novelties," then we can refer to them in domesticated animals. Examples include the twin-tail of some goldfishes, telescope and other head conformations also of some goldfishes, brachycephaly in several domestic mammals, bizarre integumental features in several pigeon breeds, cranial crests and associated brain changes in some domesticated breeds, and the peculiar horns of some goats and sheep.

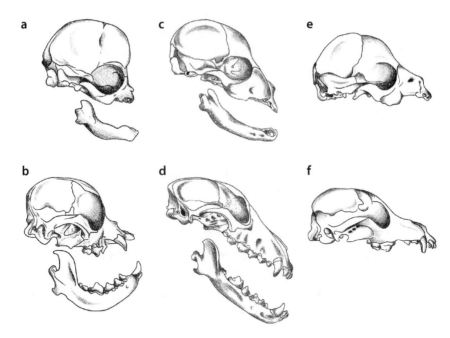

Fig 5.5. Prenatal changes are important in generating skull shape differences among breeds of dogs. Different dog breeds can be identified at the newborn stage (top row): (a–b) Pekingese, newborn and adult; (c–d) Bedlington-Terrier; (e–f) Whippet.

were associated with higher disparity values across all modules. To propose a causal relation between this adult modularity and disparity would be mere speculation.

Ontogenetic trajectories of shape before birth in mammals

Is adult disparity generated before or after birth/hatching? Although intuitively one might think that it originates postnatally, since growth can produce important changes in proportions, much of species differentiation happens during embryonic development. Quantification of ontogenetic trajectories across a small sample of eight rodents (Zelditch et al. 2016, p. 188) showed that "species differ in their post-natal ontogenies but infants are neither more nor less disparate than adults and the major dimensions of disparity distinguishing the main clades also do not change." As gestation length greatly differs across clades, even among species of similar size at birth, rates of intrauterine (or in ovo) growth must vary considerably.

Domestication can affect prenatal development, as demonstrated in the case of dogs and other mammals by the fact that newborns of different breeds can be distinguished from each other (Starck 1962; Rosenberg 1966; Fig. 5.5). Distinguishable newborns, based on several cranial metrics, were also reported for pigs of a short-snouted breed, the German Large White (Deutsches Edelschwein), common and generic domestic German breeds, and wild boar (Kim 1933).

Prenatal growth has rarely been quantified (McLaren 1976; Goswami and Pro-chel 2007). Growth rates can be quantified in many species, using simple measure-

Table 5.1. Average size (crown-rump length, CRL) of embryos of different ages

Days	Comparative information CRL (mm)						
	Sheep	Goat	Pig	Dog	Cat	Cow	Horse
20	10		12	10	10		
25	15		20	15	19		
30	20		25	20	31	8–22	30
35	30		30	36	45		
40	40		50	55	70		
45	55	43	65	82	85		
50	80	57	85	118	108		
55	90		106	145	125		
60	110	85	125	165	140	53	60
70	160	120	160				
80	195	149	210				
90	230	197	240			130	110
100	280		270				
110	320	290					
120	370					245	165
130	420	355					
140	460						
150						325	300
180						450	405
210						560	490
240						690	630
270						810	740
300							840
330							1100

Sources: Data from Evans and Sack (1973), Habermehl (1975), and Richter and Götze (1978); summary in tables presented in Schnorr and Kressin (2006).

ments and average gestation length compiled from several sources (Table 5.1). These references could serve as bases for an extensive revision of organogenesis with new data and analysis. Recent works on the development of domestic mammals are descriptive in nature but otherwise illustrate the possibilities for exploring rich data sources in collections (e.g., Bactrian camel: Kinne et al. 2010; horse: Barreto et al. 2016).

The head, neck, trunk, and tail become recognizable at different times of embryonic development in different species of domesticated mammals (Werneburg et al. 2016; Lawrence et al. 2012). It is currently unknown the extent to which this timing is affected by domestication, and, if so, whether it serves to explain adult morphological variation. Comparative studies of embryos could address this subject.

It is important to realize that characterizing growth patterns is fundamental to understanding how adult anatomy forms, but the mechanistic explanation of those

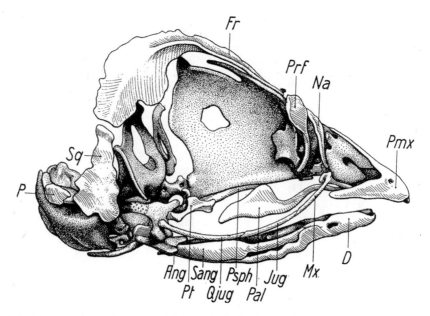

Fig 5.6. A late embryonic stage chicken skull. Embryo of total length 65 mm. Dermal bones are hatched and endochondral bones are dotted. Parts of the visceral skeleton are also depicted. Ang Angular, D Dental, Fr Frontal, Jug Jugal, Mx Maxilla, Na Nasal, P Parietal, Pal Palatine, Pmr Premaxilla, Prf Prefrontal, Psph Parasphenoid, Pt Pterygoid, Qjug Quadratojugal, Sang Surangular or Supra-angular, Sq Squamosal.

patterns is another matter. "Heterochrony," as discussed above, can refer to an observed pattern and not necessarily to a mechanism. Many factors can affect growth, and each of them and their interactions can be involved in alternative mechanisms behind patterns of growth. Among those factors are the parathyroid, thyroid, and growth hormones, vitamins A, C, and D, calcitonin, corticosteroids, testosterone, oestrogen, and insulin-like growth factor or IGF (Lawrence et al. 2012; Hall 2015).

Comparative studies of skeletogenesis are also potentially a rich source for insights on evolution. Data are available from descriptive studies of individual species (e.g., *Cavia*: Petri 1935), and from comprehensive comparative studies, including groups of placental mammals (Koyabu et al. 2014). Studies on birds also exist, some including domesticated forms (e.g., Maxwell 2008; Mitgutsch et al. 2011; Carril and Tambussi 2017; Arnaout et al. 2021).

The adult skull is an amalgamation of elements of different developmental (neural crest versus mesodermal) and ossification (dermal versus endochondral bones) origins. The embryonic development leading to the skull includes the formation of the cartilaginous anlage of much of the anterior and base portions of the skull—the chondrocranium. As one of the major components of the skull, it is partially of neural crest and mesoderm origin (Wada et al. 2011). Numerous individual chondrification centers appear at different times and fuse to form the chondrocranium. Many parts of the chondrocranium ossify during growth; others persist cartilaginous into the adult stage (e.g., nasal cartilages).

Plate 1. Top: Rock dove and examples of breeds of pigeon. Bottom: Red junglefowl and examples of breeds of chickens.

Credit: see image credits for Figures 2.13 and 2.14.

Plate 2. Top: Mallard and domestic duck breeds. Bottom: Wild and domestic geese.
Credit: see image credits for Figures 2.16 and 2.17.

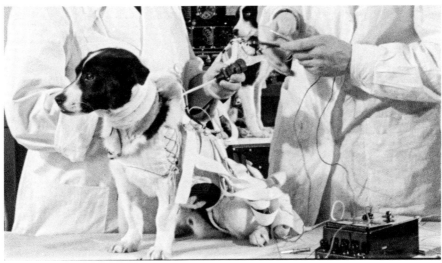

Plate 3. Top: Wolf and a sample of dog breeds. Bottom: A dog participating in the Space Dogs program in the Soviet Union.

Plate 4. Top: Examples of cat breeds. Bottom: A cat catching a small bird, a common phenomenon in many urban and wild environments around the world.

Credit: Top: see image credits for Figure 2.2. Bottom: 4 Vogeljäger-Katze_Losonsky_shutterstock_519787123.

Plate 5. Top: Examples of breeds of domestic cavy. Bottom: Wild rabbit and examples of breeds of the domestic rabbit.

Credit: See image credits for Figures 2.11 and 2.12.

Plate 6. Top: Examples of domestic goat breeds. Bottom: Moufflon and examples of sheep breeds.

Credit: Top: see image credits for Figures 2.4 and 2.5.

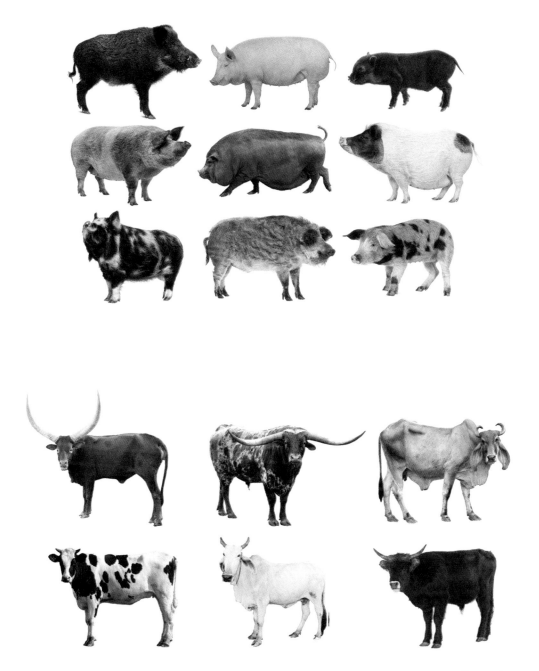

Plate 7. Top: Wild boar, *Sus scrofa*, and domestic pig, *Sus domesticus*. Bottom: Examples of domestic cattle breeds.

Credit: Top: see image credits for Figure 2.6. Bottom: clockwise starting top left: AnkoleWatusi_Anan Kaewkhammul_shutterstock_1562191900; Texas Longhorn_Charles Lemar Brown_shutterstock_1354403639; Zebu_Stepnex_shutterstock_1284626725; Schwarz-weisse_Kuh_VanderWolf Images_shutterstock_460450084; ox-bull-looking-straight-ahead-600w-1544610287; heck-cattle-called-reconstructed-aurochs-600w-20396650.

Plate 8. Examples of nonexistent domestic mammals and birds: no brachycephalic horses, no long-limbed cavies, no silkie ducks, no long-snouted cats, no long-necked chickens, and no duck-beaked pigeons.

Credit: Artwork Jaime Chirinos.

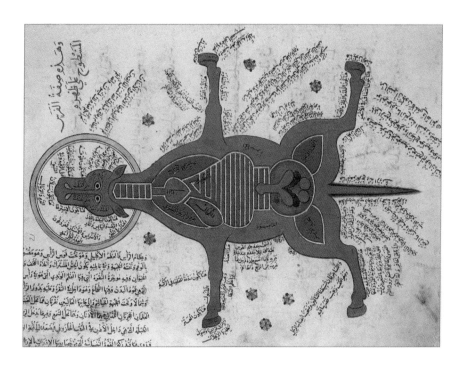

Plate 9. Top: Fifteenth-century depiction of horse anatomy in an Egyptian manuscript. Bottom: Eighteenth-century depictions of three kinds of domestic goldfish.

Plate 10. Top: Goats in an argan tree, Morocco. Bottom: Fat-tailed sheep for sale at a live-stock market, Kashgar, China.

Plate 11. Top: Dromedary camels in Petra, Jordan. Bottom: A Quechua woman herds her pack of Alpacas through the Andes. Ausangate, Cusco, Peru.

Credit: Top: camels-in-Petra_Algirdas-Gelazius_shutterstock_1308698119. Bottom. Alamy Image ID: 2A3EK3B.

Plate 12. Top: Cooking cavies in a street market in Cuenca, Ecuador. Bottom: Cavies for sale in a food street market in Ecuador.

Plate 13. Top: Yak and owner, Namtso Lake in Tibet. Bottom: Indian farmer and his cow.

Plate 14. Top: Industrial cattle farm in Para, Brazil. Bottom: Industrial cow milking facility.

Credit: Top: industrail-cattle-farm_PARALAXIS_shutterstock_1550122904. Bottom: cow-milking-facility_Official _shutterstock_95517259.

Plate 15. Top: Woman working in a traditional silk factory, Vietnam. The silkworm cocoons are soaked in hot water. Bottom: Eri Silkworms, the caterpillar of *Samia cynthia*, in the hands of a Karbi tribeswoman, northeast India. Eri are the only other completely domesticated silkworm besides *Bombyx mori*.

Credit: Top: woman-silk-factory_Angelina-Pilarinos_shutterstock_546533401. Bottom: Shutterstock B9HAH8.

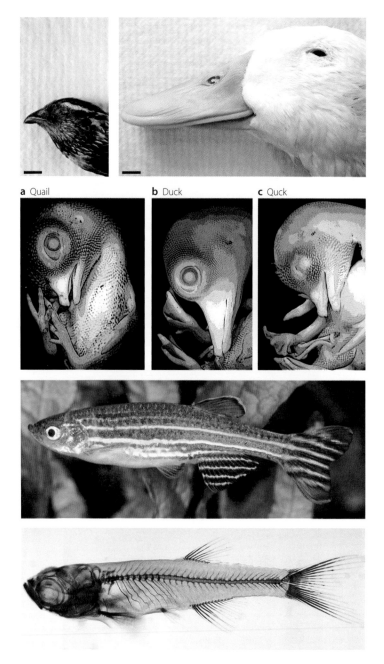

a Quail   b Duck   c Quck

Plate 16. Top: Images of adult quail and duck heads, used in experimental embryology to study neural crest development in the laboratory of Richard A. Schneider. Second row: Quail embryos have short, blunt beaks whereas duck embryos have long, broad bills. Transplanting neural crest cells, destined to form the beak, from quail donors to duck hosts produces chimeric "quck" embryos with quail-like beaks. Note that the quail-like quck has webbed feet indicative of the duck host. Reprinted from Schneider (2005). Third row: Zebrafish in aquarium. Fourth row: Whole-mount cleared and stained domesticated zebrafish, 15 mm standard length, of the AB line laboratory strain, commonly refereed to as "Wild Type." The animal was reared under laboratory conditions, and the vertebral column displays various types of malformations (Martini et al. 2020).

Credit: Top: modified from Eames and Schneider (2005), with permission from the Company of Biologists (via CCC). Second row: Courtesy of Richard Schneider, from Schneider (2005). Third row: Shutterstock zebrafish-danio-rerio-planted-aquarium-600w-718879114. Fourth row: Picture by Arianna Martini, courtesy of Eckhard Witten. See Martini et al. (2020).

Work on the chondrocranium of mammals and birds have rarely examined domesticated forms, and little is known about intraspecific variation (de Beer 1937; Maier 1993; Sánchez-Villagra and Forasiepi 2017). Much of the highly descriptive literature is from the first part of the twentieth century and is in German, though it includes a plethora of terms that tend to describe only a single or few stages.

To my knowledge, works on the chondrocranium of domestic mammals are limited and purely descriptive. They include the description of a few individuals of the Whippet breed at three embryonic "stages" (Schliemann 1966), and a 27 mm-long dog embryo of an unspecified breed (Olmstead 1911). There are also anatomical works on cat (Terry 1917), rabbit (Voit 1909, Mess 1999), and pig (Mead 1909), as well as an unpublished master's thesis from the University of Tübingen about a cow (Kuhn 2001; see also Fawcett 1918). The laboratory rat has been extensively studied by Ruf (2020). An image of the chondrocranium of a 26-day-old ferret embryo is available (Noden and de la Hunta 1985).

The few available descriptions of the chondrocranium in chickens pertain to single stages and use undetermined breeds (Hüppi et al. 2019; Fig. 5.6). Other descriptive works concerning domesticated birds concern duck, turkey, Japanese quail, and parakeet (Hüppi et al. 2021).

So far, detailed anatomical studies of the chondrocranium have necessitated histological serial sections followed by three-dimensional reconstructions, since not even the most sophisticated noninvasive imaging together with optimized chemical treatment of fresh embryos can serve to document cartilaginous structures (Metscher 2009). Perhaps a new framework for comparative work (Werneburg and Yaryhin 2018) and new methods of staining and imaging will open new avenues of comparative anatomical research in this area.

Developmental repatterning in domestic mammal skull growth

Skulls are commonly used markers of growth changes in mammals (e.g., Aeschbach et al. 2016; Flores et al. 2018), given their complexity and availability in museum collections. Such studies are usually restricted to postnatal growth, with exceptions (Koyabu et al. 2014). Most studies comparing wild versus domestic counterparts also concentrate on postnatal growth, because preserved skeletal material is more easily accessible. Examples include dog (Rosenberg 1966; Wayne 1986; Drake 2011; Geiger et al. 2017), cat (Kratochvil 1973), pig (Kelm 1938; Evin et al. 2017), dromedary camel (Al-Sagair and ElMougy 2002), horse (Radinsky 1984; Heck et al. 2019), cavy (Kruska and Steffen 2013), rabbit (Fiorello and German 1997), gerbil (Stuermer et al. 2003), and mink (Rempe 1970; Tamlin et al. 2009).

The differences in adult skull shape in some domestic forms (e.g., dog) is greater relative to adults than to juveniles of the wild forms (e.g., wolf). This phenomenon has led to the question of whether pedomorphosis (evolutionary juvenilization) has played a role in domestication (Wayne 1986; Drake 2011; Geiger et al. 2017).

Rütimeyer (1867) mentioned that shape differences between small and large forms of cattle mirror those in younger and older forms of the same species. He tried to explain differences among breeds within this framework. Theophil Studer

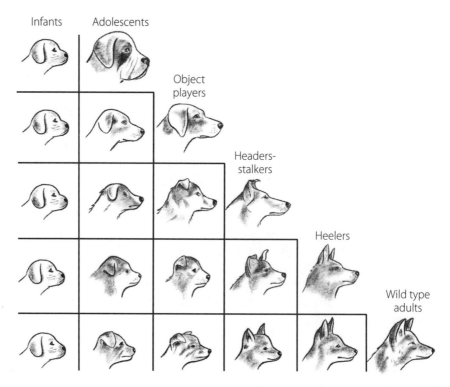

Fig 5.7. Raymond Coppinger's simplification of the pedomorphosis hypothesis of dog diversification. Ideas on juvenilization inspired much work in the early 1980s to study changes in developmental timing in evolution.

(1845–1922), the founder of a unique and extensive kynological collection in Bern (the source of specimens for many studies of dogs for decades) interpreted special features of the skulls of different kinds and breeds of dogs as retained juvenile features (Studer 1901). Hilzheimer (1877–1946), a prominent mammalogist who cofounded the German Mammalogical Society, proposed "juvenalization" as a common pattern across species resulting from domestication (Hilzheimer 1912, 1926). The simplest scenario in the case of dogs would argue for cessation of the ancestral growth trajectory, as depicted in a wonderful but sadly unrealistic illustration by Raymond Coppinger in 1982 (Figure 5.7). Klatt (1913) had already discussed and criticized these ideas at length using methods available at the time. Starck (1962) also criticized a naive juvenalization process as explanation for human and domestic animal features. He emphasized the mosaic nature of development, accelerations, retardations, and other deviations from the ancestral pattern of development that occur in different organismal modules in organic and artificial evolution. The Anglo-American discussions on ontogenetic evolution intensified in the late 1970s, predated by the influential works of de Beer (1930) and Huxley (1932).

A 3D geometric morphometrics approach can properly capture shape changes and thus test these classical ideas. A comprehensive study of growth in wolves and many nonspecialized or generalized breeds of dogs with relatively ancestral morphology did just that (Geiger et al. 2017). Dog skulls show unique features shortly after birth, and these features persist throughout postnatal ontogeny. Furthermore, at any given age, juvenile dogs exhibit skull shapes that resemble those of consistently younger wolves. The cranial morphology of adult dogs cannot be described simply as either neomorphic or pedomorphic with respect to wolves. This study was not concerned with the strong selection for some breeds and instead investigated dog breeds with skull morphology relatively close to that of wolf and also included premodern and archaeological domestic morphotypes. Geiger et al. (2017) further concluded that neomorphosis (related to heterotopy, a change in spatial/structural aspects of development) and pedomorphosis are not mutually exclusive and can occur in concert for different modules or at different time windows of ontogeny, as has also been demonstrated for human evolution (Zollikofer and Ponce de León 2010). To further understand skull changes in dog domestication, one would need a close comparative examination of prenatal growth.

Also using 3D landmark-based geometric morphometric analysis of cranial shape, a pattern analogous to that of dogs (Geiger et al. 2017) was reported for pigs (Evin et al. 2017). By comparing three growth stages, Evin et al. (2017) demonstrated that wild boar and domestic pigs follow distinct ontogenetic trajectories. Pedomorphism is not the primary pattern describing the observed differences, with the exception of the size ratio between facial and neurocranial regions.

Horses are not particularly known for their morphological diversity, which is certainly less than that of pigs and dogs; however, size variation is significant (chapter 7). There is variation in skull shape across horse breeds, and morphometric studies have shown this variation can be largely explained by size (Heck et al. 2019). These allometric effects characterize deep-time evolution of horses going back to the Eocene (Radinsky 1984). Morphometric comparison of postnatal growth series revealed that the ontogenetic trajectories of horses and ponies vary in slope and length, resulting in ponies having a skull shape that resembles horses of the same size but at a younger age (Heck et al. 2019).

Falabella horses do exhibit somewhat enlarged orbits for their size, and the back of the skull is rounder and reminiscent of a juvenile horse (Fig. 5.4). But it would be a mistake to call the Falabella "pedomorphic." It is remarkable that the snout is not proportionally shorter as the pedomorphism hypothesis would predict. Instead, a long snout may be a response to functional and metabolic constraints related to a low-nutrient and abrasive diet, which in the case of horses requires not only high-crowned teeth but also a long line of them in a long snout (Heck et al. 2019).

A study of skull growth patterns across 13 domesticated species in comparison with their wild counterparts found no universal pattern of change. This study questioned the existence of a general heterochronic trend, specifically the simple pedomorphosis hypothesis (Sánchez-Villagra et al. 2017). The domesticated mammals were represented by breeds or landraces that were not very specialized and thus approximated the first phases of domestication, as opposed to intense selective breeding. The study included members of all major clades in which domestication has occurred: dog, ferret, American mink, cat, horse, goat, sheep, pig, Bactrian camel, llama, alpaca, rabbit, and cavy. Fourteen measurements were taken to characterize skull shape, and both multivariate and bivariate analyses were conducted.

These showed that there is no single universal growth pattern; species have distinct trajectories. Furthermore, differences in growth patterns between populations of wild and domesticated animals were not the same across phylogenies. The wolf, the boar, and the wild rabbit exhibit the highest proportion of allometric growth among the wild forms (Sánchez-Villagra et al. 2017).

The classic comparison of dogs versus cats is instructive. Most variables share the same slopes in wild and domesticated cats, with some differences mainly in the intercepts (i.e., pre- and post-displacement) and extension or shortening of the trajectory. In dogs, in contrast, more variables showed different slopes (associated with neomorphism, as discussed below). In wolf and dog, most variables exhibit an allometric growth pattern in contrast to the isometric growth of cats. The latter has been cited as explaining the relative conservatism in this species when compared with dogs (Wayne 1986; Fondon and Garner 2004; Sears et al. 2007). Through minor changes in size, allometric growth produces different proportions and thus disparity. In contrast, isometric growth implies that two individuals of different size are basically much alike in terms of proportions. The large amount of allometric change in canids suggests an intrinsic propensity for change.

Body shape, axial skeleton, and developmental repatterning in mammals

A qualitative but useful comparison of overall body shape reported that a four-month-old lamb of the Suffolk sheep breed is most similar to an adult mouflon ewe. Likewise, proportions of a juvenile individual of a Middle White pig are somewhat similar to those of an adult wild boar (Hammond 1962). Notice in these cases how the two domestic forms are not juvenilized versions of the wild form, but in a general and imprecise way, are the opposite (Fig. 5.8).

Selective breeding has brought about changes in muscular and skeletal features of the limbs and axial skeleton of domestic mammals (Nickel et al. 1986). Concerning the latter, mammals are morphologically diverse but conservative in their vertebral count. Not much deviation is expected in domesticated forms compared with wild ones. *Hox* genes are involved in axial patterning, and their study in combination with ossification (Hautier et al. 2010) and morphometric data serve to increase understanding of the occurrence of evolutionary transformations and variation in populations, as reported for the laboratory mouse (Böhmer 2017).

Almost all mammals have seven cervical vertebrae, irrespective of their neck length (exceptions are sloths and the manatee). The combined number of thoracic and lumbar vertebrae (TLV) tends to be 19 in many mammalian groups, but there is variation. In some groups it is exceptional, ranging from 14 to 31, in particular among xenarthran and afrotherian mammals, two major groups lacking any domesticated species (Sánchez-Villagra et al. 2007). Indirect selection due to pleiotropy is argued as a cause of the conservation of this aspect of the mammalian "body plan." The evolution in mammals of a diaphragm muscle separating the thorax from the lumbar region may also have biased evolution in these regions (Galis 1999; Buchholtz et al. 2012; Hirasawa and Kuratani 2013).

Fig 5.8. Body shape changes and variation in sheep, pig, and their wild forms. Three growth stages of (*left*) a Suffolk breed of sheep and (*center*) a mouflon, the wild form of the sheep. *Right*: a juvenile and adult of a Middle White pig and adult wild boar. The overall body shape of a four-month-old lamb of the Suffolk sheep breed (*left, top*) is similar to an adult mouflon ewe (*center, middle*). A juvenile individual of a Middle White pig has somewhat similar proportions to an adult wild boar. In this case, the domestic forms are not juvenilized versions of the wild form, but in a general and imprecise way, the opposite.

Skeletons of mammals with an atypical number of cervical vertebrae show anomalies, such as asymmetric ribs, defective production of cartilage, fusion of vertebrae, ossification of sternum and pelvic girdle, and abnormal fibrous bands connected to rudimentary ribs (Galis 1999; Ten Broek et al. 2012; van der Geer and Galis 2017). Anomalous transitional vertebrae are located at the boundary between segments and possess characteristics of both, in some cases expressing bilateral asymmetry.

Reports on several species of domestic mammals reveal some degree of intraspecific variation in vertebral count in all regions. It is challenging if not impossible at this point to gain any major evolutionary insights from these data. They add to a body of work that attempts to discern the evolvability of this trait across mammalian clades (Asher et al. 2011).

In the cervical region of domestics, as in mammals in general (Slijper 1946; Arnold et al. 2017; Randau and Goswami 2017), variation in morphology and dimensions, and not in number, is the main source of functional specialization. Peculiar cases of variation in the cervical region include those of two rabbit specimens with eight cervical vertebrae (Sawin 1937), and that of pigs with five or six cervicals (Berge 1948; King and Roberts 1960). A predominance of six cervical vertebrae in South Down sheep from South Africa was also reported (De Boom 1965), as well as the occurrence of six cervicals in some individuals of the Japanese Bobtail cat breed (Pollard et al. 2015).

A comprehensive survey across multiple kinds of dog breeds found that the occurrence of cervical ribs was more common in the Pug than in any other breed, at 46% (Brocal et al. 2018). Pug dogs with cervical ribs were more likely to have a presacral vertebral count of 26 and a transitional thoracolumbar vertebra. The same survey found no association between the presence of cervical ribs and abnormal sacrum, vertebral body formation defect, bifid spinous process, or caudal articular process hypoplasia.

Concerning the thoracolumbar region, most interest and most data exist for livestock, as their vertebral number is related to body length and carcass traits, making it economically important. The loin muscles, *longissimus thoracis et lumborum*, are valued in the livestock industry as sources of quality cuts of meat and are commercially important. These muscles run across the thoracolumbar spine, and as such are affected by the number or length of vertebrae. Thus the interest in studying the rib and vertebrae number, and carcass length in pigs, for which some variation is recorded (Berge 1948; King and Roberts 1960; Borchers et al. 2004). In goats, most variation is recorded in lumbar and sacral numbers (Simoens et al. 1978). A study of thoracic vertebrae in fetal sheep reported that the thoracic vertebral number did not change between weeks 6 and 20 of fetal development, whereas total length of the thoracic vertebral segment increased (Nourinezhad et al. 2013).

Variations were also found in a sample of thoroughbred racehorses, with 39% of 36 individuals studied having deviations from the expected number of six lumbar and five sacral vertebrae (Haussler et al 1997). Comprehensive studies of ponies and small breeds are lacking. The Falabella, the smallest of breeds, reportedly lacks one or two lumbar vertebrae in comparison with the standard horse (Hendricks 1995).

Anatomical deviations are studied to understand clinical matters, as in the lumbosacral vertebrae of the German Shepherd (Wigger et al. 2009). Lumbosacral transitional vertebrae were found in 29% of a sample of 4,386 individuals, with an isolated spinous process of the first sacral vertebrae (78%) as the most common condition. A correlation was found between lumbosacral transitional vertebrae and cauda equina syndrome, the damaging of the bundle of nerves at the end of the spinal cord. In contrast, no correlation was found between those lumbosacral transitional vertebrae and canine hip dysplasia (Wigger et al. 2009).

In the ferret the thoracolumbar region is variable. Aside from seven cervicals, 14–15 thoracic and 5–7 lumbar vertebrae are reported (Proks et al. 2015).

In mammals, intraspecific variation in the number of caudal vertebrae is much more common than in any other region of the axial skeleton. Domestic forms display peculiar extremes, such as in the Japanese Bobtail cat, where an outstanding reduction in number of caudal vertebrae is associated with aberrant vertebral formulae (Pollard et al. 2015).

Charles Darwin and domestication

Domestication was central in the development of Darwin's ideas on evolution (Browne 1995). Darwin was born and raised in rural Britain, where he became well acquainted with artificial selection for livestock and other domestic forms. He could experience first hand that selection was key to organic transformation (Ruse 2010). After his voyage around the world, Darwin settled down in an area outside London where he was in contact with local breeders and studied many kinds of domestic animals and plants, most prominently pigeons. The Galapagos finches are not mentioned at all in Darwin's *On the Origin of Species* (1859); domesticated pigeons, on the other hand, are discussed at length as examples of how selection can lead to rapid and significant behavioral and morphological divergence from an ancestral form, and to subsequent variation.

The first chapter of Darwin's *On the Origin of Species* concerns variation under domestication. The second chapter is dedicated to variation in nature. In the fourth chapter, Darwin starts by comparing variation resulting from the breeding of "domestic productions" by humans in short periods of time with the variation occurring in nature in geological time. Darwin presented a central analogy of the book: just as humans select their specimens to improve the stock in selective breeding, so does natural selection favor the best parents for the next generation (Ruse 1975; Bartley 1992). Several domestic animals figured prominently and repeatedly in Darwin's writings. An example is the now extinct *vaca ñata* of Uruguay and Argentina, a brachycephalic form (chapter 7).

Darwin's (1868) *The Variation of Animals and Plants under Domestication* is a landmark summary of the subject. In this book, he tried to decipher the mechanisms of heredity and proposed the idea of pangenesis, suggesting that each part of the body continually emits small organic particles that aggregate in the gonads. This idea became obsolete later when results of genetic studies, often conducted on domesticated animals and plants, became available. Darwin failed to discover mechanisms, but provided natural history documentation and insights.

Domestication in Darwin's thought can be seen in relation to the ideas of how science works promoted by two philosophers that shaped the foundation of Darwin's work. A consilience of inductions—different classes of facts coinciding, providing the basis of a general principle—was seen as the ultimate test of a theory's truth by the philosopher William Whewell (1840). He greatly influenced Darwin when Darwin was a student in Cambridge and afterward (Ruse 2010). Darwin was also influenced by another Cambridge philosopher, Herschel (1831), who stressed the importance of observational evidence in finding what he called vera causes for the phenomena to be explained.

Darwin attempted to merge or rather consider these two positions in his formulation of the theory of evolution by natural selection (Ruse 2010). Herschel's vera cause was found in selection, with domestication as the starting point. The consilience of inductions was based on Darwin's examination of data from diverse sources, including paleontology and embryology.

The first figure in Darwin's (1871) *The Descent of Man* compares a human embryo with a dog embryo (Figure 5.9). In the same book, Darwin (1871, p. 181) drew a parallel between the origin of what he uncritically accepted as human "races" and the origin of breeds. He mentioned parallels, but also stated that the absence of any deliberate selection of traits during human evolution, as is typical in animal husbandry, spoke against a direct comparison with domestication (Brüne 2007).

Some of the major contemporary critics of Darwin explicitly referred to his treatment of domestication. Louis Agassiz (1860), in his review of *On the Origin of Species*, stated that Darwin's analogy of selection applied to domesticated animals and plants cannot be compared with the "chance influences which may affect animals and plants in the state of nature." Agassiz stated that selections imply design, and this is not found in nature but is found in domestication. Criticisms on the subject are also found in George Mivart's (1871) anti-Darwinian *The Genesis of Species*.

It is ironic that during Darwin's time, and for many decades afterward, the mechanism for which domestication was supposed to serve as analogue, and proof of its potential, natural selection, was not well accepted, while evolution was an increasingly established concept (Bowler 1992).

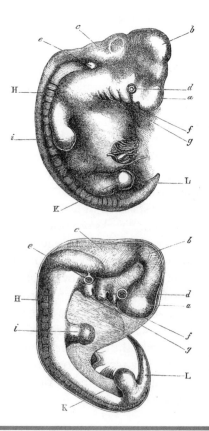

Fig 5.9. The first figure in Darwin's (1871) *The Descent of Man* is one comparing a human embryo with a dog embryo. In a letter to his colleague Asa Gray, C. Darwin stated "embryology is to me by far the strongest single class of facts in favor of a change of forms" (Darwin 1887, vol 1. p. 72).

Human variation and human self-domestication

In the first half of the twentieth century a zoological perspective on human variation was presented by some authors alluding to domesticated species, such as the one by Adolf Remane in his 1927 paper on "Species and Race" published in a journal on physical anthropology. This paper was not explicitly ideological, a stance later taken by others (Rieppel 2017). Some views of the time on humans as a domesticated species implied that some human populations are more domesticated than others, a concept used to justify racist and eugenic ideas (Brüne 2007).

Not much can be salvaged from several early works focused on very general and nonsensical comparisons with dogs, as the nature of variation in dogs and humans is quite different (Norton et al. 2019). There is a larger global genetic variation in humans than in dogs. Large human variation within local populations is behind this difference, and not the variation between regional groups. In dogs, in contrast, low levels of heterozygosity occur within breeds; instead, the large amount of genetic variation corresponds to breed differences (Parker et al. 2004). It is tragic that the loaded history of studies of variation in human populations (Gould 1981; Lewis et al. 2011) has prevented an enlightened examination of human diversity when it comes to the potential identification of natural, historical (phylogenetic) groups, beyond a crude and misplaced characterization of races (Leroi 2003). Perhaps a new term should replace the socially constructed one of "race." I propose the Greek word Φυλή—Phyle in our alphabet—for such groups. It spells similarly in its Greek version to the word Φιλί, which means "kiss," a nice coincidence reflecting the intertwined nature of Phyles. The intricate history of exchange and migration would present a huge challenge for researchers trying to identify the Phyle, which after all, may not exist beyond ethnolinguistic units embedded in millennia of human evolutionary and ecological history. Such research could be a celebration of diversity free of the colonial and depicable past history of studies of human diversity. Comparison of the range of human morphological variation with that of other animals could be another rich subject of investigation (McKellar and Hendry 2009). Here the topic of the effects of climate on skeletal morphology offers an ample area of research (Harvati and Weaver 2006).

Some perceived similarities in diversity patterns between domesticated species and humans have led to suggestions of domestication-like or a "self-domestication" phase in human evolution. There have been different versions of this idea. Most are of historical relevance but otherwise are of little significance for current scientific understanding of ourselves. The similarities are superficial or refer to some aspect of domestication, when in fact the domestication process is diverse, as are its phenotypic and genomic consequences. Konrad Lorenz's (1940) notion of "Verhausschweinung" suggested that urban life makes Western people similar to domesticated pigs, a crude analogy that rendered this idea a mere curiosity. Most of the recent ideas I summarize below are surely like any other embedded in a sociocultural context in which they came to be, but without the problematic background of some of their nineteenth- and twentieth-century predecessors (Sánchez-Villagra and van Schaik 2019).

It has been hypothesized that anthropogenic environments that resulted from sedentism resulted in a selection process analogous to domestication. Self-domestication resulted from "acclimatization to life in a household and . . . its buildings, yards, gardens, and orchards" (Leach 2003).

The leading idea of human self-domestication proposes that selection for tameness in animals undergoing domestication is a model of selection for behaviors that characterized a certain phase of human evolution in which seemingly independent features also arose as a result of developmental correlation. Starting in the Middle Pleistocene, reduced aggression and enhanced social tolerance, physiologically caused by a reduction in average androgen reactivity, were coupled with craniofacial feminization (Cieri et al. 2014). Many of the features that distinguish modern humans from our Late Pleistocene ancestors reportedly include reduced tooth and brain size and shorter rostrum. As many of these features derive from the neural crest, "selection for intragroup prosociality over aggression" (Hare 2017) or "selection against reactive aggression" (Wrangham 2019) is seen as analogous to the mechanism involved in selection for tameness and its morphological correlates (chapter 4). It has been suggested that the morphological by-products of self-domestication could have acted as honest signals of reduced xenophobia, leading to a reduction of aggressive behavior (Sánchez-Villagra and van Schaik 2019). The cultural explosion experienced by *Homo sapiens* starting around 100,000 years ago, as documented in the archaeological record, may have resulted not only from a change in kinds of aggression (Wrangham 2018) and from prosociality, but also from demographics, with more humans contributing to increased cultural accumulation and transmission. The human skeletal gracilization that occurred during the late Pleistocene and Holocene also characterizes various mammalian species and may be correlated with interglacial warming.

The current human self-domestication hypothesis was very much informed by studies of the contrast between chimpanzees, *Pan troglodytes*, and bonobos, *Pan paniscus*, with the latter exhibiting morphological, physiological, and behavioral features akin to domesticates (Hare 2017). It should be possible to link neural crest hypotheses to ideas of extended juvenilization in development before sexual maturity, as driven by altered concentration of thyroid hormones (Crockford 2002; Wilkins 2017a).

These ideas on human self-domestication (Wrangham 2019) have productively promoted research in genomics, cognition, and behavior. Genes under positive selection in modern humans and domesticates have been identified (Theofanopoulou et al. 2017). Experimental data on patients with special conditions, together with genetic comparisons between modern and archaic humans, have led to identification of scenarios of genomic changes associated with facial and behavioral ones (Zanella et al. 2019). Scenarios of human language evolution are also informed by these self-domestication ideas (Benítez-Burraco and Kempe 2018). These studies, along with those of variation in morphological traits across the evolution of *Homo sapiens,* will surely sharpen any comparisons and scenarios of evolution based on our knowledge of the domestication process.

A common critique of the concept of "self-domestication" by humans or other species is the lack of an external agent, as it seems improbable if not impossible that a species would be its own agent of domestication. This seems at first justifiable, but upon further consideration this argument falls apart in the presence of evident examples such as the domestication of house mice. Standard literature on domestication treats the house mouse as self-domesticated (Weissbrod et al. 2017).

Summary: Variation in developmental repatterning

Different kinds of developmental repatterning can occur that lead to morphological change of the adult phenotype. A pattern of change does not necessarily imply the same kind of mechanism behind it. A heterochronic pattern of change in a growth trajectory may not show a developmental change in timing in the molecular and developmental mechanisms that produce it.

Studies of skull growth patterns across domesticated mammals in comparison with their wild counterparts have found no universal pattern of change. They have rejected the hypothesis of global heterochrony, and specifically of simple pedomorphosis. In mammals, studies of prenatal ontogeny are missing, and this is regrettable as they would be fundamental to understand how under the same gestation lengths, different breeds produce neonates that vary so much in size and morphology. Studies of the chondrocranium and its intraspecific variation are also an area of relevance for those concerned with the origin of cranial differences among breeds.

Given the ontogenetic changes and with that the palimpsest nature of integration patterns, more ontogenetic studies of the subject are needed. Domesticated animals offer a great subject given their availability and possibility of comparisons across species. It is doubtful that a simple and universal relationship between the degree of module integration and morphological diversity exists.

Whereas mammals are rather conservative in evolvability of the number of vertebrae, birds are more plastic, as reflected in the variation recorded among breeds or individuals within populations. There are several documented cases of variation in mammals. An example is goats with a number of cervical vertebrae deviating from the almost universal seven cervicals. The challenge will be to move from these anecdotal reports to an understanding of the reasons behind the viability of forms that rarely occur in nature as far as we know.

CHAPTER 6

Life history and growth

An animal, in trying to maximize its fitness, could theoretically reproduce immediately after being born and continue to do so throughout its entire life—thus becoming a "Darwinian Demon" (Law 1979). But organisms do not have endless resources in the form of time and energy to spend, as they need to obtain that energy and also grow to reach the condition in which they can reproduce, namely sexual maturity. There are trade-offs, and the pattern in which the organisms operate is their life history. Among its variables are gestation time, the condition at birth or hatching, the age at reproduction, longevity, interbirth interval, neonatal size, litter or clutch size, age and body mass at first reproduction, and growth patterns. The multivariate "life history profile" of a species includes developmental, reproductive, and demographic variables. Theories around life history revolve around optimization, but the many deviations from it recorded in field biology speak against that null assumption (Diogo 2017).

Domestication, in particular selective breeding, would seem to provide the conditions under which Darwinian Demons could exist, as in many cases domesticates are devoid of predators and of competitors, have unlimited food supply, and are provided with an environment particularly fitting to live and reproduce. However, no domesticated animal reproduces directly after being born, nor do domesticates produce infinitely many offspring, nor do they live indefinitely. There are limits to the extent to which domestication alters life history traits (Kihlström 1972; Herre and Röhrs 1990; Geiger et al. 2016; Heck et al. 2017, 2018a). However, seemingly improbable features have been "achieved" in cases of artificial and intense selection. In fact, this is at the core of industrial animal production. Some chicken breeds now produce 10 times as many eggs as their wild counterpart, and Holstein cattle can produce >30,000 kg of milk in a year, almost 90 kg/d on average (VandeHaar and St-Pierre 2006). There are trade-offs in life history modifications, so that no chicken can be both a broiler and layer; no chicken can be optimized for both meat and egg production (Bennett et al. 2018). More importantly, the welfare of these animals is poor, as the exaggeration of traits has led to an imbalance. This statement is not a reflection of a naive idea on the pureness and goodness of nature but a conclusion reached from empirical data.

There are several ways in which energy budgets and life history vary (Stearns 1992), including the ability of the young to assimilate the milk yield in the case of mammals, the allocation of energy to pre-birth and post-birth development of different organs (e.g., brain) and the timing of birth / hatching between conception and attainment of physical maturity (Gaillard et al. 1997).

Mammals and birds feature diverse levels of development of their organs at birth or hatching: altricial forms are somatically immature at birth, whereas precocial mammals are relatively mature. The concepts of altricial and precocial are extremes of a continuum, and different organs may be at different stages of development, in a mosaic pattern of strategies for the condition at birth. These and other life history traits are treated in exemplary form, first for mammals, then for birds, in this chapter. A section is devoted to the dog, since much is known on this species.

Mammals

THE CONDITION AT BIRTH: THE ALTRICIAL / PRECOCIAL CONTINUUM

In general, altricial mammals invest mostly in postnatal growth and have short gestation, whereas precocial mammals invest less in postnatal growth and have longer gestation periods. All studies of the condition at birth coincide in hypothesizing that the newborn of the last common ancestor of placentals was altricial; it probably had closed eyes and an almost naked skin, and its limbs were evenly developed (Werneburg and Spiekmann 2018). There were surely multiple independent origins of precocity and altriciality within the group, as quantitative studies across phylogeny of diverse organs at birth demonstrate (Werneburg et al. 2016).

Different developmental events have been used to characterize the condition at birth, for example "eyelid closure" (Müller 1972). The degree of development of the brain would seem another good marker, in particular given its close connection to life history. In fact, classical studies of precocity and altriciality consider the degree of cerebralization at birth (e.g., Portmann 1951). However, there are different parts or modules of the brain with different functions and phylogenetic history, so that a separate consideration of each would be more appropriate. The cerebellum, for example, is a region that plays a crucial role in visual sensorimotor control: movement of an animal in space is monitored by visual perception. A wide diversity of conditions at birth are seen in the histological development of the cerebellum, as measured in the differentiation of its layers of cells and tissue types. A parsimony-based phylogenetic reconstruction suggested that the last common ancestor of placentals was born with an altricial cerebellum, in which the molecular layer was just present between the external granular layer and the prospective Purkinje cell layer. The species with the most mature cerebelli are those with their eyes open at birth (Sánchez-Villagra and Sultan 2002).

The degree of independence and of development in several anatomical and physiological aspects can simultaneously provide a better characterization of mammalian neonates. Derrickson (1992) categorized neonates in nine life-history traits related to thermoregulation, sensory organs, locomotion, and nutrition. She defined precocial development based on independence in all four traits, altricial by dependence in all four, with intermediate stages. Derrickson's survey showed that mammalian neonatal patterns have not evolved along a single continuous axis of diversification and that the altricial and precocial conditions probably arose independently in response to more than one kind of selection pressure.

Based on her comparative study, Derrickson (1992) concluded that neonate mass varies according to developmental category, whereas litter mass and age of first parturition show no significant variation among developmental categories. She also found that growth rate and litter size vary in different proportions along the altricial-precocial spectrum.

Starck and Ricklefs (1998) hypothesized that "In mammals, the developmental mode is highly correlated with neonate size and gestation period, suggesting that the developmental advancement of precocial species reflects a prolongation of the embryonic developmental period." Starck and Sutter (2000) suggested that this pattern is also highly characteristic of Galliformes within birds.

GROWTH: SIZE, MODULARITY, DISPARITY AND THE ALTRICIAL-PRECOCITY CONTINUUM

The important role of size in evolvability has been shown in studies of the skull (Marroig and Cheverud 2010; Cardini and Polly 2013; Porto et al. 2013) and the postcranium (López-Aguirre et al. 2019). Changes in size may serve as a most parsimonious pathway of least evolutionary resistance (Zelditch et al. 2004; Marroig and Cheverud 2010; López-Aguirre et al. 2019). It has been hypothesized that in mammals a link exists between life history strategy and evolutionary flexibility, as achieved by size-related variation (Porto et al. 2013). Clades with altricial neonates invest more of the energy budget in postnatal growth and have relatively larger amounts of size variation, whereas the precocial strategy allocates fewer resources to postnatal growth, associated with lower amounts of size variation (Wilson 2018).

In a comprehensive study of integration and modularity in skull shape using 3D geometric morphometrics as well as simulations examining the influence of size, Porto et al. (2013) concluded that integration magnitudes are closely correlated with the amount of size variation present in an evolutionary group. They found that more size variation corresponds with more highly integrated skull traits—and reduced evolutionary flexibility, that is, reduced capacity to respond in the direction of selection. Porto et al. (2013) found that altricial mammals show more size variation and lower evolutionary flexibility than precocial mammals. One might then expect altricial mammals to show lower magnitudes of disparity compared with precocial mammals. Furthermore, differences in disparity may exist among domesticated and wild pairs of altricial mammals compared with similar pairs for precocial ones (Wilson 2018).

Comparative studies of prenatal development among domesticated mammals, including altricial and precocial species, could provide important data to investigate the relation of life history strategy to evolvability and disparity. The condition at birth may impact the magnitude of response to ontogenetic modifications in evolution, as shown by a study of allometric disparity in postnatal skull growth series of 12 pairs of wild versus domestic forms of mammals (Wilson 2018). A comparison of altricial and precocial forms found support for the hypothesis of Porto et al. (2013) of lower evolvability of altricial forms. This stems from the fact that the majority of directional change in slope and slope length between domesticated and wild pairs was for precocial forms, four out of five pairs (Wilson 2018). Another support for this hypothesis is that most of the domesticated wild pairs that displayed heterochronic shifts (three of four pairs) were also precocial forms (Wilson 2018). This study also found greater allometric disparity among precocial compared with altricial forms.

In contrast to the idea above, morphological evolvability has been hypothesized to be linked in the opposite way to the relative timing of birth, at least in carnivorans, based on embryonic staging in domesticated dogs and cats (Werneburg and Geiger 2017). The authors described a series of more than 80 specimens of domestic dog developmental stages by synthesizing many single depictions of domestic dog organogenesis (e.g., Bischoff 1845; Bonnet 1897, 1901, 1902; Keibel 1906; Evans and Sack 1973; Rüsse and Sinowatz 1991; Miglino et al. 2006; Evans and Lahunta 2012) and compared the timing of external developmental characters with that of homologous features in domestic cat and ferret development. They recorded differences in the timing of external characters and discovered that dogs exhibit fewer well-developed organs at birth than cats. Furthermore, in the cat, the period from birth until the onset of hearing, eye opening, and weaning is much shorter than in the dog, although both have practically the same gestation length (Werneburg and Geiger 2017). Dogs were thus hypothesized to be capable of a greater magnitude of response to human selection (directed at postnatal stages in dogs) than cats (Werneburg and Geiger 2017).

SEASONALITY OF REPRODUCTION AND FECUNDITY

In the wild forms of domesticated animals, sexual activity, mating, and parturition tend to occur at regular intervals. This seasonality probably responds in part to ecological context (e.g., photosynthetic production) produced by changes in the distribution of sunlight, which affect pituitary activity. Parturition tends to coincide with seasonal blooms of food. In domesticated animals, reproduction is less dependent on the environment. Both sexes of domestic animals are more fertile than their ancestral forms in all species in which this could be studied (Bradford et al. 1991).

In domestic animals, the reproduction rhythm is different from that of wild forms. In many cases, females are in heat not only more often per year, but also regardless of the season. There are exceptions to this of course, and the degree of divergence from the ancestral, wild form, can vary. In domesticated forms, births tend to succeed each other in much shorter intervals than in the ancestral forms. In the latter, females are in heat at a certain season of the year, synchronized so that their offspring are born in a suitable season conducive to their growth. In female cattle of many breeds, the rhythm of sexual processes is independent of the season (Hammond 1947), a pattern reported for other domesticated species (Price 2002a). In the case of cattle, as in the horse, the extinction of the wild form prevents exact comparisons. Reports on the aurochs from the sixteenth century, before it went extinct, may provide some information on its social and reproductive behavior, but these reports are rare and limited in scope (Van Vuure 2005).

FECUNDITY: LITTER SIZE VARIATION

Many species of domesticated mammals demonstrate increased fecundity. The litter size, or number of young per parturition, is the result of ovulation rate and prenatal survival. These two factors can be breed-specific, and they vary according to genetics, not only because of environmental influence (Bradford et al. 1991).

Wild rabbits have an average litter size of 4.5, whereas an average of 10.1 was reported for the Flemish Giant rabbit (Herre and Röhrs 1990). In the wild boar, litter size is between 4 and 6, whereas the litter size of a domestic pig can exceed 20. Wild sheep and wild goat generally give birth to one young per year. Some breeds of domestic sheep and goats are characterized by regularly giving birth to twins or triplets (Herre and Röhrs 1990).

Litter size can be affected by the living conditions. A study documenting stereotypic behavior, stress activity, and levels of health condition in farmed minks (*Neovison vison*) showed that decreasing inactivity correlated with increased reproductive output in captivity (Meagher et al. 2012).

The heritability of litter size in domesticated forms can be low, as exemplified by studies of cattle, sheep, and pigs, with values around 5%–10% (Bradford et al. 1991). Environmental factors, including nutritional and health status and temperature, affect fecundity. Breed differences, however, do have a strong genetic basis. There is much genetic variation in ovulation rate, with breed differences in mean numbers of corpora lutea, ranging for example from 1.2 to 4.5 in sheep and from about 10 to 20 in pigs (Bradford et al. 1991; Lamberson et al. 1991).

Experiments on domestication of rats clearly showed the effect of domestication on litter size and other life history variables. King (1939), in her experiment with wild gray rats in captivity (chapter 8), found that after 25 generations, the average length of the reproductive period was nearly eight months longer than in the wild rats she collected in the vicinity of Philadelphia. She reported earlier breeding and the persistence of reproduction to a more advanced age in the captive rats. Fertility, as measured by litter production, also increased, reaching its maximum at the nineteenth generation, when females produced an average of 10.2 litters each, in contrast to an average of 3.5 litters each for generation 1. On the other hand, there was no significant change in the size of individual litters, which remained at an average of 6.1 throughout generations 2–26. In another experiment, after 19 generations of captivity, lab rats increased in average litter size from 3.2 to 10.2, whereby an increase of body size also took place (Castle 1947). Richter (1952) reported that after 30 years of divergence, domestic rats show faster sexual maturity compared with wild ones. In a comparison between captive wild and domestic rats, Clark and Price (1981) found that domestic females reached sexual maturity earlier, produced slightly larger litters, and raised a higher percentage of their young to weaning age, and males achieved first copulation earlier.

As with many other variables, litter size as well as weight at birth are studied within the context of animal production. For example, it has been quantified that low weight at birth negatively affects carcass and meat quality in pig production (Vázquez-Gómez et al. 2020). Birth (at term) is a critical set point for permanent programming in mammals, in particular in precocial ones. This has been shown in sheep, showing the impairment of a wide range of endocrine functions regulating growth following prenatal undernutrition, which cannot be reversed by dietary correction later in life (Khanal and Nielsen 2017).

GESTATION LENGTH

Gestation length has been hypothesized to be constant within species (Promislow and Harvey 1990; Heck et al. 2018a). In general, clades of mammals tend to be conservative in gestation length (Clauss et al. 2014). This is reflected in relatively little variation produced by domestication. For example, different-sized domestic dogs have very similar gestation length (Kirkwood 1985; Clauss et al. 2014). Moreover, the length is almost invariant between the domestic dog (64–66 days; Concannon 2000) and the wolf (60–65 days; Seal et al 1979). However, some variation in gestation length is recorded among individuals of the same species. Gestation length is not likely to have been a direct target of selection, and the phenotypic deviants recorded in domesticated species may simply reflect naturally occurring variation within species or the correlated effect of selection for other life history traits.

A minor shortening of gestation length can be achieved by hormonal treatments, but these are too labor intensive and expensive for regular usage (Bristol 2000). Performance pressure in production animals, such as cattle, sheep, and pigs, favors selection for fast reproduction, and this can include shorter gestation periods (Figure 6.1). One would then expect production animals to be less variable in gestation time than animals bred for nonproduction purposes, such as dogs or most horse breeds. This is indeed what we found in our study of 192 gestation lengths for eight mammalian species (Heck et al. 2018a). In general, selective breeding, as described below, can greatly influence the time of birth and the frequency of reproduction, but can hardly alter the gestation length itself.

PARTURITION IN MAMMALS: LABOR AND BIRTH

Parturition is the process of delivery of a fully-grown fetus upon completion of the normal pregnancy period. At parturition, the uterus starts contracting and the cervix relaxes sufficiently to allow passage of the young to outside the mother's womb, passing through the birth canal, formed by the uterus, cervix, and vagina. Many domestic animals are prone to maximum injuries and infections, some of them immediately endangering the life of the fetus and the newborn, and some affecting the future productive and reproductive life of the mother. Thus, different methods of assistance have been developed for domesticated species (Purohit 2010). Research is conducted on the mechanisms that trigger the birth process, the role of cytokines and hormones during ripening of the cervix, the effects of fetal body movements in preparation for parturition, and timing in the transportation of an individual fetus from its intrauterine location, via the pelvic canal, to the outside world (Taverne and van der Weijden 2008).

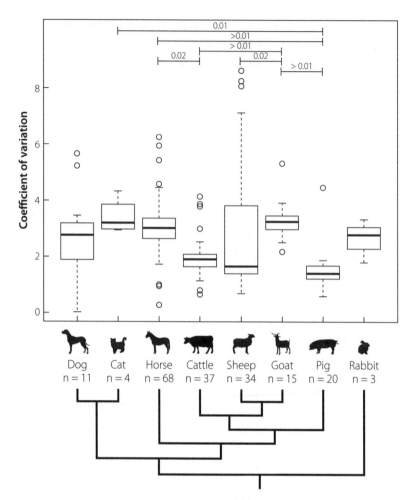

Fig 6.1. Production animals exhibit less variability in gestation length than dogs and most horse breeds. In production animals such as cattle, sheep, and pigs, economic pressures may have favored the selection of short gestation periods, that is, fast reproduction. This pattern fits the expectation generated by this idea from Darwin (1859): "If a trait is of little importance, then its variability is probably high, whereas if the trait is important, then variability should be low."

The increases in size in domesticated forms and other morphological changes that probably occurred in modular form led in some cases to conflicts in the relationship between the progeny and the mother at birth. Data on our own species show environmental changes (e.g., dietary) can lead in short periods of time, within a generation, to complications at birth because of a larger fetal head. To begin with, our species may confront special challenges given our bipedality, resulting in the hypothesized "obstetrical dilemma" (Washburn 1960), a trade-off between a narrow and biomechanically efficient pelvis (males and females), as well as a wide pelvis for a large birth canal to accommodate a large head. Difficulties at parturition are well documented and a subject of much study in cattle and other domesticated

Fig 6.2. Vietnamese Pot-bellied pig.

forms (Mekonnen and Moges 2016). In Texel sheep, difficult births are frequent. At birth, the lambs are already so large that they pass through the birth canal with difficulty (Naaktgeboren et al. 1971). Some double-muscled breeds require caesarian section at every birth, while others hardly ever have problems on their own.

ACCELERATION OF SEXUAL MATURITY

In domestic mammals many cases of acceleration of the onset of sexual maturity in relation to what is considered adult, usually established in terms of skeletal maturity, have been reported. Among some breeds of domestic pigs this acceleration is noticeable, such as in the Vietnamese Pot-bellied pig, a traditional breed from Vietnam (Figure 6.2). These domestic pigs come into heat after only a few months and can mate successfully. The ensuing births are normal. At that point their body weight is still low; the animals still grow substantially following their first births (Herre and Röhrs 1990).

As in the experiment with captive wild and domestic rats demonstrating the acceleration of sexual maturity in the latter (Clark and Price 1981), the experiment with silver foxes also showed an acceleration in the domestic population (chapter 9). External conditions are used by breeders to influence reproductive time. The reproduction of racehorses is influenced by breeders using hormones and artificial light (Bristol 2000; Nolan et al. 2017). The breeding season is advanced to mid-February by inducing ovulation with artificial light.

DENTAL ERUPTION PATTERNS AND SKELETAL MATURITY

Dental maturity is reached when all teeth of the adult generation are erupted. In mammals there is a species-specific sequence of dental eruption, and variation in this pattern across species is at least to some extent tied to life history features. According to "Schultz's rule," there is a trend of an increasing number of replacement

teeth (incisors, canines, premolars) erupting before or simultaneously with the molars in species with slow growth, late sexual maturity, long gestation, and extended
longevity (Schultz 1956, 1960; Smith 2000; Godfrey et al. 2005). A prolonged juvenile
phase in dental development characterizes slow-growing species, with the deciduous
teeth gradually wearing out before the permanent teeth become functional. Jaw size
may also play a role in this equation: when not large enough to contain all molars,
the replacement teeth erupt relatively early. Support for Schultz's rule was reported
in some studies on primates and "ungulates" (Smith 2000; Henderson 2007). On
the other hand, deviations from the expectations and ambiguous support was reported for several groups of artiodactyls, primates, and hyraxes (Monson and Hlusko
2016; Veitschegger and Sánchez-Villagra 2016; Asher et al. 2017; Geiger et al. 2018a;
Geiger and Asher 2019).

The absolute timing of tooth eruption may be influenced by body size, brain size,
period of learning, resource availability, and diet and its protein content (Godfrey
et al. 2001). In environments where the risk of starvation is high, slow dental growth
could be beneficial because it reduces caloric need per unit time (Godfrey et al.
2001; Geiger et al. 2018a). As fecundity increases due to domestication, an earlier
eruption of molars relative to premolars would result if the prediction of Schultz's
rule is fulfilled (Geiger et al. 2018a). An early eruption of molars could indeed
happen in absolute terms, but what does not necessarily change is the sequence of
dental, sexual, and skeletal maturity (dogs: Geiger et al. 2016) or the sequence of
dental eruption (sheep: Geiger et al. 2018a).

The order of eruption is similar in wild, domestic, and Soay (feral) sheep, despite
the differences among them in life history (Geiger et al. 2018a). Contrary to domestic sheep breeds, Soay sheep erupt their teeth at an absolute older age and also tend
to grow more slowly, as estimated based on jaw measurements. Concerning the
latter, they resemble the extinct *Myotragus* from the Balearic Islands Majorca and
Menorca in the Mediterranean, which has been hypothesized to have evolved a
"slow" life history strategy (Köhler and Moyà-Solà 2009).

The timing of dental eruption was described for five breeds of cattle (Glaus
1932). These and other data surely available from veterinary works may open the
possibility of further examination of the evolution of this important developmental
trait in life history variation within domesticated species.

In the case of the postcranium, a temporal sequence of events also occurs with
increasing age in growth plate fusion, such as fusion of the epiphyses to the main
shaft or diaphysis of many bones, including the long bones of the limbs (Geiger
et al. 2016). The quantification of this sequence in populations of wild versus domestic forms provides useful comparisons. These data provide information on age
and sex that can be used to generate demographic profiles in the zooarchaeological
record, used to infer kinds of interactions, such as intensity of management, and
domestication of populations. Age profiling also concerns dental remains (Ruscillo
2006; Klevezal 1996), or a combination of postcranial and dental data, as studied in
domestic pigs and wild boar (Bull and Payne 1982; Bridault et al. 2000), as well as in
feral and domestic goats (Noddle 1974). Examples abound.

Based on a study of Asian and comparisons with European wild boar populations, eleven different age classes based on a regular pattern in the sequence of fusion

of postcranial elements were defined (Zeder et al. 2015). These ages corresponded well with dentition-based age classes (Lemoine et al. 2014). In some archaeological sites, postcranial fusion-based harvest profiles can be a valuable tool for understanding ancient exploitation strategies, and much stronger when used in tandem with dentition-based profiles.

A comprehensive study of demographic profiles of pigs based on dental eruption pattern, dental wear stages, and long bone fusion (Lemoine et al. 2014) was conducted for three Near Eastern archaeological assemblages from three distinct time periods. By considering different levels of resolution, finer and broader differences in pig exploitation strategies could be identified.

In goats, the ages of epiphyseal closure varied according to the breed and the sex of the animal. A study found that some feral goats are about a year behind domestic goats, and castrate males are up to four years behind females. In sheep, in contrast, maturation is much earlier than in the goats (Noddle 1974).

GROWTH IN DOMESTICATED MAMMALS

A comparative study across placental mammals (more than half being rodents and lagomorphs) of growth curves based on body weight data found the following general pattern: altricial mammals have their peak growth rate later in the growth trajectory process than precocial mammals (Gaillard et al. 1997). There is no pattern of this kind within genera, perhaps indicating the establishment of patterns early in the differentiation of the major groups. Concerning the shape of the growth curve, there is much variation, with some species having a decelerating growth rate from birth to maturity, and others a sigmoid curve, with a peak after birth (Gaillard et al. 1997).

The shape of the growth curve of domestic mammals and the approaches to fitting nonlinear models to growth data have been studied for many species and populations (Lawrence et al. 2012; Figure 6.3a). Examples include studies of rabbits (Fiorello and German 1997), goats (Parés-Casanova and Kucherova 2014), and sheep (Pittroff et al. 2008). The correspondence between inflection points in the growth curves and life history events (e.g., the onset of puberty) is a major issue (Brody 1945), with an interest in establishing the tempo and mode of fat gain for production purposes as a major goal (e.g., Pittroff et al. 2008 on sheep).

There is much information on growth in farmed animals, with documentation of weight changes of the whole animal and of individual organs, muscles, bones, and fat. Body composition is measured in terms of relative proportion of fat and protein in the carcass. Much of this information comes from management reports of companies and other organizations concerned with animal production. The aim has been to establish how the proportions of the body change and where to find an optimum at which to slaughter animals of a special breed or under some specific treatment for consumption. In particular, many breeds of pigs and cattle have been selected for some combination of increased growth rate, decreased fattiness, and increased muscularity. Selection for growth rate is coupled to changes in digestive, respiratory, circulatory, and excretory systems—related to increased metabolic rate.

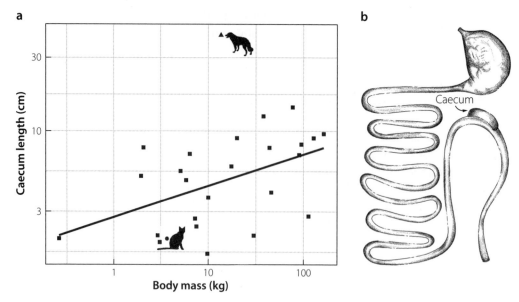

Fig 6.3. (a) Caecum length in dogs (*triangle*) and cats (*circle*) as compared to other carnivores (*squares*). (b) gastrointestinal tract of the dog.

An early example is the thorough study by Wilson (1958) of dwarf goats. This study provided percentages of edible portions, changed proportions of organ systems (skin) through growth, and many other figures that provide a complete picture for the animal producer aiming to optimize consumption of some specific organ or overall outcome. Data from the USA National Pork Producers Council Terminal Sire Line National Evaluation Program exist for nine breeds and "sire lines" of pigs (Lawrence et al. 2012, table 8.8.). Here, feed conversion efficiency is calculated, with quantification of measurements such as daily lean tissue growth and carcass lean weight. In some pig breeds there has been a drastic increase in pig muscle growth and a corresponding decrease in fat deposition due to consumer demand for lean meat. The genetics behind some changes have been identified, as in the *RN* locus affecting glycogen content in skeletal muscle, and the *IGF2* locus affecting muscle growth (Andersson 2016).

Changes in diet in domestic in comparison with wild forms may lead to changes in digestive anatomy. Dogs have a more plant-based diet than wolves, as reflected in genetic changes related to the production of some enzymes associated with starch digestion and fat metabolism (Axelsson et al. 2013). This change may explain the longer caecum of dogs compared with their wild, closest relatives (McGrosky et al. 2016; Figure 6.3). Comparative data from a greater number of wild canids, and from domestic dog breeds, would serve to test this hypothesis.

Gut length can vary between wild and domestic forms, tending to be longer in the former, as reported for minks in a detailed study (Drescher 1975) and also for rabbits (Balon 1995). Darwin (1868) cited a report showing the intestines of domestic cats to be wider and longer than in wild cats of the same size, presumably resulting from the less strictly carnivorous diet in the domestic form. The latter is peculiar, as

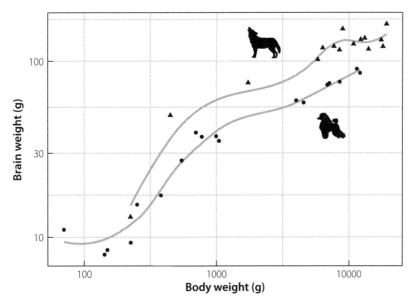

Fig 6.4. Relation of brain to body weight during growth in wolves and poodles.

indeed domestic cats are generally considered hypercarnivorous. Here, again, wild versus domestic may be too simplistic a dichotomy, and phenotypic plasticity as determined by the conditions of life as opposed to genetic determinism may play a role. The study of the evolution of the digestive tract in relation to domestication should reveal differences and patterns worth exploring.

The studies of variation in growth trajectories across species and within species—for example, among breeds of very different size but with equal gestation length (Brody 1945)—represent examples of the evolution and plasticity of ontogeny. Furthermore, they show the nonlinear relation between individual organ growth and animal body weight. This is evident in the case of brain growth in dogs (Figure 6.4).

The significant effect of fetal nutrition to health and growth in postnatal life is another subject that is quantified for many domesticated forms, as has been done in sheep (Khanal and Nielsen 2017). It was found that undernutrition in late gestation negatively affects endocrine functions that regulate growth. Whereas the negative effects on health of a high-fat diet in early postnatal life can be reversed by dietary correction, those of malnutrition before birth are more difficult to circumvent.

LONGEVITY

The one aspect of life history that is prone to more variation, and is thus difficult to characterize precisely is longevity. Likewise, gathering data on longevity is difficult, and thus much data come from studies of animals in captivity. Mammals in zoos generally live longer than in the wild, as demonstrated for 84% of more than 50 species in a comprehensive study (Tidière et al. 2016) that compared four survival metrics: longevity, baseline mortality, onset of senescence, and rate of senescence between free-ranging and zoo populations. Zoos offer protection against predation, intraspecific competition, and diseases. Species with a faster pace of life (i.e., a short life span, high reproductive rate, and high mortality in the wild) showed significant changes in increased longevity. Species with a slower pace of life (i.e., a long life span, low reproductive rate, and low mortality in the wild) may have less potential for a reduction in mortality: they benefit less from captivity in terms of longevity (Tidière et al. 2016). An analogous situation may occur in domestication. The altered selection regime, with access to food, protection from predators, and assisted birth in some cases, may lead to longer life. On the other hand, some selective breeding has led to physical and life history traits which may have the opposite effect.

Hypotheses to explain why some species live longer than others concern their metabolic rate and size. Aging has been linked with energy expenditure, because more oxidative metabolism leads to more free radicals as by-products, which in turn produce more cellular damage. Given the higher mass-specific metabolic rate of small species compared with large ones, a positive interspecific relationship between size and longevity would be expected. In general, there is a negative interspecific relationship between energy use and longevity in mammals (Speakman et al. 2002). In general, small mammalian species have relatively higher metabolic rates compared with large species. We thus expect that species longevity increases with body size. The comparison between mice and elephants fulfills this expectation. *Mus musculus*, weights 20–60 g, has a life span of 2–3 years and a heart rate of 310–840 beats per minute. In contrast, *Loxodonta africana* weighs 2,500–5,000 kg, has a life span of up to 70 years, and a heart rate of 25–35 beats per minute. Traditionally it has been expected that, within species, small individuals also have higher metabolic rates than large individuals. But exceptions abound, as shown in a study of mice with high metabolism and longer life (Speakman et al. 2004), or when considering some species such as bats or rodents, that are small but long-lived.

Other hypotheses concerning longevity have been proposed. The "mutation accumulation" theory suggested that selection has little power to remove late-acting mutations from the gene pool in populations that experience high mortality rates early in life, so they accumulate deleterious mutations that would otherwise reduce fitness (Medawar 1952). High mortality rates favor selection for earlier maturity and a higher rate of investment in reproduction early in life, which may incur a cost later in life (Williams 1957). In other words, late-acting mutations tend to "accumulate" (they are consistently transmitted from generation to generation) because they are shielded from selection—high mortality rates early in life preclude the effects of late-acting mutations on the phenotype during the reproductive phase. Anti-aging mechanisms may be costly (Kirkwood 1977), and selective breeding may have resulted in dramatic changes in some life history variables that affect longevity.

As studied in mice and humans, telomere shortening to a critical length results in aging and shorter life spans. The mechanism for this shortening involves induction of a persistent DNA damage response at chromosome ends and loss of cellular viability. A study across eight species of mammals and birds, including goats and

Life history and growth   157

mice, found that telomere length is a determinant of species longevity across these species. The study concluded that critical telomere shortening and consequent onset of DNA damage and cellular senescence greatly influence life span (Whittemore et al. 2019). A study of Soay sheep, using longitudinal life history and telomere data, did not find an association between telomere shortening and mortality risk, but reported a correlation between telomere length and longevity and hypothesized a genetic basis for that correlation (Froy et al. 2021).

Dogs

SEASONALITY OF REPRODUCTION AND FECUNDITY

The mating season of wolves is set between January and March, and the pups are born in April or May. The yearly periodicity of reproduction physiology is not limited to females. Herre and Röhrs (1990) reported that the sexual behavior of male wolves kept at the Institute of Kiel displayed a seasonal dependence. Haase (2000) established that testicle weights in wolves reached peak values during the breeding season. Only when the females are in heat are spermatozoa produced.

Wolves are monoestrous, whereas domestic dogs are generally in heat multiple times per year (Herre and Röhrs 1990). Some domestic dogs are reportedly in heat more than twice per year if they are not fertilized (Lesbouryies 1949).

Regardless of whether they live under direct human control or are free-living, domestic dogs are not seasonal in their reproduction. Males are capable of reproducing at any season (Haase 2000), and females come into estrus every seven months on average; pups can be born year-round, at any time (Macdonald and Carr 1995; Boitani et al. 2007). Some seasonality has been recorded in some parts of the world, likely influenced by the environment. For example, tied to the late monsoon season and winter, free-living pups in India tend to be born between October through March (Oppenheimer and Oppenheimer 1975; Pal 2001; Chawla and Reece 2002). A comparative study of seasonality between domestic dogs, captive dingoes, and captive wolves demonstrated year-round conception in dogs, one between January and July in dingoes, and a more restricted one in wolves, between late January and early March (Thomson 1992).

In terms of reproductive seasonality, age of first reproduction, and pair-bonding, dogs are unique among canids, in that they are not seasonal. Females tend to breed at a younger age, and they do not pair-bond; rather, both males and females are promiscuous (Lord et al. 2013). This behavior could be the result of "relaxed selection" caused by direct care and protection by humans, or the result of direct selection for an increase in fecundity by humans (Fox 1978; Packard et al. 1985; Haase 2000; Boitani et al. 2007), or alternatively, it could be adaptations to a new niche, created by permanent settlement and stationary human refuse (Lord et al. 2013). As argued by Lord et al. (2013, p. 140), "this waste niche releases dogs from the high transportation and acquisition cost of finding and processing live prey typical of other members of the genus. The relative ease of finding and processing food year-round makes extended parental care and seasonality maladaptive. Reducing parental care and breeding multiple times/year increases their fecundity."

LITTER SIZE VARIATION

Wolves have an average litter size of 5.8, with 8 being the maximum (Schneider 1950). In the domestic dog, the range of variation in litter size is considerable, ranging mainly from 1 to 22 pups (Table 6.1). In addition to the effects of cohabitation with excess of resources, in dogs and cats the administration of ovulation-inducing hormones (GnRH, hCG) can also lead to above-average litter size (Günzel-Apel and Bostedt 2016).

The relation of body size to the number of pups per litter in domestic dogs has been investigated for decades (Sierts-Roth 1953; Kaiser 1971; Kirkwood 1985), but so far not in the context of the current breed diversity (Parker et al. 2017). Female Great Danes with an average body weight of 53 kg have an average litter size of 10.2 (2–19). Giant Poodles of roughly 20 kg body weight have an average litter size of 6.6 (1–13), whereas female Pomeranians that weigh only about 3 kg give birth to an average of 3.3 pups (1–8) per litter (Herre and Röhrs 1990). Kirkwood (1985) reported statistics significantly correlating litter size with body weight of the breed, in an allometric fashion, in disproportion to size. There are, however domestic breeds in which this relation of litter size and body size does not hold, such as in the Chow-Chow. Weighing approximately 20 kg, they give birth to an average of only 4.3 pups (1–9) per litter. Therefore, there are size-independent differences in litter size in domestic dogs. When using the height at withers as the reference basis, the results remain generally the same (Kaiser 1971). Kirkwood (1985) suggested that the covariation in litter size and weight among breeds may be due to variation in ovulation rate and not to fetal absorption. Across species of mammals there is no simple correlation between body and litter size (Kirkwood 1985), but this subject deserves more scrutiny (Sikes and Ylönen 1998).

In domestic dogs, the weight of young in percentage of mother's body weight is quite variable and allometric: 1.03% in Great Danes, 6.2 % in Pomeranians. Neonate size increases less-than-linearly with body mass in dog breeds, as is the case across mammals. Smaller breeds have higher values for individual gestation products than large breeds (Kaiser 1971; more relevant figure, though, is the weight of the entire litter in percentage of mother's body weight, the maternal output, and this percentage has the same values in most breeds and also corresponds to those of the ancestral form. In the wolf and the domestic dog this value corresponds to an average of 11%–12% (range 10%–15%), independent of body size (Herre and Röhrs 1990; Wehrend 2013).

SEQUENCE OF DENTAL, SKELETAL, AND SEXUAL MATURITY

In contrast to all the data clearly supporting significant changes in life history features brought about by wolf domestication, one aspect that remains unchanged is the relative sequence of dental, skeletal, and sexual maturity—as shown by a quanti-

Table 6.1. Number of newborns (pups) as well as average weight at birth from dogs of different breeds

Breed	Average body mass*	Average number of pups	Range of variation	Weight at birth of a single neonate (g)
Miniature Poodle (Zwergpudel)	ca. 5–7 kg (b)	3.5	1–7	200
King Charles Spaniel	ca. 3.6–6.3 kg (a)	4.6	1–9	250
Teckel	ca. 9 kg (a); 7–15 kg (b)	4.2	1–11	200
Chow-Chow	ca. 20–32 kg (b)	4.6	1–10	350
Whippet	ca. 11–18 kg (b)	5.7	1–12	340
Small Münsterländer	ca. 18–27 kg (b)	6.0	–	300
Dalmatian	ca. 24–32 kg (a); 20–32 kg (b)	8.2	1–15	360
German Longhaired Pointer	ca. 30 kg (a); 25–36 kg (b)	6.8	1–13	460
Pointer	ca. 20–34 kg (b)	8.2	6–10	440
Boxer	ca. 25 kg in females, more than 30 kg in males (a); 29–36 kg in males, about 7 kg less in females (b)	6.8	1–14	405
Gordon Setter	ca. 25.5–29.5 kg (a); 20–36 kg (b)	7.0	1–14	380
German Shorthaired Pointer	ca. 20–32 kg (b)	6.7	1–13	400
Braque Français	ca. 18–25 kg; Pyrenean type (b)	8.0	4–13	400
German Shepherd	ca. 22–40 kg (a); 23–41 kg (b)	4.9	1–12	440
Leonberger	ca. 41–77 kg (b)	8.2	1–20	450
Hovawart	ca. 29–41 kg (b)	8.0	2–12	485
Newfoundland dog	ca. 54–68 kg (a); 45–68 kg (b)	6.2	3–10	656
Rottweiler	ca. 42–50 kg (a); 36–61 kg (b)	6.5	1–12	620

* (a) from Krämer (2009); (b) from American Kennel Club (https://www.akc.org/; retrieved on 18 June 2020)
Source: Data from Günzel-Apel and Bostedt (2016, table 3–1)

tative study with a comprehensive sample of domestic dogs (Geiger et al. 2016). The sample included 137 domestic dogs of 15 breeds, a great range of body sizes, including both sexes and excluding neutered individuals, compared with 64 wolf individuals, with data obtained from radiographs and examination of macerated bones. In an ontogenetic series, the authors documented the timing to complete permanent dentition from eruption to occlusion, as a proxy for dental maturity, as well as the timing of growth plate closure at the proximal humerus (as a proxy for skeletal maturity). Their comparisons showed that neither the domestication process nor breed formation changed the sequence of dental, skeletal, and sexual maturity of the wild wolf, although the absolute values of life history variables in domestic dogs do have a greater range of variation than in the wild wolf. The reports of timing variation in tooth eruption among breeds (Huidekoper 1891; Scott and Fuller 1965) require quantification, but the conclusion concerning comparison of wolves versus domestic dogs is in any event valid (Geiger et al. 2016).

LONGEVITY

Galis et al. (2007) studied longevity in dogs, based on two large data sets: (1) a veterinary medical database (VMDB), with weight (in classes) as a size measure of 44,363 dogs from 134 different breeds, and (2) data from a collection at the Natural History Museum Bern (NMBE), including length of skull as a size measure of 859 dogs comprising 42 breeds. They found a negative correlation of life span and size or weight among breeds, but a slightly positive correlation between life span and size or weight within breeds. This result means that, within breeds, smaller dogs die younger, as expected from oxidative stress theory, given the higher mass-specific metabolic rate in small individuals. On the other hand, dogs from large breeds die young. This finding is not explained by oxidative damage due to size-related energy expenditure, as dogs from large breeds have lower mass-specific metabolic rate than small breeds (Burger and Johnson 1991; Speakman et al. 2003). Galis et al. (2007) argued that the selection for large adult size led to developmental diseases that diminish longevity in large breeds. Large breeds suffer from high rates of bone cancer, hip dysplasia, and osteochondrosis, which are all linked to high growth rate. Elevated mortality is thus caused by high growth rates, coupled with high oxidative damage in early life (Rollo 2002).

It would be important to investigate the extent to which larger individuals in wild species are shorter lived than smaller ones (in contrast to the pattern across species, in which large-bodied mammals live longer), to put the dog pattern in context. Here, as in all cases, it would be important to consider that there might be clade-specific patterns, for example, canids as opposed to felids or to other non-carnivoran mammals.

The case of dogs, though, does seem remarkable, as some breeds diverge in body size by almost an order of magnitude and in average life expectancy by a factor of 2. Early onset of senescence, elevated baseline mortality, higher early mortality, or increased rate of aging may all be aspects that characterize large size dogs. Kraus et al. (2013) studied these different mortality components based on data from 56,637 dogs representing 74 breeds and found that a key aspect in the trade-off between size and life span is a strong positive relationship between size and aging rate. They found no clear relationship between age at onset of senescence and size and concluded that large dogs die young mainly because they age quickly.

Information on life history and demographic data of particular breeds and dog populations are available in databases of specific organizations and countries (McGreevy et al. 2018). This kind of data can be used to assist prioritization of health issues, prophylactic breed-specific attention, common disorders (e.g., overweight/obesity, otitis externa, degenerative joint diseases), and wellness checks by owners.

Comparative bone histology and life history

Bone microstructure preserves a record of life history. Bone growth marks record cyclic variation in bone tissue deposition as zones, lines of arrested growth (LAGs), and annuli, the latter two indicating an interruption or decrease in cortical bone apposition. The number of cyclical growth marks within a bone cortex can provide insight into age at maturity, growth rate, and longevity. Bone histology is used to study fossils, as has been done for mammals on islands, with their specializations in life history traits (Kolb et al. 2015). The study of living species, including domestic forms (Firth 2006; Zedda et al. 2008, 2020; Cambra-Moo et al. 2015), can also greatly benefit from the bone histological approach (Marín-Moratalla et al. 2013). Differences in growth rates among breeds of different sizes could be estimated by comparing the distances between LAGs. However, the interpretation of differences in bone histology is non-trivial given confounding factors of variation and preservation (chapter 1).

Differences in microstructure in parts of the bone are related to biomechanics. High mechanical stress correlates with a higher number of osteons in some regions of the bone (Nacarino-Meneses et al. 2016). Bone remodeling can complicate the identification of LAGs, as it erases much of the growth record of the individual. In some bone portions, the primary bone is completely replaced by Haversian tissue (Straehl et al. 2013). Horses, for example, experience rapid growth in the first two years of their lives (Stover et al. 1992), but in some cases this phase is erased later in life because of remodeling).

A standard procedure is to use histological sections from the mid-shaft of long limb bones, degreasing them, embedding them in epoxy resin, cutting and grinding to produce thin sections that are studied under special microscopy (Padian and Lamm 2013). This series of steps means that the method is invasive, involving so-called destructive sampling.

The study of bone microstructure in other tetrapods besides mammals is also widespread in paleontology, in particular in dinosaurs, including birds (Chinsamy-Turan 2005). A study of ducks led to the discovery of a rare (or rather unique to date) occurrence of chondroid bone in postembryonic periosteal bone formation of any tetrapod limb (Prondvai et al. 2020). Chondroid bone, which in the case of the ducks is nonpathological, is a kind of skeletal tissue intermediate in many features between bone and cartilage (Witten et al. 2010). This discovery in ducks may lead to further examinations of other birds, a group in which this skeletal specialization may be coupled with the extreme growth rates in their limbs (Prondvai et al. 2020).

The effects of nutrition and husbandry practices on bone mechanical performance are not uniform across bones or portions of them, as shown in a study of femora and humeri of broiler chickens (Bonser and Casinos 2003).

Fig 6.5. Precocial pigeon (*left*) and altricial Muscovy duckling.

Birds

THE CONDITION AT HATCHING: THE ALTRICIAL / PRECOCIAL CONTINUUM IN BIRDS

Precocial chicks are very active, have to search for food on the ground, and do not receive much parental care, showing lower growth rates than altricial chicks (Gaillard et al. 1997). In contrast, fully altricial birds hatch with their eyes closed, are relatively immobile, and are dependent on the parents for feeding and thermoregulation (Starck and Ricklefs, 1998). All studies of the condition at hatching coincide in hypothesizing that the last common ancestor of birds was precocial. Precocial birds are born with their eyes open, are mobile, and some are capable of feeding on their own (Figure 6.5).

Different features can be used to characterize the altricial-precocial spectrum for birds. For example, a combination of lipid-free dry/wet weight ratio of tissues and neonate mass is a way of characterizing the condition at hatching (Starck and Ricklefs 1998). External features and patterns of activity and behavior have been used in a "typological" approach that is used to create categories with different grades of maturity (Nice 1962). In fact, these categories do not imply any sequential, stepwise evolution among them. The qualitative scoring includes eight categories that refer to the eyes, plumage, activity, and behavior (Table 6.2).

The great majority of domesticated birds are reportedly "precocial 2" (Starck and Ricklefs 1998), including chicken (*Gallus gallus*), Japanese quail (*Coturnix coturnix*), duck (*Anas platyrhynchos*), Muscovy duck (*Cairina moschata*), and geese (*Anser anser* and *Anser cygnoides*). The turkey (*Meleagris gallopavo*) is precocial 3. In contrast, the pigeon (*Columba livia*) is semialtricial 2, and the zebra finch (*Taeniopygia guttata*) is altricial (cf. Murray et al. 2013; Ikebuchi et al. 2017). The canary (*Serinus canaria*) and the Bengalese finch (*Lonchura striata*; cf. Yamasaki and Tonosaki 1988) are best characterized as altricial.

A comparison of the developmental trajectories of Galliformes with diverse degrees of precocity at hatching indicated that they are species-specific, so they can be compared to reconstruct the pattern of heterochronic change underlying different degrees of precocity (Starck and Sutter 2000). The brush turkey, peacock, and pheas-

Table 6.2. A qualitative scoring framework from altricial (1) to precocial 1 (8) in hatchlings

Type	Quality	1 altricial	2 semialtricial 2	3 semialtricial 1	4 semiprecocial	5 precocial 4	6 precocial 3	7 precocial 2	8 precocial 1
eye	closed	+	+	−	−	−	−	−	−
plumage	naked	+	−	−	−	−	−	−	−
	down	−	+	+	+	+	+	+	+
	contour	−	−	−	−	−	−	±	±
activity	motor	−	±	+	+	+	+	+	+
	locomotor	−	−	±	+	+	+	+	+
behavior	stay in nest	+	+	+	−	−	−	−	−
	fed by parents	+	+	+	+	±	−	−	−
	follow parents	−	−	−	±	+	+	+	−
	search alone	−	−	−	−	−	±	+	+
	no interaction	−	−	−	−	−	−	−	+

Sources: From altricial (1) to precocial 1 (8) in hatchlings, based on Nice (1962); from superprecocial to altricial, according to Starck (1993; see also Starck 2018).

ant follow the same developmental trajectory, but the brush turkey hatches late on that trajectory, becoming superprecocial (Starck and Sutter 2000). The patterns of development of wing and leg do not vary independently of each other. This integration may not be universal for birds, and modularity in other groups besides Galliformes may explain different degrees of evolvability in limb proportions (Faux and Field 2017).

LIFE HISTORY STRATEGIES AND EVOLVABILITY IN BIRDS

A relation was found between the condition at hatching and the evolvability of toe arrangements in birds (Botelho et al. 2015b). The groups that independently evolved super-altriciality are those exhibiting transformations of toe arrangements. No precocial or semi-precocial birds exhibit toe specializations, which include zygodactyly, semi-zygodactyly, heterodactyly, pamprodactyly, and syndactyly. It was hypothesized that in altricial species, with their delayed skeletal maturation, there is more epigenetic influence of embryonic muscular activity over developing toes, facilitating the repeated evolution of morphological innovations (Botelho et al. 2015a).

The condition at hatching may limit and facilitate specific morphological changes or evolvability of traits, including brain features. It has been suggested that altriciality facilitates the evolution of telencephalic expansion by delaying telencephalic neurogenesis. Furthermore, delays in telencephalic neurogenesis may generate delays in brain maturation, resulting in neural adaptations that facilitate learning (Charvet and Striedter 2011). Delayed telencephalic neurogenesis may facilitate the evolution of neural circuits that allow songbirds and parrots—including the canary among domesticates—to produce learned vocalizations.

FECUNDITY IN BIRDS: CLUTCH SIZE VARIATION

The red junglefowl generally lays 10–30 eggs per year, whereas some breeds of domestic chicken lay up to 300 eggs or more. One report is 63 for fowl and 361 eggs for domestic chickens (Clayton 1972). Diet and environmental conditions affect these figures (Burt 2007). Social conditions can affect the timing of sexual maturation. A comprehensive experiment on chickens showed that rearing with males accelerates the onset of sexual maturity in females—although this did not lead to a higher egg production (Widowski et al. 1998).

In domestic ducks, an egg-laying capacity exceeding 200 per year is not rare, whereas the clutch of the wild form consists of 10–12 eggs. Clayton (1972) stated 146 as the maximal egg number per year in wild ducks and 418 for domestic ducks. In the wild, the Muscovy duck (*Cairina moschata*) has a normal clutch size of 12–14 eggs. Under conditions of expert husbandry these birds can be much more prolific: two 20-week laying cycles separated by a 10-week moult and rest period can achieve production of 120–150 hatching eggs per duck (Clayton 1984a). The Japanese quail (*Coturnix japonica*) has increased its egg-laying capacity from 7–14 in the wild to 200–300 (Chang et al. 2009; Hubrecht and Kirkwood 2010). Even the domesticated budgerigar or parakeet (*Melopsittacus undulatus*), a long-tailed, seed-eating parrot, can reach an egg-laying capacity of 300 per year if the eggs are consistently removed (Herre and Röhrs 1990).

Geese have also been affected by domestication in this life history trait. The clutch size of many European goose breeds has increased by 600%–1200% compared with the greylag goose, which lays about 5–6 eggs. For example, the Chinese goose breed lays about 875%–1200% more eggs than its wild counterpart the swan goose, which lays 5–8 eggs (Kozák 2019).

In birds, gonad maturation and sexual behavior can respond to changes in day length. Given this, manipulation can lead to reproductive changes. As an example, ducks can be made to lay at any season of the year using artificial light. Domestic ducks may be brought into breeding by six months of age if exposure to light is conveniently regulated, whereas in the wild, on average the mallard starts breeding 12 months after hatching (Clayton 1984b).

In the wild, the guinea fowl lays about 15 eggs in a laying period lasting two months. Free-range guinea fowl can produce up to 100 eggs in their first laying season, extending from April to September in the Northern Hemisphere. Whereas individuals subjected to normal seasonal variations in light and temperature lay their first egg at 18 weeks of age, climate-controlled facilities and artificial insemination can lead to significant changes: sexual maturity can be reached at 32 weeks of age (in order to avoid the production of eggs too small to be incubated), and 170 eggs in 43 consecutive weeks can be obtained. From these 170 eggs, around 110 chicks were recorded hatching. During the second year of laying, an additional 20–30 offspring can result (Mongin and Plouzeau 1984).

Numerous studies have been conducted to select for increased egg production in different species. For example, in turkeys a line of 28 generations was selected for increased 180-day egg production and compared with the randomly bred control population from which it originated. The selected line exhibited linear increases in egg production, rate of lay, clutch length (average and maximum), number of effective days of production, and percentage hatch of fertile eggs 12 weeks into production (Anthony et al. 1991). This study and numerous other works in the poultry industry examine changes in body weight and life history variables associated with egg production (Zuidhof et al. 2014).

GROWTH IN CHICKENS AND OTHER DOMESTICATED BIRDS

As in domestic mammals, the growth pattern of domestic birds has been studied using different mathematical functions (e.g., Gompertz, logistic, Bertalanffy) for different species of domesticated birds (Starck and Sutter 2000; Nahashon et al. 2006; Narinc et al. 2010; Eleroğlu et al. 2018). Most studies have been conducted on specific breeds and treatments in chickens, and have aimed at characterizing the time and mode of growth and carcass traits (Goliomytis et al. 2003; Murawska et al. 2005, 2011; Schmidt et al. 2009; Zuidhof et al. 2014; Soares et al. 2015). These studies illustrate developmental plasticity and the intensity with which humans have modified growth parameters for management and production purposes. There are numerous studies of other species besides chickens, including these on ducks: Bochno et al. (2005), Murawska (2012), Murawska et al. (2016), Svobodová et al. (2020); geese:

Strain	1957	1977	2005
0 d	–	–	–

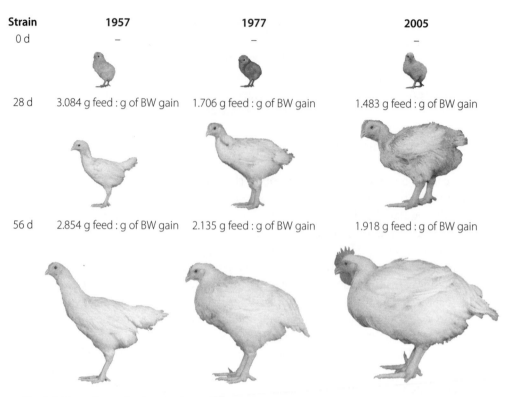

| 28 d | 3.084 g feed : g of BW gain | 1.706 g feed : g of BW gain | 1.483 g feed : g of BW gain |

| 56 d | 2.854 g feed : g of BW gain | 2.135 g feed : g of BW gain | 1.918 g feed : g of BW gain |

Fig 6.6. The effect of selective breeding in broiler chickens over five decades. Illustrated are age-related changes in size for three ages in individuals of University of Alberta Meat Control strains.

Bochno et al. (2006), Kozák (2019); turkey: Murawska et al. (2015); and guinea fowl: Voroshilova (1974).

There is a trade-off between growth rate and functional maturity in organs, although there is no universal rule applicable to all species to understand that relation (Starck 2018). The intense selection for higher posthatching growth rates has led to a remarkable increase in body mass (Figure 6.6). The trade-off is disrupted, but at the expense of meat quality and bone properties.

LONGEVITY IN CHICKENS

"Matilda" (1990–February 11, 2006), a chicken from Alabama, reportedly received the title of World's Oldest Living Chicken from Guinness World Records. Besides curious cases like this one, domesticated animals can be the subject of more systematic studies on longevity. Chickens have been used to study the biochemistry of the accumulation of modified proteins in aging (Gracy et al. 1990). Artificial selection for high growth rates in chickens, and in fact in many other domesticated species, can produce different kinds of health issues, some of which result in reduced longevity.

Domestic animal production and the Western World Social Paradigm

One of the dominant beliefs in Western societies is that economic prosperity is tied to growth, and that growth will resolve most social dissatisfaction and conflicts. This belief is coupled to another core belief: that all major issues can be resolved with continued pursuit of industrial and technological advancement, which will result in the development of new solutions to new issues that arise. Thus the belief that the main solution to a pandemic is a vaccine, as opposed to it being a temporary solution for one specific pandemic, without addressing the roots of the problem that will lead to new pandemics soon to follow.

Domestic animal production is at the core of these beliefs. In this view, world hunger can be solved by increased production (growth), which can be accomplished by genetic manipulation of crops and animals (technology). Empirical evidence speaks against this view, as its implementation has paid a big price: increased incidence of zoonoses (Recht et al. 2020), environmental destruction (Naess 1989), disregard for animal welfare (Jensen 2017), and decrease in the nutritional value of food (Pawlick 2006).

Large-scale animal production has resulted from a growing global demand for animal protein. Supplying this protein at low cost and in an industrial fashion has been enabled by increased antibiotic consumption in domestic animals, which has led to increased antimicrobial resistance, affecting the health of both humans and animals (Van Boeckel et al. 2019; Moore 2019). Industrial livestock farming has led to disruption of nitrogen and phosphorous cycles and greenhouse gas emissions from enteric fermentation and manure. Intense overgrazing and land-use change have negatively impacted ecosystems. Addressing these problems requires international collaboration; thus the World Health Organization formulated a combined "One Health" approach regarding humans, animals, plants, and their shared environment. A transition to sustainable animal farming practices can be accomplished only by changes in consumer behavior; otherwise regulations are needed that are unlikely to be established without demand from the population and decisions at the leadership level.

Establishing a new Ecological Paradigm has been a response to this crisis, based on the idea that limitations on industrial and population growth can make sustainable and healthy human life and environmental protection possible. The Deep Ecology movement led by the late Arne Naess (1989) recognized the intrinsic value of all living beings, seeing humans as part of and linked to nature. In this view, money is not considered wealth (as resulting, for example, from "efficient" breeds of chickens), but instead wealth is based on human skills, plants, forests, soil, and animals. Those human skills can be put to good use in the cultural management of the environment and in our relations with plants and with other animals. Comparative ethnological examination of the concept of nature and the views of domestic animals is central to understanding human variation and possibilities in this context (Davis 2001; Descola 2013).

Summary: Variation in life history evolution

Increased fecundity is a common result of domestication, but the extent of change and the trade-offs involved are species-specific. Environmental factors, including nutritional, health status, and temperature, affect fecundity. Breed differences, however, do have a strong genetic basis. On the other hand, the heritability of litter size in at least some domesticated forms can be low, as exemplified by studies of cattle, sheep, and pigs

There is no universal similarity or matching of the patterns and perhaps mechanisms of relations between life history variables (e.g., size versus longevity) reported from interspecific, broad comparisons to those within species of domesticates.

The relation between size and metabolic rate is not linear, and the expected pattern of small individuals having higher metabolic rates than large individuals is not universal. Neonate size increases less-than-linearly with body mass in dog breeds; in assessments of the relationship for a total litter, the relationship is rather linear.

Schultz's rule predicts that relatively later eruption of molars keeps the dentition functional during a prolonged life span, but there are many exceptions to this general pattern, valid only for some clades.

Different developmental landmarks have been used to characterize condition at birth, for example "eyelid closure" and degree of development of the brain, and depending on the parameter chosen, a different categorization results, although these features tend to be correlated in their timing of occurrence.

Life history traits are variable across species, and these traits influence the morphological outcome of their domestication. As an example, dogs are less mature than cats at birth, and arguably more evolvable under a domestication/selective breeding process.

Growth rate and litter size vary in different proportions along the altricial-precocial spectrum. Comparative studies of prenatal development among domesticated mammals, including altricial and precocial species, could provide data to inform the relationships among life-history strategy, evolvability, and disparity. There is a reported pattern of mammalian clades with altricial neonates investing more of the energy budget into postnatal growth and having relatively higher levels of size variation. This strategy may lead to more adult disparity than the alternative, precocial strategy. The latter allocates fewer resources to postnatal growth, associated with lower amounts of size variation. Morphological evolvability has on the other hand been hypothesized to be linked in the opposite way to the relative timing of birth in carnivorans. It has been argued that precocial forms are much exposed to artificial selection at postnatal stages, resulting in higher diversity than in altricial species.

CHAPTER 7

Morphological diversification

Exploring the morphological diversity of domesticated animals shows us that in many cases the range of variation within a species, at least in certain organ systems or modules, exceeds the standing variation expressed in the wild form (Darwin 1868). This is particularly the case when considering intensive artificial selection (van Grouw 2018). Characterizing the variation in the wild form is fundamental to this comparison. Only then can one establish whether the assumed pattern of new traits and extended variation in the domestic forms is truly novel or simply a set of rare conditions becoming increasingly more frequent in a given population (Bateson 1894, p. 266).

In this chapter I will examine several morphological novelties and trends that result from selective breeding. Examples are given from a few of the many organ systems that could be compared. The extent of morphological diversity generated by domestication is reminiscent in many cases of what is recorded across species in macroevolution. The term "disparity" is widely used in paleontology to refer to such diversity. I do not use it here given the intraspecific nature of the variation in domestication.

Changes brought about by the initial phase or nonintensive domestication can be more subtle. They concern the so-called domestication syndrome (chapters 3 and 4).

Divergence from the wild

Domestic animals differ significantly in morphology, physiology and behavior from their stem forms, but there is also considerable variation in the extent and composition of those differences. In some features, domestication is not about the odd innovation but about preservation of the odd (Lord et al. 2020b). This may be the case when certain wild individuals express some of the variation observed in domesticated forms.

The study of morphological variation has a long tradition in biology, and some examples concern domestic animals (Bateson 1894). However, much is unknown, and some organ systems are more intensively studied than others. Studies of the variation within wild forms are uncommon (e.g., Eda 2020 on red jungle fowl), although they are important for evaluating the divergence that has taken place, and of particular importance in zooarchaeology. Concerning the latter, an example is the evaluation of Late Pleistocene samples of either wolves or dogs.

Variation in the environments in which Pleistocene gray wolf populations lived may have resulted in a range of wolf ecotypes that were morphologically distinct (Perri 2016), as suggested by skull shape and tooth wear of extinct Late Pleistocene populations that were specialized hunters and scavengers of extinct megafauna (Leonard et al. 2007). Considering the high mobility of wolves when tracking migratory prey or when searching formates and territory, variation in archaeological *Canis* material may result from this dispersion or migration (Perri 2016).

Another example of the characterization of wild-type variation is the range of coloration in horses. Coat variations painted in the 25,000-year-old cave of Pech-Merle in France included a range of black, bay (a brown or reddish-brown body color, with black tail, mane, ear edges, and lower legs) and leopard spotting. At first, it was not known whether these were unrealistic or truthful depictions of the variation in those wild horses. However, we now know these predomestic horse populations did possess three coat color variations, as determined by ancient DNA studies of single nucleotide polymorphisms present in six genes linked to specific coat colors (Pruvost et al. 2011). In fact, there was a significant increase in coat color variation that coincided with the beginning of horse domestication at least 5,000 years ago, as shown by allelic variation in genes related to coat color (Ludwig et al. 2009).

A study of horse teeth quantifying premolar shape variation in Pleistocene wild populations determined that this feature remained unaltered over thousands of years across Eurasia based on comparisons with modern Icelandic and Thoroughbred domestic horses, recent Przewalski's wild horses, and domestic horses of different ages throughout the Holocene. In contrast, in the two millennia after the Iron Age, an explosive increase in the scale and pace of dental variation has resulted in a twofold increase in the magnitude of shape divergence among modern breeds (Seetah et al. 2016).

Tempo and mode of major patterns of morphological diversification in domesticated animals

The tempo and mode of morphological diversification occurring in the domestication process can vary, and this variation can be tied to phases or transitions in human history (e.g., migration to the Americas, industrialization, origin of kennel clubs). Analogous to models discussed in the context of mammalian diversification before or around the Cretaceous/Paleogene boundary (Springer et al. 2003), one can define explosive, soft explosive, long fuse, and short fuse patterns of diversification (Fig. 7.1). In the long fuse model, an initial lengthy period of invariance is followed by a great increase of change, as mentioned above for dental variation in horses (Seetah et al. 2016). The short fuse model illustrates an old and fast diversification. Perhaps this is the case of integumental patterns in carp domestication, an old domestication process for which records of great variation exist since domestication was first recorded (chapter 9). The explosive model implies a recent and fast diversification, following a more recent event of domestication. Probably a case of an explosive model is the canary, because diverse forms started appearing soon after domestication of this bird began in historical times (chapter 2). The chinchillas may be another example, after their domestication process was initiated in California in the 1920s, leading to most farm chinchillas, although the diversification in this species has been rather modest (Zimmermann 1961). A third example is that of the golden hamster, which domesticated around the same time as the chinchilla and soon exhibited great variation of fur color and length. The soft explosive model could be illustrated by rabbit domestication, already recorded in Roman times, but with phenotypic diversity starting to be recorded centuries later. These are oversimplifications if meant to represent the history of a species, given regional bursts of diversification or other changes in time and space that do not fit any of these overall patterns. For example, there could be regional patterns of diversification at different times than in other parts of the world, as with dog diversification in pre-Columbian America (Segura et al. 2022).

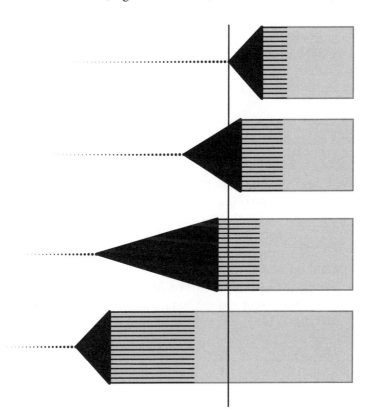

Fig 7.1. Patterns of morphological or genetic diversification following domestication. From top to bottom: explosive, soft explosive, long fuse, and short fuse patterns of diversification following domestication.

Fig 7.2. Creatures of the paintings of Hieronymus Bosch are examples of animals that do not exist. Although fantastical, they are still limited: the creatures consist of parts we know. As in many monsters of Greek mythology, "the human mind is not able to invent something that is not derivative of human experience" (Syropoulos 2018, p. 6). Forms are combined, or somewhat distorted; nothing is created anew. The creatures are coherent and formed with ordinary elements that are exaggerated or distorted.

Unoccupied Morphospace

A universe of potential phenotypes is called a "morphospace". Some of it is realized, occupied, as in the examples in this chapter. Some of it is not, as a result of developmental evolution and contingency (chapters 4 and 5). The creatures of the paintings of Hieronymus Bosch are examples of this second case (Figure 7.2). Although fantastical, the forms of these creatures are limited because they consist of shapes we know. As in many monsters of Greek mythology (Syropoulos 2018, p. 6), "the human mind is not able to invent something that is not derivative of human experience." Forms are altered, modified or distorted, nothing is entirely new. Bosch's creatures are coherent and formed from ordinary elements, albeit exaggerated or distorted (Jacob 1981).

In the real world we can find many examples of unoccupied morphospace. Despite the impressive number of species and variation in color, lip shape, and feeding apparatus of cichlids in African lakes (Kratochwil et al. 2018; Salzburger 2018), none are "snake-like" (angilliform), nor are there flatfish cichlids. Golden moles, talpids, and fossorial South American rodents all have limbs; there are no limbless burrowing mammals, whereas this is a common phenomenon in squamates (Greer 1991). These patterns may result from ecological aspects, but internal, structural, and developmental aspects surely play a role as well. In the case of domesticated animals, some concrete examples of unoccupied morphospace are illustrated by the absence of long-snouted cats, brachycephalic horses, silkie ducks, long-limbed cavies, and long-necked chickens (Fig. 7.3).

Fig 7.3. Examples of unoccupied morphospace among domestic mammals and birds. (a) no brachycephalic horses, (b) no long-limbed cavies, (c) no long-snouted cats, (d) no Silkie ducks, (e) no duck-beaked pigeon, and (f) no long-necked chicken.

The Quantification of Morphospace

Different methods have been developed to quantify morphospace, particularly in three dimensions. Comparison of form can be approached by recording homologous points across individuals—landmarks—and applying the statistical approaches of geometric morphometrics. Unlike results in traditional morphometrics based on linear measurements, results are reported as distances between 3D coordinates, and deformations or deviations from a mean form (or shape) (Mitteroecker and Huttegger 2009). Identification of the same set of biologically homologous landmarks in every sampled individual is easily accomplished when comparing wild with domesticated forms. Historically, this approach goes back in its conception to D'Arcy Thompson (1917), who introduced the concept of geometric transformations, showing how one can mathematically transform organismal forms by warping a geometric grid (Fig. 7.4). A system of coordinates can be elongated, flattened, or partially distorted. These transformations can then describe and quantify shape differences among organisms.

The ability to identify and describe shape patterns mathematically has been immensely beneficial to the evaluation of morphological change brought about by domestication. However, mathematically complex methods, along with big data, can make replication testing tedious if not impossible and give the impression that the method is more important than the question. The big data approach often focuses on the search for universals where there are few, or none. As more scientists worldwide apply morphometric methods, a more inclusive and collaborative community makes discoveries. Many unusual domestic populations are localized to different corners of the world, absent from collections in the main European and

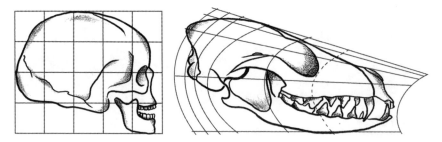

Fig 7.4. Human and "dog" skulls in a Cartesian grid in the work of D'Arcy Thompson (1917) *On Growth and Form*. This work was a precursor of the geometric morphometric approach commonly used today to quantify shape differences across populations and taxa. The reference grid of the human can be modified to produce a shape of the dog, in this simplified depiction resembling more a generic wild canid.

North American museums, which are the ones usually accessed for this kind of research.

Another approach to capturing patterns of morphological disparity is using discrete character-taxon matrices (Lloyd 2016). These matrices are commonly generated in phylogenetic analyses in which characters are defined to represent individual hypotheses of biological homology and to conduct comparisons across species (Hetherington et al. 2015). The same can be done for breeds within a species. Integumental variation in chickens was studied using this approach (Núñez-León et al. 2019). Traits included were comb, earlobe and wattle shape and color, several feather features, and colors of eggshell, beak, eyes, skin, and shanks. Principal components analysis was used to reveal clusters and patterns in accordance with the hypothesis of high hybridization among chicken breeds and fit with genetic analyses (Rubin et al. 2010). Another noteworthy finding was the grouping of the Araucana breed with Asiatic breeds, perhaps supporting the Polynesian origin hypothesis (Storey et al. 2012; but see Thomson et al. 2014). The latter could suggest pre-Columbian exchange, an important subject that should be approached with new empirical data and analyses, and with neither a dogmatic anti-diffusionist position nor an inclination to accept as evidence any hint of potential ancient transpacific voyages.

Imaging and digital approaches and information

Some differences among populations or species are not apparent, as seen in internal structures visible only through digital imaging (Billet et al. 2017). Besides veterinary studies conducted mostly on dogs (Carrera et al. 2009; Pérez et al. 2010; Mielke et al 2017), there is little of this kind of work done for domestic species with a comparative and evolutionary perspective (Marchant et al. 2017).

Digital imaging techniques can be used in morphospace studies for capturing gross morphological data (Davies et al. 2017). Techniques include X-ray computed tomography (CT) and magnetic resonance imaging (MRI). An example of an application of these methods together with appropriate staining techniques is the superb description of the head anatomy of a rock dove (Jones et al. 2019; see also Bonsmann et al. 2016 for a study of the ear of the quail, *Coturnix coturnix*).

Zooarchaeological studies of domesticated animals can also benefit from these techniques. An example is the digital documentation of petrosal bones, a favorite subject of destructive sampling involved in ancient DNA extraction (Evin et al. 2020).

The ethmoidal or nasal region has been investigated using CT in dogs, but rarely in other domestic animals. Several studies reveal how the delicate turbinals are different in their proportions and geometry among breeds (Craven et al. 2007; Wagner and Ruf 2020). These are in many cases broken during maceration and in any event invisible even in intact skulls as they are located inside the nasal cavity. Turbinals are a potentially rich subject of study of functional anatomy of respiration and olfaction. But in terms of archaeological research, preservational artifacts make turbinals a hopeless subject. Instead, imaging techniques are better suited for documenting the size and morphology of the cribiform plate, a more robust area of the skull that reflects innervation related to olfactory function (Bird et al. 2018).

Through digitization, great amounts of data are now produced in the form of images accessible from a distance, so data sharing becomes possible, in principle. However, ideas of ownership are often a hindrance to accessability, so it has been properly claimed that psychology, not technology, is a great impediment to the widespread sharing of digital data (Hipsley and Sherratt 2019). Another issue is access to hardware and software, which in many cases becomes very expensive and inaccessible to many researchers around the world. If a part of scientific excellence is equal access to opportunities, digital imaging development and suggested standards (Davies et al. 2017) should consider how people in all parts of the world can produce and gain access to those data.

In studies of geometric morphometrics of skull shape, the use of CT scanning has become standard. This technology is excellent for many purposes, but it is expensive and inaccessible to most research institutions around the world, consumes much energy, and may not be necessary for some of the research for which it is used. A comparative study of the reliability and precision between CT scanning and surface scanning in a small rodent revealed that compared with μCT scans, even low-resolution 3D surface scans of very small specimens are sufficiently accurate to investigate intraspecific differences (Marcy et al. 2018).

Close-range photographic techniques such as photogrammetry are a great option in technical possibilities and cost when dealing with the surface of organic structures. By recording and combining photographic images, researchers can construct 3D models of skulls and other kinds of material evidence associated with domestication in a cost-effective and accessible way. A photogrammetric approach applied to wolf crania produced 3D models that accurately describe the original cranium in terms of geometry, and even coloration and texture (Evin et al. 2016).

An example of the great (but expensive) technological possibilities of digitalization is R2OBBIE-3D, which integrates 3D-reconstruction approaches combining a robotic arm, a high-resolution digital color camera, and an illumination basket of high-intensity light-emitting diodes (Martins et al. 2015). This system can generate highly accurate 3D models of biological objects between 1 and 100 cm in size, based on multiview photometric stereo scanning, making it possible to capture color-texture and geometric resolutions better than 15 μm; thus R2OBBIE could be used in detailed studies of the integument.

Coat variation in mammals

In mammals the first visible deviations between domestic and wild populations are in coat color, recorded as well in island populations of mammals (van der Geer 2019). Experiments in selection for tameness (chapter 8) have indeed shown that changes in pigmentation are the first to appear, although the role of pleiotropic effects of selection for tameness in this context has been questioned (Linderholm and Larson 2013). It is often hypothesized that humans selected for individuals with color variants in founding populations of domesticates, perhaps simply for aesthetic reasons, or in order to avoid camouflage, which facilitated husbandry. Selection for color variation among domestic animals also allowed humans to distinguish prized domesticated animals from their wild forms (Linderholm and Larson 2013; Andersson 2016).

While most wild species show very modest variation, domestic animals exhibit remarkable coat color diversity (Herre and Röhrs 1990; Cieslak et al. 2011; Wiener and Wilkinson 2011). Color variants in domestic animals have been recorded for millennia, as in written records from the Ur III dynasty in Mesopotamia, dated to about 5,000 ybp (Zeder 1994).

Changes in body size in mammals

Size reduction has been recognized as one of the first indicators of domestication in large mammals (Boessneck and von den Driesch 1978). Linear measurements of modern and ancient bones and teeth have been used to identify the wild or domestic status of mammals in the archaeological record (Vigne et al. 2005a). In fact, a decrease in size during domestication has been reported for many species, including dogs, cows, sheep, goats, and pigs (e.g., Davis 1981; Tchernov and Horwitz 1991; Albarella et al. 2009; Zeder 2006; Hongo et al. 2009), but not in South American camelids (Arbuckle 2005; Balcarcel et al. 2021a). As with many other features, size changes remain questionable for investigating the early stages of the domestication process (Meadow 1989; Steele 2015; chapter 1).

Changes in body size, if any, must reflect the effects of specific husbandry factors, including disease, nutrition, and crowding. The exception of domestic South American camelids may be related to the nature of their husbandry, in which exploitation strategies did not subject animals to nutritional constraints or to a highly anthropogenic environment (Arbuckle 2005; Balcarcel et al. 2021a). Small mammals reportedly experience an increase in size with captivity or domestication, as in the cases of the Norway rat, house mouse, gerbil, and cavy (Arbuckle 2005). This pattern has not been thoroughly tested, though. Further investigations of the plasticity and the immediate environmental effects on body size of individuals have been suggested (Evin 2020).

As for selective breeding, changes in size can be significant, as can the range across breeds. This pattern is well exemplified by standard surveys of livestock species. Cattle (mature bull figures) range from the very small at about 450 kg to large, reaching 700 kg. Goats (mature buck figures), sheep (mature ram figures), and pigs (mature boar figures) range from 45 to 68 kg, 64 to 170 kg, and 180 to 360 kg, respectively. In these species, frame size, reported in terms of small, medium, or large, is used as a descriptor in commodity meat markets (Ekarius 2008). There are outliers to these standard ranges, some listed below.

A Chianina bull, which averages 1,360 kg and stands 1.8 m tall at the withers, is a very large breed, while a Dexter bull, which averages around 430 kg and 1.1 m at the withers, represents a small breed. There is the odd record, such as that of "Knickers," a steer (a neutered male) of the Holstein-Friesian breed from a small town south of Perth in western Australia that weighed 1.4 tons and was 193 cm tall. Even larger individuals are documented in diverse media, although mature steers of this breed typically weigh half of Knicker's weight (Ekarius 2008).

There are about 200 breeds of domestic goats, with much size variation among them. Nigerian dwarf goats are one of the smallest breeds, weighing around 9 kg; pygmy goats weigh between 24 and 39 kg; Nubian goats weigh up to 113 kg, and Boer goats up to 135 kg (Porter 1996). Within a region, there can be great variation, as in Punjab in Pakistan with forms ranging from the Barbari goat at 30 kg to the Pahari goat at 67 kg (Saif et al. 2020).

The range of adult pig body mass including most breeds is between 30 and 350 kg, excluding unusual individual outliers (Porter 1993).

The body size differences among horse breeds are remarkable, ranging from 70 cm to around 2 m wither height. The weight of a Falabella, the smallest of breeds, can be as low as 30 kg, while Shire horses from England and Percherons from France weigh up to around 1,200 kg (Hendricks 1995; Brooks et al. 2010). There are records that go beyond these figures.

Rabbits became particularly diverse morphologically (chapter 2), and their weight ranges mostly from 1 to 7 kg (Herre and Röhrs 1990). They are an excellent subject for study of growth allometry and the generation of morphological diversity among breeds (Fiorello and German 1997).

Size variation in dogs is remarkable. Breed differences in stature (Rimbault et al. 2013; Stone et al. 2016) are far greater than across humans (Norton et al. 2019). Mastiffs can be 50 times heavier than Chihuahuas. The median shoulder height in Great Danes (76 cm) is four times greater than in Pekingese (19 cm). When scaled to humans, this would be equivalent to a difference of almost 5 meters. Differences within breeds can also be large. A toy poodle weighs 2.5 kg, while the large, standard ("king") poodle weighs 33 kg (Herre and Röhrs 1990).

Rensch's (1950) rule predicts that sexual size dimorphism across species is less among smaller species and greater among larger species if males are the larger sex. In domestic dog breeds, the proportional size difference between males and females of large and small breeds is essentially the same (Sutter et al. 2008); therefore, Rensch's rule, formulated to predict relative scaling of sexes across species, does not hold in dogs.

Within domestic species, examining the evolution of body size given the network of breeds is now possible thanks to methodological developments in comparative methods that consider reticulations as opposed to more traditional evolutionary trees (Bastide et al. 2018).

The examples above illustrate that in domestication, as in evolution, size is a variable that can change more easily and faster than others. Size has been described as a line of least evolutionary resistance (Marroig et al. 2009). A study of the macroevolutionary patterns of body size evolution in mammals concluded that decreases in size have occurred at a rate 10 times higher than increases in size (Evans et al. 2012; Polly 2012).

Changes in body proportions in mammals

Because of the allometric—as opposed to isometric—relation of shape and size, much morphological variation is generated by scaling. This becomes evident when examining skulls, brains, limbs, and other organ systems discussed below.

Selective breeding associated with animal production or with changing breed standards can lead to changes in body proportions. These changes have happened, among other cases, in cattle, where different body shapes have resulted, ranging from slender to robust body types, typologically characterized as the respiratory and the digestive types (Fig. 7.5; Duerst 1931). These types are also recorded in horses, where running horses are the respiratory and gaited horses the digestive type (Duerst 1931).

The diversity in body proportions among dog breeds is remarkable. As a general trend, as weight increases, dogs become relatively taller at the shoulder and narrower at the hip (Kirkwood 1985). Already in dynastic times in Egypt, different kinds of dogs of varied proportions existed (Figure 7.6). Selective breeding in pre-Columbian America also produced different kinds of dogs at different times and places (Segura et al. 2022). Leg to body size ratio is one way in which breeds within a species vary. In dogs, the Basset Hound, Dachshund, and Corgi are well-known examples. Among sheep, the Ancon breed has a mutation that causes limbs to be considerably shorter than in most sheep (Landauer and Chang 1949)—an example of discontinuous variation (Darwin 1868).

Breed formation from Victorian times onward has resulted in marked intrabreed differences in many dogs (Nussbaumer 1982; Drake and Klingenberg 2008; Geiger and Sánchez-Villagra 2018). An example of variation in proportions within a breed is that of the German Shepherd (Fig. 7.7). Present-day German Shepherd show lines have a sloping back and strongly angled hindquarters (Räber 1993), in contrast to

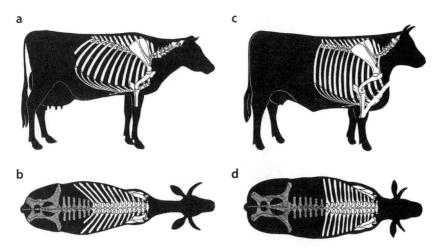

Fig 7.5. The respiratory and digestive types of cattle, two extremes of a continuum of body shape and proportions. (a) An example of the respiratory kind is the Friesian-Holstein breed, and (b) of the digestive kind, the Aberdeen-Angus breed.

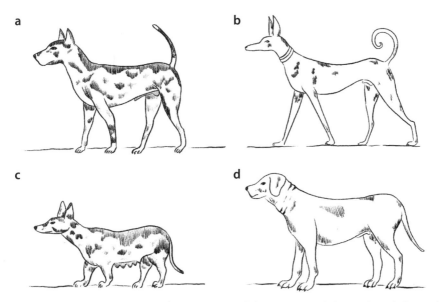

Fig 7.6. Varieties of Dynastic Egyptian dogs. (a) robust pariah or shenzi dog with pricked ears and tan back and ears, grey flanks, and white venter. (b) "Tjesem," ancestor of the Seluki Greyhound or hunting dog. (c) female black and white brindled short-legged terrier. (d) robust Mastiff hound with drooping ears (legs restored). Drawings not to scale.

Fig 7.7. Historical changes in body proportions of the German Shepherd dog breed. The top figure illustrates an individual from approximately 1899, the middle one from 1960, and the bottom one is a modern-day German Shepherd.

Fig 7.8. Head shape variation in dogs, horses, and cattle.

the first dogs awarded "best in show" at the beginning of the twentieth century (Verein für Deutsche Schäferhunde 1922).

The head shape in domesticated species shows wide variation, as illustrated here for dogs, horses, and cattle (Fig. 7.8). Faces in horse breeds vary even if there is conservation across breeds of similar allometric proportions among body length, withers height, and head length (Parés-Casanova 2018).

Changes in size and in body proportions in birds

As in mammals, body size in birds is a trait that responds readily and rapidly to selection. One can come up with adaptive explanations for changes in size in the farmyard environment. Larger individuals may become viable in an environment with unlimited food and would probably be more prolific. Smaller individuals may have better fitness if the food is limited and they have a shorter life history, or larger individuals would prosper in cases of food competition, the stronger individuals gaining access (Tchernov and Horowitz 1991). Surely, a multitude of factors play a role, and the evolvability of size may be different in the several species of domestic birds. The genetics behind body size changes has been studied for some species, including chickens (Henriksen et al. 2016) and ducks (Zhou et al. 2018).

There are no systematic studies of body size in breeds of domestic birds, but manuals for breeders and association compendia are sources of these data. Differences of weight within domesticated species of birds can exceed a 1 to 10 ratio. There are both increases and decreases relative to the wild form. Here I provide some examples.

Range in body weight in adult chickens is between 1 and 5 kg (Herre and Röhrs 1990). Examples of the two extremes are the small Bantam chicken and the Japanese Chabo chicken.

Range in body weight in adult domestic ducks is from less than 1 kg in the Call breed to the largest meat strains of Aylesbury and Pekin weighing about 6 kg; the Rouen is another large breed.

Historically noteworthy in terms of selection for weight in domesticated turkey was the case in the first decades of the twentieth century in England of a form that could reach up to 18 kg in toms (males) and 13 kg in hens (females) at 9 months of age (Crawford 1984). Subsequent introduction of this breed in North America led eventually to even larger extremes in size. The wild turkey typically weighs 5–11 kg (male) and 2.3–5.4 kg (female).

The "English Budgerigar" is more than three times the weight of the wild form, and males of this breed are reportedly unable to mount a female without assistance (Sossika 1982).

Selective breeding in geese has led to changes in body shape and proportions (Serjeantson 2002), and increases in size alone are often correlated with loss of the ability to fly (Darwin 1868). For example, compared with the weight of the swan goose (gander, the male, 3.5 kg; goose, the female, 2.8–3.5 kg), the African goose developed higher body weights (gander by 285%–362%, goose by 292%–311%). Compared with the greylag goose (gander, 2.8–4.1 kg; goose, 2.5–3.8 kg body weight), the body weight of the Toulouse gander has increased by 331%–421%, and that of the Toulouse goose by 286%–364% (Kozák 2019).

In many breeds of domestic birds, the posture changes as to be more upright, as is clearly the case in some ducks and geese. The diversity of postures in pigeon breeds includes the Pouter pigeon, with its permanently inflated neck, and the Fantail pigeon, which bends its neck back to the erect tail. A peculiar case of derived posture is seen in a variety of canary with what has been described as a more or less crippled vertebral column that causes a hunchback appearance and prevents flight (Sossinka 1982).

Given the many duck breeds differing in body posture, a gradual series from the wild condition to an erect form can be portrayed, as in the beautiful depiction of van Grouw's (2018) *Unnatural Selection*. This figure, called "The Ascent of Mallard," does not portray a true, gradual transformation of an evolutionary process. It is simply a clever play on the common (and wrong) depiction of human evolution implying linear progress to the current erect, bipedal posture of *Homo sapiens*. That kind of depiction, so common for human evolution, is highly misleading about the evolutionary process (MacFadden et al. 2012).

Domestication and changes in brain size and anatomy in mammals

Domestication leads to a reduction in overall brain size, albeit to a different degree in different species (Kruska 2005; Balcarcel et al. 2021c). Relative brain size has been measured through allometric mass comparisons requiring large samples of true brain and body masses (Kruska 2005; Balcarcel et al. 2021c). Using this approach, researchers have reported the following values: the largest change in pigs (33.6%), with high values for dogs (28%–30%) (Klatt 1955; Kruska and Stephan 1973; Ebinger 1980) and cats (24%) (Hemmer 1978) as well. The average value for sheep, derived from several breeds, is around 24% (Ebinger 1974). In horses the value is 16% (Kruska, 1973). In smaller mammals the reduction values are smaller. They are reportedly 8.7% in the case of the laboratory rat, 13.4% in the cavy, 13% in the rabbit (Ebinger 1972; Fischer 1973; Sorbe and Kruska 1975). The reality of brain reduction as a universal feature of the "domestication syndrome" has been questioned (Lord et al. 2020c). Nevertheless, there are multiple and independent reports of reduction in brain size of domesticated animals compared with some corresponding wild populations.

Multiple studies have used regressions of a variety of skull dimensions versus endocranial volume to simulate the allometry of brain and body size in several groups of mammals, including artiodactyls and carnivorans, groups containing many domesticated forms (Finarelli 2006, 2011; Balcarcel et al. 2021c). Using this approach, the following values were reported (Balcarcel et al. 2021c): cattle brains reduce by 25.6%, goats by 15.5%, pigs by 24.7%, and llamas by 14.9%. In contrast, the alpaca brain appears 16.5% larger than that of its wild ancestor the vicuna, which could be explained in part by the alpaca's hybrid status (Balcarcel et al. 2021a). In view of the vagaries of sampling and methodological issues involving, for example, body size corrections (Rogell et al. 2020), it has been suggested that brain size changes be reported as a range of values rather than as absolute numbers (Balcarcel et al. 2021c).

Brain size changes first require an explanation of the reduction, and then of differences among species.

Studies of morphology of domesticated animals in history

Some of the historically most notable and scientifically fundamental discoveries in comparative anatomy were based on the study of domestic animals. Aristotle was probably not the first, but his pioneering description of the embryology of the chicken was extensive in anatomical detail. Furthermore, it was used to reveal a general pattern based on comparison—with humans and horses—that predated the famous observation of Karl von Baer in the nineteenth century that in the process of embryonic development general features are formed first, with specific features appearing last (Leroi 2014).

Early works documented diversity of many breeds of many animals. Aldovandri (1600, in Lind 1963) on chickens is a classic reference in this context, but his approach and text are largely tied to cultural contexts and biblical references, and in some aspects it includes a mixture of imagination and real documentation (Asma 2001). Goethe discussed the vertebral nature of the back portion of the mammalian skull and the premaxilla based largely on observations of cattle, sheep, and horses (Peyer 1950).

Buffon (1707–1788), in his multivolume *Histoire naturelle, générale et particulière*, provided descriptions of varieties of horses, asses, cattle (volume IV), and sheep, goats, pigs, and dogs (volume V) and also included ideas on genealogical relationships among breeds in the case of dogs (chapter 2).

Isidore Geoffroy Saint-Hilaire (1854) produced large monographs on the "use" and diversity of domesticated mammals. Darwin (1868), in *The Variation of Animals and Plants under Domestication*, published a survey of diversity in several domesticated species.

The Swiss L. Rütimeyer in Basel and J. U. Duerst in Bern were among the first to conduct systematic studies of domesticated animals. Rütimeyer (1861) worked on past lake settlements, conducting archaeological research; he characterized and differentiated skeletal material of domestic goats, pigs, and cattle from their wild counterparts. Duerst (1931) continued this kind of research on skeletal variation in cattle (Davis 1987).

The Kiel school of zoological studies of domestic animals in the second half of the twentieth century had a long and lasting influence on the German-speaking world. The many works of diverse authors were summarized in the comprehensive book by Herre and Röhrs (1973, 1990), *Haustiere—zoologisch gesehen* (A zoological look at domesticated animals), never translated into English. Among the precursors in Germany of this school is Berthold Klatt, with his pioneering studies of dogs, chickens, and other domestic animals (Klatt 1913).

Several compendia works documented variation across species, including those of Zeuner (1963), Mason (1984), Epstein (1969, 1971), and Clutton-Brock (1987, 1999). Van Grouw (2018) documented mostly skeletal and integumentary changes in a wide variety of mammals and birds as the result of selective breeding.

The cognitive buffer hypothesis aims at linking brain size to ecology. According to it, large brains enhance cognitive capacity for feeding innovations and extractive foraging (Barrickman et al. 2008). If domestication involves less exposure to environmental stimuli or challenges, an assumption that has not been properly tested and may not apply universally across species or phases of domestication, then a reduction of brain size would be predicted from domestication.

In fact, a "decline of environmental appreciation" (Hemmer 1990), meaning fewer challenges in the behavioral/ecological context of domesticates, has been proposed as the main explanation for brain reduction related to domestication. This argument was used to explain the contrastingly small reduction in brain size in South American camelids when compared with that of other artiodactyls. Reflecting husbandry conditions, llamas and alpacas have not been subject to the same selection pressures as some other domestic animals, including exposure to predators, breeding with wild camelids, and less exposure to an anthropogenic environment than other kinds of livestock (Arbuckle 2005). Here, comparisons among breeds may provide insights, as discussed below.

A study of marsupials that controlled for several confounding variables found that brain size is not tied to behavioral complexity, measured in terms of social complexity and environmental interactions (Weisbecker et al. 2015).

Several factors may explain the different proportions of brain size changes associated with domestication among species. One is the encephalization quotient of the species. This quotient is the ratio of the observed to expected brain mass for an animal of a given size, based on a nonlinear regression of a range of living species (Jerison 1973). The degree of neonatal maturity (chapter 6) may also play a role in the evolvability of brain size (Weisbecker and Goswami 2011).

Comparison among species suffers of course from the unavoidable vagaries of sampling particular to each study. One concerns the difficulties in wild versus domestic comparisons, with any population taken as proxy for the wild form being a major and unproven assumption. They are only an approximation of the original wild form.

When making these comparisons, scaling issues need to be taken into account. Here there are methodological issues, with different approches often providing somewhat different results. Establishing the allometric function serving as a baseline is one issue (Bronson 1979; Kruska 2005). Brain size is studied in relation to body size, and since there are usually body size changes associated with domestication, questions arise as to how best to approach these comparisons. The use of body mass as independent variable is widespread and justified, but it has the disadvantage of being influenced by the (temporary) condition of the animal. Potential alternatives, such as the length of the vertebral column, are usually not available and also have disadvantages.

The reduction in brain size reported for mammalian domesticates does not affect all regions of the brain equally (Kruska 2005). The telencephalon shows the greatest quantitative loss, followed by the diencephalon and the cerebellum with similar reductions. The brain stem is less affected. The telencephalon has cortical areas and sensory perception structures that are particularly affected (Kruska and Stephan 1973; Kruska 2005).

Concerning reductions in subsections of the brain, it has been hypothesized that the size and function are tightly correlated (Bennett and Harvey 1985; Rehkämper et al. 1988; Iwaniuk and Hurd 2005; Sultan 2005) and that the brain of mammals and birds reacts with plasticity to sensor stimuli, experiences (learning), or hormonal influences. On the other hand, Finlay et al. (1998) suggested that the size of a brain region responds to the rate of neurogenesis particular to that region and species, and is not simply a reflection of function. Parts of the mammalian brain such as the cerebrum and the cerebellum evolve as influenced by selection for certain functions and as determined by developmental programs of more or less conserved neurogenetic scheduling. A detailed study of the brain in three marsupial species revealed that their adult proportional differences result from differences in growth rate and duration (Carlisle et al. 2017).

Contrasting experimental results illustrate the complexity of the matter and the difficulty in finding simple universals to predict brain size changes. The first generation of wild gerbils (*Meriones unguiculatus*) bred in the lab had no lower brain weights than their wild-caught kin, although domesticated Mongolian gerbils had lower brain weights than wild gerbils (Stuermer and Wetzel 2006). In contrast, an experiment with tree shrews (*Tupaia*) showed that captive first generation gained a significant increase in brain weight in comparison with the wild population (Frahm and Stephan 1976). The recorded areas of change include enlargement of the telencephalon, generally associated with information processing, and reduction of the amygdala, an area associated generally with emotion. The differences between studies may be due to taxa-specific features and to differences in the stimulation and complexity provided by laboratory environments.

In island evolution, which resembles domestication because of establishment of a new selection regime following drift (but being different in the usual lack of introgression from the original population or some other), there is a lack of uniform pattern in brain change across taxa (van der Geer et al. 2010). In islands, size reduction has not been accompanied by a proportional (following allometric predictions) change in brain size. In the cases of the Cretan dwarf deer *Candiacervus*, the Sardinian dog *Cynotherium sardous*, and the Sicilian pygmy elephant *Elephas falconeri*, brain size is larger than predicted. The opposite trend, size reduction accompanied by a larger than expected reduction in brain size, has been documented for *Myotragus*, the Malagasy dwarf hippopotamuses, and for *Homo floresiensis*. These patterns refer to overall brain size, a rough measure that does not consider the degree of convolution of the brain, for example, or the areas that become reduced or expanded and those that do not (van der Geer et al. 2010). There is no universal pattern on islands, and no universal pattern should be expected for domesticated forms either.

Diverse external and internal factors may be responsible for the lack of a universal pattern of brain size change resulting from domestication in all its phases and intensities (chapter 1). One aspect that may be of fundamental importance for understanding behavioral and brain changes in domesticated animals is changes in the "microbiota-gut-brain axis" (Cryan and O'Mahony 2011). The host physiology is modified by the microbiota, as shown mostly by studies of laboratory mice (Diaz Heijtz et al. 2011; Hsiao et al. 2013; Desbonnet et al. 2014). Given the proven importance of gut microbes in regulating metabolites and their association with neu-rodevelopmental disorders, it seems that domestication may have affected the brain via changes in the microbiota.

Little work has been done in systematically examining brain size and anatomy among breeds. What we do know reveals the evolutionary plasticity of the brain and the diversity of forms within a domesticated species. In a comparison across breeds of cattle, it was shown that bullfighting cattle have relatively larger brains than other breeds, while reduction in other breeds also depends on selection type (Balcarcel et al. 2021b). Bullfighting cattle are uniquely selected for aggressive behavior and reactive temper. Brain reduction is highest in dairy cattle, followed by dual-purpose (beef + dairy), beef, and bullfighting breeds (Balcarcel et al. 2021b).

Selective breeding has led to diversity in the brains of different lineages of do-mestic dogs in different ways (Ebinger 1980; Roberts et al. 2010; Hecht et al. 2019). Reported variation in the regional volumetric variation of the brain in 33 dog breeds is nonrandom (Hecht et al. 2019). There are specific networks of regions, and these correlate significantly with different tasks, including sight guarding, hunt-ing, scent hunting, and companionship. This variation occurs in the terminal branches of the genealogical tree of dog breeds, indicating strong, recent selection (Hecht et al. 2019). A study based on citizen science data on more than 7,000 pure-bred dogs from 74 breeds suggested a strong relationship between estimated abso-lute brain weight and breed differences in cognition. Reportedly, estimated brain weight and various cognitive measures varied widely, but larger-brained breeds performed better on measures of short-term memory and self-control (Horschler et al. 2019).

A seasonal plasticity in brain size has been reported for some shrew species among mammals (Dehnel 1949; Lázaro et al. 2018), a remarkable feature that might occur in other species. It seems unlikely that any of the domesticated spe-cies reflect any phenomenon of this kind, but to my knowledge seasonal plastic-ity has never been tested. This phenomenon is more widespread in bony fishes (chapter 9).

Domestication and changes in brain anatomy and size in birds

For domestic fowl, as with mammals, brain size reductions have been reported in the species studied so far. There is much variation, modularity, and many method-ological issues in the comparisons. The degree of cerebralization probably affects the extent of brain reduction, as do the husbandry conditions of the domestics.

Some comparisons have been made between the wild form and a specific breed—for example, between the red junglefowl and some specialized kind of chicken. These do not serve to characterize changes under domestication per se, but instead serve to investigate the potential of selective breeding to bring evolutionary change reveal mechanisms behind variation within species.

Reductions in size of 6.9% in the domestic pigeon and 16.1% in the domestic goose have been reported (Ebinger and Löhmer 1984, 1987). In the case of the turkey, the difference in brain size between wild and domestic toms is as much as 35.2% smaller in domestic toms, and 23.8% in domestic hens compared with wild counterparts (Ebinger et al. 1989). In general, a total brain reduction of 15.8% is assumed in domestic ducks compared with wild ones (Ebinger 1996).

In contrast to the general pattern reported for domestication, a study of sister-taxa comparisons and phylogenetic ancestral trait estimations considering more than 1,900 avian species showed that species living on islands have relatively larger brains than their mainland relatives (Sayol et al. 2018). These differences are the result of in situ evolution rather than reflecting colonization success of large-brained individuals (Sayol et al. 2018). Here an examination of the selective pressures involved would be important.

The sexual dimorphism in brain size of some wild forms is lost in the domesticated form in at least some species (Ebinger et al. 1989). Even in the domestic duck, the sexual dimorphism present in wild ducks can hardly be detected in brain size (Ebinger and Löhmer 1985).

Some authors have emphasized the idea of degeneration related to a less demanding environment to explain brain size reduction in birds, as in mammals. Given the differential changes of parts, with some for example even increasing in size, the functional correlations are of a more complex nature (Ebinger 1996; Rehkämper et al. 2008; Katajamaa et al. 2021). Here metabolic considerations may also play a role.

Studies of chickens, ducks, and pigeons have shown that there are breed-specific brain proportions. Much variation has been generated through selective breeding (Rehkämper et al. 1988, 2003; Frahm and Rehkämper 1998; Frahm et al. 2001; Cnotka 2012), as described below.

In general, chickens, domesticated pigeons, geese, and turkeys have a proportionately larger cerebellum than their wild counterparts (Ebinger and Löhmer 1984; Ebinger and Röhrs 1995; Henriksen et al. 2018). But there are exceptions to this general pattern, as in crested ducks, which have a proportionately smaller cerebellum (Cnotka 2012).

There is differentiation and variation in the brain of domestic pigeons compared with rock doves, their wild form (Rehkämper et al. 1988, 2008). Total brain, cerebellar, and telencephalic volumes are significantly smaller in five breeds of domestic pigeons compared with the rock dove (Rehkämper et al. 2008). In contrast, the hippocampus is significantly larger in the domestic forms, particularly in homing pigeons (Rehkämper et al. 2008). The homing pigeons also have significantly larger olfactory bulbs, perhaps a sensory adaptation to homing (Rehkämper et al. 2008).

Concerning adult brain neurogenesis in pigeons, two breeds—the Racing Homer and utility Carneau—were compared using endogenous immunohistochemical markers for proliferating, and for immature and migrating neurons, with no differences found (Mazengenya et al. 2017). It was concluded that adult neurogenesis is a conserved trait irrespective of body size.

In domestic pigeons the hyperstriatum ventrale and the tectum opticum show the greatest decrease (21% and 12%, respectively), and an increased volume was

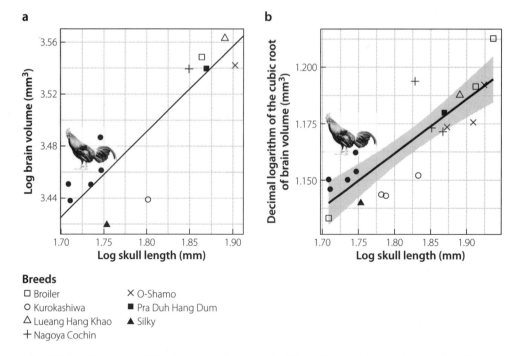

Fig 7.9. Scatter plots of the log-transformed skull length and brain volume of red jun-glefowl and domestic fowl. (a) Individual red junglefowl individuals and averages of seven chicken breeds. (b) Same red junglefowl data and individual data for breeds. Reduced major axis (RMA) regression line (solid) and 95% prediction intervals (shaded) drawn for chickens. Reanalysis of data provided in Kawabe et al. (2017).

even observed for the paleostriatal complex (Ebinger and Löhmer 1984). Rehkämper et al. (1988, 2008) also noted a partial increase in volume of the olfactory bulb.

The brain of chickens

There is great diversity in shape of the brain among chicken breeds and between chickens and red junglefowl, in part at least independent of size of this organ or of total body size (Kawabe et al. 2017). A study of chickens reported that genes controlling variation in brain mass and body mass have separate genetic architectures and are therefore not pleiotropically linked (Henriksen et al. 2016).

In contrast to the red junglefowl, the common layer breed, the White Leghorn, has an estimated body size around 85% larger, almost doubling it, and a brain mass around 15% larger. Another study reported that brain volume relative to skull length is distinctly larger in red junglefowl compared with broiler chicken, but an exploration of the original data provided reveals much variation within breeds (Fig. 7.9). The red junglefowl has larger, rounder cerebral hemispheres, and large telencephalon and optic tectum (Kawabe et al. 2017). This study traced ontogenetic changes in brain shape in broiler chickens and reported that, scaling for size, the shape of the red junglefowl brain is intermediate between that of young and adult broilers (Kawabe et al. 2017).

a

b

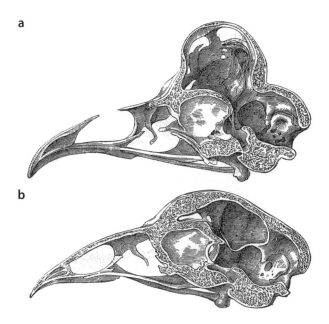

Fig 7.10. Sagittal sections of chicken skulls. The dorsal area, greatly modified in comparison with other chicken breeds, accommodates the "cerebral hernia" of the Polish breed (*top*), in contrast to the Cochin breed (*bottom*).

Similar brain size and shape were reported for chicken breeds clearly differing in body shape and skull morphology, namely the O-Shamo and Nagoya Cochin, the former a game breed and the latter an egg-type breed (Kawabe et al. 2017; Fig. 7.9). The area of the cerebellum adjacent to the cerebral hemispheres is a particularly variable part of the chicken brain (Kawabe et al. 2013, 2015, 2017).

Some chicken breeds have a feather crest on the head, and below it a cerebral hernia formed by the telencephalon located in a frontal bone protuberance (Klatt 1910, 1911). The cerebral hernia is controlled by a single autosomal recessive gene and is associated with crest formation (Fisher 1934; Yoshimura et al. 2012).

These skull and brain peculiarities are present in several breeds, such as the Houdan, Paduan, and White Crested Polish chickens (Darwin 1868; Yoshimura et al. 2012). There is an archaeological record of chickens with cerebral herniae and crests going back at least 1,500 years (Brothwell 1979; Gál et al. 2010). Historical records of crested chickens go back to the seventeenth century (Yoshimura et al 2012).

The White Crested Polish chicken has been studied in detail since Darwin (1868; Fig. 7.10). The size of different portions of the brain is significantly altered in this breed compared with those of other, uncrested domestic chicken strains (Frahm and Rehkämper 1998).

Considering the allometric relation with size, the brain is relatively larger in crested chickens, but not all regions change equally (Frahm and Rehkämper 1998). The optic tract, diencephalon, telencephalon, accessory hyperstriatum, dorsal and ventral hyperstriatum, and neostriatum are significantly enlarged in White Crested Polish chickens. On the other hand, the tegmentum, cerebellum, tectum, paleostriatum, hippocampus, septum, and olfactory bulb are not significantly larger in White Crested Polish chickens compared with uncrested chicken breeds (Frahm and Rehkämper 1998).

The brain of ducks

There is also much variation among breeds of ducks (Cnotka 2012). The crested forms are variants of several duck breeds (Cnotka 2012). In them, the olfactory bulbs and the cerebellum are significantly smaller than in uncrested forms (Frahm et al. 2001). Contrary to the chicken condition in which the crest sits atop a bony protuberance, the duck crest is located atop a pad of fat and connective tissue in the parietal region of the skull (Darwin 1868; Requate 1959; Frahm and Rehkämper 1998; Bartels et al. 2001; Frahm et al. 2001). This pad is nourished by brain vessels via small holes in the skull (Frahm et al. 2001). Crested ducks have been depicted by Dutch painters of the seventeenth century and are mentioned in the stories of Wilhelm Busch (1867). Darwin (1868) wrote about them as well.

An intracranial body of fat or lipoma in the cranium, associated with the brain, is present in many individuals of several breeds of crested ducks. This feature may lead to neurological disorders, with clinical signs similar to those of infectious and noninfectious diseases (Bartels et al. 2001). Lipomas in crested ducks are composed of a collection of highly differentiated univacuolar adipocytes surrounded by a rich capillary network (Bartels et al. 2001). A similar pattern of fatty acids in lipomas and crest pads suggests that they may derive from the same anlage in development (Bartels et al. 2001; Cnotka 2012).

The occurrence of intracranial lipomas has been studied in humans and in the laboratory rat (Brander and Perentes 1995). Intracranial lipomas are rare congenital malformations, reported as the result of abnormal persistence and maldifferentiation of the meninx primitive (the loose mesenchymal covering of the brain from which the dura mater, pia mater, and arachnoid mater are formed). These malformations are generally asymptomatic and benign, although symptoms can occasionally be observed, depending on the location of the lipoma (Jabot et al. 2009).

A thorough study of "Landente" duck breed showed that lipomas, when present, are highly variable in size (Cnotka 2012) and mostly localized in the region of the tentorium cerebelli or between the telencephalon and the tectum opticum (Krautwald 1910; Requate 1959, Bartels et al. 2001; Frahm et al. 2001; Cnotka 2012).

Much of the variation recorded in duck brain size is actually generated by the presence of lipomas (Cnotka 2012). Allometric comparison of the entire brain size in relation to body mass including body of fat, if present, shows much more diversity deviation from the norm in Landenten than in other breeds, and is above the expectation for body weight. The allometric comparison without the lipoma, that is, the net brain volume, shows how the Landenten is on average smaller than the reference group, also reducing its variation. With the data so treated, brain size in the Abacot Ranger is then significantly larger by allometric comparison. Quantification of different parts of the bain revealed that the cerebellum, tegmentum, hyperpallium apicale, and olfactory bulb of the Landenten are, after correction for allometry, significantly smaller in comparison to several other breeds (Cnotka 2012).

Domestication and cranial changes in mammals

In mammals the adult skull has been the most widely used marker of morphological diversity, as it is complex and correlates with feeding and sensory function. On the practical side, its study benefits from the fact that museum collections keep such specimens available for research.

Anatomical comparisons of domestic species skulls (Hanot and Bochaton 2018) have become rarer as geometric morphometrics became more commonly used. Variable features have been found by investigating suture closure (Rager et al. 2014; Esteve-Altava and Rasskin-Gutman 2015) and bone contacts and arrangements, which can vary among closely related species and could well be variation generated by selective breeding (Julian et al. 1957; Geiger and Haussman 2016).

Skull shape has been quantified for populations of many domestic species. As with much else on domestication, dogs are the most studied species, while many others have been barely studied. Some works concern only local breeds and are of restricted scope. Many of these works are listed here. Many species, for example domesticated bovids, have not been investigated in this regard, or there are few studies about them.

Dog: Dahr 1941; Rosenberg 1966; Lüps 1974; Nussbaumer 1982; Wayne 1986; Regodón et al. 1993; Drake and Klingenberg 2010; Schmitt and Wallace, 2014; Drake et al. 2017; Curth et al. 2017; Ameen et al. 2017; Selba et al. 2020
Ferret: Rempe 1970
Mink: Pohle 1969; Kruska and Sidorovich 2003; Tamlin et al. 2009
Cat: Künzel et al. 2003
Common cattle: Veitschegger et al. 2018
Sheep: Baranowski 2017; Gündemir et al. 2020
Goat: Sarma 2006
Bactrian camel: Martini et al. 2018
Llama and alpaca: Balcarcel et al. 2021a
Pig: Owen et al. 2014.
Horse and donkey: Evans and McGreevy 2006; Zhu et al. 2014; Hanot et al. 2017; Heck et al. 2018b; Parés-Casanova et al. 2020
Rabbit: Böhmer and Böhmer 2017
Cavy: Kruska and Steffen 2013

These works show how skull shape has changed in subtle ways in the first phases of domestication, or has been greatly modified in many cases with selective breeding. Most domestic forms thoroughly studied exhibit greater skull shape diversity than their wild counterparts. Llamas and alpacas do not, and an explanation may reside in the kind of husbandry and living conditions they have, not dramatically different from the wild, as discussed above for brain size (Balcarcel et al. 2021a)

There is a conserved allometric pattern relating size to skull shape across placental phylogeny that results in larger mammals having longer faces than smaller ones (Cardini and Polly 2013). This predicts, then, that simple changes in size, as occurring in the first phases of domestication, can easily lead to traceable changes in proportions. It has been claimed that a shortening of the rostrum results from domestication. This pattern is indeed recorded for many, but not all species (Sánchez-Villagra et al. 2016; Lord et al. 2020b; Balcarcel et al. 2021a).

Much of the diversity of skull shapes recorded in domesticated forms concerns variation between short and elongate skulls, as predicted by the allometric relation just discussed (Klatt in 1913 recorded this for dogs), and variation in the angle between face and the neurocranium. Different clades may vary in terms of the features correlated with such changes. Anatomical and functional features such as visual field, locomotor and feeding behavior, bullar size, and the dimensions of other structures and areas such as neck muscle attachment regions in the skull can thus be impacted by cranial changes associated with domestication. The angle between face and the neurocranium has been investigated in the clade of rabbits in relation to the ecology of various species of the group (Kraatz et al. 2015). The ontogenetic changes that occur in this angle during postnatal life have also been documented in rabbits (Moore and Spence 1969).

In some cases, there is overlap in a feature or proportions between the wild and the domestic form, but extremes in this distribution of variation are found only among domesticates. Only in domestic dogs does one see extremely wide orbital angles above 60 degrees, whereas wolves have a much narrower angle, often under 35 degrees. These figures were obtained with three-dimensional CT scan images; other measuring methods cannot be compared directly (Janssens et al. 2016). Because of the overlap in values, the orbital angle is of limited use to discern wolves from recent and archaeological dogs, but the extreme values do serve to distinguish between wolves and recent dogs.

The changes in skull proportions recorded in domestic forms have biomechanical costs and benefits. Basic consideration of lever arms suggests that a longer face can result in a reduction in skull strength (Ellis et al. 2009). In wild felids this has been shown to be compensated by changes in bone thickness: larger felid species, with their longer faces, have relatively thicker skull bones than smaller ones (Chamoli and Wroe 2011). This subject has not been thoroughly considered in other groups.

Skull shape can change throughout the history of breed. This has been documented in some dogs, which show significant change in a short time period (Drake and Klingenberg 2008; Geiger and Sánchez-Villagra 2018). The angle between face and the neurocranium has greatly changed in purebred Bull Terriers in just four decades (Nussbaumer 1982; chapter 8).

Horns in bovids and caprids

The horns of domesticated bovids exhibit much variation (Duerst 1926; Fig. 7.11). The reduced competition over mates or resources which characterizes domestication surely results in this differentiation from the wild form. Domesticated cattle usually have smaller horns, in some cases even having lost them. Selective breeding has led to large and diverse horns in some cases, in others it led to their elimination, with concurrent welfare issues in both conditions. Horns reach exceptional sizes in some cases, as in the Ankole-Watusi and Texas Longhorn breeds. One may think

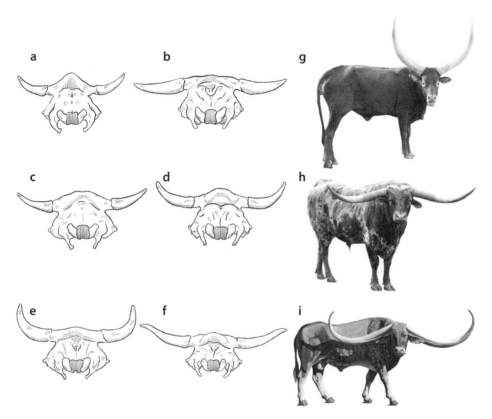

Fig 7.11. Austrian Blondvieh cattle horn cores and extreme sizes in domestic and extinct bovids. (a–f) variation in the Austrian Blondvieh cattle horn cores. Extremes of horn size are found in (g) Ankole-watusi and (h) Texas Longhorn cattle breeds. (i) *Pelorovis antiquus,* an extinct relative of the living Cape buffalo. *Pelorovis* was a massive animal, possibly weighing as much as 1,200–2,000 kg (2,400–4,400 lb), and the tips of its enormous horns may have spanned 2.5 m (8 ft).

these extremes are impossible to achieve in nature, but the fossil record tells us otherwise (Fig. 7.11), The extinct *Palerovis antiquus,* a relative of the living African buffalo *Syncerus caffer,* also had bone cores in the horns, some extending more than 2 meters (MacPhee 2018).

The variation in horn shape in goats and sheep is remarkable (Duerst 1926; Riedl 1975; Benecke 1994b). In the wild, fighting success may be determined not only by size, but also by form, so sexual selection acts on the latter. Relaxed selection may imply that more variation becomes possible, as it is not selected against. In many breeds, sexual dimorphism in horns is lost. Among Swiss goat breeds, the Appenzeller lack horns, whereas in the "capra grigia," both males and females have long and conspicuous horns (Ammann et al. 2012). The power of selection in changing the size and shape of horns is shown by the recorded consequences of trophy hunting, with hunters shooting animals with larger horns. The analysis of a 30-year record of the hunting in Alberta of bighorn sheep, *Ovis canadensis,* revealed how in a short time period a population responded to selec-

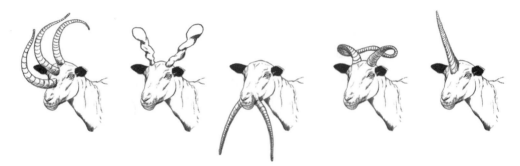

Fig 7.12. Examples of nonexistent horn shapes and positions in caprids, here illustrated in imaginary goats.

tive shooting of big-horned big rams (Coltman et al. 2003): mean ram weight declined from 85 to 65 kg, a 24% drop, mean horn length had a 28% drop. For all the variation recorded in caprid horns, there are many forms that simply do not exist—surely determined by the developmental process that leads to horn position and early growth (chapter 4; Riedl 1975; Fig. 7.12).

Sexual size dimorphism in sheep and goats

The changes in sexual dimorphism associated with domestication have been thoroughly studied in goats and sheep. Domestic goats and sheep are much less dimorphic than most of their wild relatives, which are highly dimorphic animals (Polák and Frynta 2009). The marked dimorphism of wild caprids correlates with a social system of sexually segregated groups in which male combat is a regular part of male–male competition (Schaller 1977). The sexual selection hypothesis provides the most intuitive explanation for larger male size in caprids. The mechanism behind this pattern is reportedly the delayed sexual maturation of males in wild populations (Côté and Festa-Bianchet, 2001), whereby differences in size among sexes and populations are affected by food conditions and individual life history (LeBlanc et al. 2001).

The reduction in sexual size dimorphism (SSD) during domestication may be explained by: (1) relaxed intrasexual selection as a result of the female-biased operational sex ratio and reduced role of male combat in captive herds; (2) artificial selection against males attaining large size; and/or (3) the effect of reduced sexual segregation in captivity where bucks/rams usually graze together with females in mixed herds. Although none of these factors are well documented throughout the domestication process and their relationship to SSD is more or less hypothetical, these explanations are plausible. Regional distribution of SSD among goat and sheep breeds supports the view that large male size is less advantageous or simply difficult to attain in poor environmental conditions (Mbayahaga et al. 1998).

Skull variation: the case of brachycephaly

Brachycephaly, a short skull relative to its width and a dorsally rotated rostrum, is a condition common in many breeds of domestic species. The condition refers to an extensive shortening of the rostrum as an inherited trait, occurring consistently in individuals of a population, in some cases being breed-defining.

Among domestic carnivorans, there are brachycephalic varieties and breeds of both dogs and cats, whereas in the ferret, brachycephaly occurs occasionally. At least 20 dog breeds have been described as brachycephalic (Geiger et al. 2021). Persians and Exotic shorthairs are the most extreme examples of facial shortening in cats and are usually categorized as brachycephalic breeds.

The shortening of the snout in brachycephalic dog breeds such as the Pug and the Bulldog, has led to a remarkable decrease of surface area and surface density in the turbinal skeleton. However, in these breeds the olfactory epithelium and other soft tissues are not much reduced—so the dermal skeleton changes more rapidly in response to selection (Wagner and Ruf 2020).

The study of brachycephaly in cats and dogs has been largely focused on associations of the condition with a range of diseases that pose a considerable welfare issue in extreme varieties (Bessant et al. 2018). Snout shortness is expressed of course not only in the skeletal elements, but also in the associated soft tissues (Diogo et al. 2019). A plethora of anatomical aspects are associated with brachycephaly, with clinical implications (Geiger et al. 2021). Here I refer to one of them: brachycephalic syndrome (BS) of short-nosed dog breeds, characterized by severe respiratory distress due to obstruction of the nasal, pharyngeal and laryngeal lumen (Ravn-Mølby et al. 2019). It is caused by excessive soft tissue within the upper airways. A narrowing of the nasal inlet is part of this condition, which can have different degrees of expression (Fig. 7.13).

Given the known ontogenetic trajectory of snout elongation in mammals, the question arises whether a brachycephalic form is not simply one exhibiting retention of a juvenile stage (Fig. 7.14). Brachycephaly does not simply result from the retention of a juvenile character—a shorter face—and in the case of most brachycephalic forms across species, they are not the smallest breeds or varieties (Geiger et al. 2017; 2021). Prominent examples include the extinct ñata cattle from Argentina and Uruguay, of average size compared with other taurine cattle (Veitschegger et al. 2018); brachycephalic goats being large animals (Acharaya, 1982); and also brachycephalic pigs of different sizes (Porter 1993). The brachycephylic pig breed Middle White derives its name from its average body size compared to that of related breeds (Porter 1993).

In some cases, brachycephaly occurs in association with chondrodysplasia (chapter 4), as in some cattle (e.g., Dexter, horned Hereford dwarf, Aberdeen Angus) and sheep breeds (e.g., Cabugi, Texel, Cheviot, Suffolk, Hampshire, and Merino).

The domestic rabbit breeds exhibiting short snouts relative to the braincase differ from the norm just described. Based on their proportions they could be classified as brachycephalic, but here this condition is highly associated with changes in proportion that result from the allometric—as opposed to isometric—relation of skull shape and size (Klatt 1913; Fiorello and German 1997).

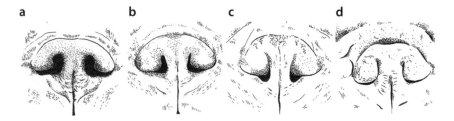

Fig 7.13. Different degrees of nostril stenosis, narrowing of the nasal inlet, in French Bulldogs. (a) In the normal condition, the nostrils are wide open. (b) In the mildly stenotic nostrils, these are slightly narrowed and the lateral nostril wall does not touch the medial nostril wall. (c) In the moderately stenotic nostrils, these are open only at the bottom, and the lateral nostril wall touches the medial nostril wall at the dorsal part of the nostril. (d) In the most severe condition, the severely stenotic nostrils are almost closed. In the last, the dog tends to switch from nasal to oral breathing after very gentle exercise or when stressed.

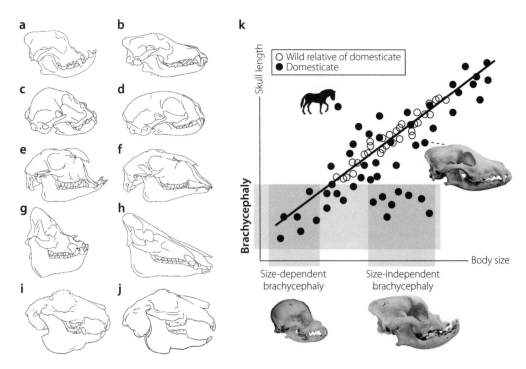

Fig 7.14. Examples of the diversity of brachycephalic forms across domesticated mammals, and relation to size. (a–j) Lateral view of the skull in ancestral (*right*) and brachycephalic (*left*) conditions. (a, b) wolf, bulldog. (c, d) cat, Siamese cat. (e, f) goat, Egyptian goat. (g, h) wild boar, brachycephalic pig breed. (i, j) non-brachycephalic, short-snouted rabbit. (k) A short snout can correlate with small size or can be disproportionate—a significant shortening of facial bones. Illustrated is the scaling relationship between body size and skull length in any wild animal (white dots) and its domestic counterpart (black dots). Skulls not to scale.

Brachycephaly does not occur in horses. Perhaps the possibility of this evolving in a ruminant bovid like the cow and not in the horse has to do with their digestive physiology and associated oral and dental anatomy. Grazing animals eat earth and dust along with the plants. Mechanisms to deal with the excessive abrasion of the teeth are different in ruminants, including cows, goats, and sheep, on the one hand, and perissodactyls such as horses on the other hand. Horses have very high premolar and molar teeth starting from their roots; with them they compensate for the abrasion caused by dust and sand. Cows and other ruminants manage with somewhat shorter teeth. They have a stomach system with multiple chambers that use bacteria to digest the plant material they eat. The food is sorted into material already small enough to digest, and larger pieces that are regurgitated to be chewed again. Using computer tomography, it was discovered in goats that the sand ingested with the plants was not equally distributed within the stomach system. The digestive physiology of ruminants includes a "washing system" (Hatt et al. 2019). There is less sand in the place where the material to be ruminated is regurgitated, than in the ingested food itself. The sand sinks down in the rumen and is collected in a chamber called the abomasum, passing through the bowel to then be expelled in the feces. This explains why ruminants chew their food much less thoroughly the first time they ingest it than they do afterward, when they ruminate cleaner material (Hatt et al. 2019). Horses have to cope with dust wearing down their teeth instead.

Domestication and dental changes in mammals

Tooth shape has been used in zooarchaeological research to differentiate wild from domestic forms in pigs (Evin et al. 2013, 2015a) and horses (Seetah et al. 2014; Cucchi et al. 2017). Domestication produced completely new dental phenotypes not found in wild boar, although the amount of variation among domestic pigs does not exceed that found in the wild (Evin et al. 2015a). The distinct domestic dental phenotypes probably resulted from differing types and intensity of husbandry practices. Captivity also appears to impact tooth shape.

There are differences in tooth shape among horse breeds and in comparison to their wild ancestor (Seetah et al. 2014, 2016). This divergence from the wild did not arise at the start of the domestication process. After a long phase of little change in premolar shape following initial domestication, there was a dramatic increase in the amount and pace of variation at the end of the Iron Age (Seetah et al. 2016). Other changes yet to be studied in horses, in particular among extant and ancient breeds in archaeological samples of preserved skulls, are the proportion and position of the tooth row in relation to the rest of the skull. The evolutionary change in these features and their allometry are well documented in horse evolution (Radinsky 1983; 1984). However, no comparison has been made with the patterns resulting from domestication.

Many studies of dental changes associated with domestication have mostly used geometric morphometrics to quantify them. Another approach would be the codification of discrete characters. For example, species of *Equus* can be differentiated based on dental features (Bennett 1980). These features could be examined in horse breeds. Enamel folding is a good taxonomic marker across species of *Equus* (Cucchi et al.

2017); it remains to be studied whether it can be used to differentiate between wild and domestic horses, or whether variation in this feature ever arose in some breeds.

It is in most cases difficult to discern the evolutionary processes leading to subtle differences in dental anatomy. Drift can be one of them. When selection results in significant cranial changes, then issues of tooth crowding or changes in proportions naturally occur, as recorded in dogs (chapter 1).

Anomalies such as supernumerary teeth, missing teeth, uneven tooth wear, and malpositioned teeth—either rooted outside the normal tooth row or rooted in the tooth row but deflected either labially or lingually—have been found in many domestic species (Nieberle and Cohrs 1967; Miles and Grigson 1990). These may be more prevalent in inbred populations, such as in some feral populations of goats (Vanvuren and Coblentz 1988). Dental anomalies may also be more common in domestic forms with a foreshortening of the maxillary and dentary bones, as in the American mink, *Neovison vison* (Pohle 1969).

In brachycephalic dogs, the maxillary bone offers less space for dental alveoli, leading to reduction of teeth, crowding, and abnormal rotation of the alveoli to accommodate teeth in a smaller space (McKeown 1975; Harvey 1985; Kupczyńska et al. 2009; Schlueter et al. 2009; Lobprise and Dodd 2019). This results in problems with occlusion (Geiger et al. 2021). Neither felids nor artiodactyls with extremely short faces were so affected in their dental function, given their dental reduction and presence of diastemata (toothless space), respectively. In domestic rabbits, malocclusion does occur in connection with the rostral shortening (Korn et al. 2016; Lobprise and Dodd 2019).

Postcranial functional anatomy and locomotion in dogs

The different sizes and kinds of dogs may be reflected in the variation in their postcranial anatomy. This matter has been extensively investigated regarding elbow shape, in which disparity among domestic dogs was shown to exceed that in wolves. In fact, this disparity is comparable with that of wild Caninae, including species of wild dogs, wolves, jackals, coyotes, and the California island fox (Figueirido 2018). This expansion of morphospace has not meant, however, a morphological correspondence of dog kinds with their analogues among wild species; for example, fast-running and fighting breeds have not converged with extreme pursuit and ambush wild carnivores, respectively. Nor has it meant an expansion surpassing the disparity of wild forms of carnivorans, as recorded for skull shape (chapter 7).

The limbs of dogs cannot be distinguished in general proportions from those of many wolflike canids of equivalent size (Wayne 1986). However, dogs are consistently separated from fox-sized, wild canids by subtle but evolutionarily significant differences in scapular, metapodial, and olecranon morphology (Wayne 1986).

The allometric proportions of the limbs across dog breeds and in comparison with wolves are largely conserved, with the femur showing more deviation than other bones (Casinos et al. 1986).

Some geometric properties of bone are different in dogs of different sizes. Smaller bones are not an exact reduction of larger bones. The cross sectional properties of the radius and ulna, as can be established by a survey using computer tomographic scanning, are not equal (Brianza et al. 2006). Given similarly normalized loading, toy dog moments of inertia are significantly lower, exposing this group to higher local stresses and eventual failure at lower peak loads (Brianza et al. 2006). These differences make the antibrachii of toy breeds more susceptible to fracture than those of large breeds.

The sarcomeres can be seen as the smallest unit of force production in muscle function. Sarcomere length depends on actin filament lengths—the latter is quite conserved among at least a sample of diverse breeds and mixed dogs (Dries et al. 2019).

The exceptional athletic peformance of some dog breeds, such as the running speed of Greyhounds, can be understood based on considerations of the muscle relations to the skeleton. As with cycling humans, Greyhounds power locomotion by torque about the hips (Usherwood and Wilson 2005).

The kinematics of dog locomotion has been the focus of numerous studies (Maes et al. 2008; Saunders et al. 2013). In spite of the variation in the gait parameters of dogs, as documented in a comprehensive study involving 32 different dog breeds (Fischer and Lilje 2014), the basic locomotory pattern is conserved. But there are subtle differences that may be important for clinicians. For example, a high precision 3D in vivo hind limb kinematic study of four breeds of French Bulldog showed peculiarities, including an unusually large femoral abduction (Fischer et al. 2018). This condition may make them susceptible to a higher long-term loading of the cruciate ligaments.

Domestication and cranial changes in birds

Anatomical comparisons have shown much innovation in domestic forms in birds (van Grouw 2018). The changes in bone proportions and in sutures have been particularly singular in crested forms, with their associated brain features discussed in this chapter. Morphometric studies of domestic bird skulls have been less extensive than those of mammals. Those on pigeons (Young et al. 2017) and chickens (Stange et al. 2018) have also shown the wider occupation of morphospace than in the wild counterparts.

The remarkable cranial diversity of pigeons became known with Darwin's (1868) study. Domestic pigeons exhibit a morphological diversity that greatly exceeds that of the evolutionary clade of almost 300 species of pigeons and doves as a whole (Young et al. 2017). There are superficial and less superficial resemblances to diverse forms (Fig. 7.15). There is much variation in beak size and shape. Extremely short-billed breeds include owl pigeons and short-faced tumblers. In the latter, the beaks of the offspring are too small for them to break the eggshell, so the breeder must do this for them (Reznick 2010).

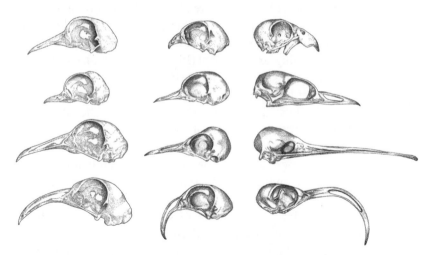

Fig 7.15. The skull of the rock dove compared with that of diverse pigeon breeds, and with other bird species. *Left*, top to bottom: pigeon breeds: wild rock pigeon *Columbia livia*; Short-faced Tumbler, English Carrier, Bagadotten Carrier. *Center:* pigeon breeds with extreme craniofacial shapes qualitatively converging on distantly related avians: Blondinette, Domestic show flight, Budapest long face, Scandaroon. *Right:* avian crania with shape similar to corresponding domesticated pigeon breeds: Parakeet, Grebe, Kiwi, Sickelbill. Individual crania are scaled to similar braincase length.

In chickens the most variable part of the skull is the cranial vault (Stange et al. 2018), surely related to brain shape variation, especially in crested breeds (Verdiglione and Rizzi 2018). Neural crest-derived portions of the skull exhibit a higher amount of variation than mesoderm-derived portions. Beak deformities might be more common in some breeds than in the wild form. An example is the crossed beak, in which the maxilla and the mandibula were affected, while other parts of the skull are normal (Joller et al. 2018).

The simple examination of skulls in the case of geese reveals obvious differences contrasting with the wild form, here with the added subject of the dual species origin and the different degrees of hybridization of the different breeds (Van Gouw 2018).

In many domesticated birds there are forms with a pronounced crest, including chickens, ducks, and pigeons, as discussed above in the section on brains, but also in geese, budgerigar, canary, Zebra finch, and Bengalese finch (Sossinka 1982).

There is a common cranial evolutionary allometric (CREA) pattern in birds, as in mammals, in which smaller species have relatively bigger braincases and smaller faces than larger species, and larger species have disproportionately larger faces (Bright et al. 2016). A study of Galliformes, the bird clade including chickens, examined this idea and provided unexpected results (Linde-Medina 2016). Looking at the allometric relation with a multivariate regression between shape and centroid size supported the CREA hypothesis. A bivariate regression analysis comparing the two portions of the skull revealed a negative allometry for beak size and braincase size, contrary to expectation. This study is important in showing the vagaries of choosing a method and measures to study shape and proportions.

Summary: Variation in the morphological diversification of domesticates

Domestic animals have been somewhat underexploited as subjects of studies of morphological diversification. The sheer cataloging of the seemingly endless variation provides the fundamental basis of our understanding and appreciation of diversity. Because anatomical studies have a long tradition, it is often assumed that there is not much to do in that area, but this is far from true. Several aspects associated with domestication have not been thoroughly studied in a comparative and evolutionary context in spite of the great amount of variation and the insights into homology, patterns and mechanisms of modification, and functional anatomy that could be gained.

Different clades exhibit different sets of modifications resulting from selection for tameness; thus the "domestication syndrome" does not apply as a universal set of characters to all domesticates.

The remarkable diversity within domesticated species, with the notable examples of dogs and pigeons, shows that selective breeding can lead to modifications of the phenotype in a species far beyond the range of variation of the ancestral population. But there are limits to what artificial (and natural) selection can produce, as determined by the kind of variation that is generated during development.

Morphometric studies of skulls of domestic versus wild populations mostly show an expansion of morphospace occupation in the former. The extent of that expansion varies, as does the amount of variation in individual modules of the skull.

The pace or rate of phenotypic change in a domestication process may vary even within the same structure, suggesting various resulting patterns of tempo and mode of evolution. An example is an initial lengthy period of invariance followed by an explosive increase in phenotypic change.

Comparisons among domestic animals have provided major insights into the evolution of digestive strategies and differences among groups. For example, cows and other ruminants have evolved a mechanism to wash abrasive elements from their diet before intense chewing, resulting in different adaptations and evolvability in cranio-dental features when compared to horses and zebras, which lack that digestive specialization.

Reduction of brain size after selective breeding has been reported for many species but in different degrees and with variations in the proportions of the parts becoming reduced. Since the reference data for the calculations have usually been at best approximations of both the wild and the domestic forms, the validity of the reported values is worth revisiting.

The kind and intensity of domestication, and if relevant the kind of husbandry, influence the degree and kind of brain size changes resulting from the domestication process. The size and encephalization quotient of the original species must also affect the degree of change in brain size following domestication. Decreased fighting for mates and resources, as well as reduced competition, are likely selective forces underlying a reduction in the first place. The potential of using domesticated models for understanding brain plasticity and evolution has not been fully realized, including comparisons across breeds, as opposed to the always problematic wild/domestic comparison using some derived breed as proxy for the original domesticated process.

Breeds of domestic animals provide the opportunity to analyze populations under conditions in which some selective pressures are relaxed (e.g., sexual selection), whereas others are enhanced (e.g., fecundity selection). Not all sexually dimorphic domestic animals become monomorphic.

There is almost a universal trend in the perception that big data and high-resolution technology are desirable approaches in any ambitious research project, the scientific value of which is also measured in terms of its financial cost. Here the environmental cost of data collection and storage is overseen. The approach is often seen as a call for large and unique responses to what is framed as a large question. Low-resolution 3D scans are suitable for many morphological studies that currently use high-resolution technology.

CHAPTER 8

Feralization and experimental domestication

Examining phenotypic plasticity or genetic changes resulting from controlled selection regimes or the control of environmental variables, and testing for correlations of traits, can provide insights into the evolutionary patterns and mechanisms of the domestication process. Likewise, placement of populations of domesticated animals in natural ecosystems results in changes in the selection regime, and with that a "natural" experiment results from feralization.

Feralization

Populations of many species of domesticated animals have broken free of captivity or human control and direct care and live instead in a wide and diverse range of places and environments; ferals are free-living animals in these populations (Price 2002a). As with so many categorizations in biology (Levins and Lewontin 1985), the use of the term feral, as well as its value judgment or significance, is not universal across species. Feralization has conveyed different degrees and consequences of changes in the form of interactions with humans (Baker and Manwell 1981; Wilson et al. 2017). Although there is considerable interest in how specific feral populations react and respond to natural conditions, researchers are usually looking for universals, for general rules. These, however, as with much else concerning domesticated animals, are hard to find.

The focus on feral animals is often tied to conservation—how do those populations affect native species, and how can we control them? Predation by feral cats is a leading cause of bird mortality in the United States (Loss et al. 2013) and in many cities around the world (Marra and Santella 2016), and cats have been implicated in the decline and extinction of many species worldwide (Medina et al. 2011). Reportedly at least 268 species (including 247 plant and 8 bird species) in the United States have experienced population declines caused by feral pigs (Gurevitch and Padilla 2004). Some feral populations have become invasive species and have caused a cascade of effects in ecosystems (Doherty et al. 2016). In Hawaii, feral pigs alter ecosystems in a variety of ways: they damage and consume endemic plant species, are dispersers of invasive plants, alter soil fertility, prey on the eggs of native ground-nesting birds, and create breeding grounds of invasive mosquitoes carrying avian malaria by virtue of the microhabitats they produce (Nogueira et al 2009; Cole et al. 2012). On the largest island of the Aldabra Atoll, Grande Terre (Seychelles), goats altered species composition and slowed the regeneration of natural vegetation, reducing shade cover and forage for the giant tortoises—this habitat loss has led to coordinated action for the eradication of feral goats (Bunbury et al. 2018). Horses are also reported to cause habitat damage when feral (Cherubin et al. 2019). Feral pigeons can carry diseases that affect both humans and livestock (Haag-Wackernagel and Moch 2004). Feral populations are therefore in many cases a problem, and control or extermination has been conducted. On the other hand, in some cases feral populations are considered a reservoir of genetic diversity and are thus conserved, as in the case of the sheep of Santa Cruz Island in California (Wilson et al. 2017).

The study of some feral forms has had some historical significance given the people involved in their study. Darwin (1868) discussed at length the feral rabbits of Porto-Santo, an island in the Madeira group of Portuguese islands in the Atlantic Ocean. He pointed out several unique features of this animal in the integument and skeleton, and its small size; later Ernst Haeckel named a new species for this population, an unjustified and rejected taxonomic decision (Herre and Röhrs 1990). The Porto-Santo rabbit is much smaller than the continental one, weighing around 500 grams. It was introduced in the middle of the fifteenth century and multiplied. Like introduced rabbits on other islands, most prominently in the Australasian region, the Porto-Santo populations became a pest.

Phenotypic variation has been intentionally reduced in the case of many specialized breeds selected for divergent environments and/or purposes, and this suboptimal trait variance reduces population fitness. This has been listed as a kind of "maladaptation" in feral animals (Gering et al. 2019). This subject is particularly important in aquaculture (chapter 9). An example among mammals is the case of captive-bred oldfield mice *Peromyscus polionotus* that exhibit reduced responsiveness to predators compared with wild mice, reducing fitness during feralization (McPhee and Carlstead 2010).

Genetics and feralization

Feral animals usually hybridize with both domestic and wild forms, which allows selection to draw from disparate gene pools and result in a temporally optimal phenotype (Henry 2016). The genetic "contamination" produced by feralization can be either detrimental or favorable to the animals in question (Henriksen et al. 2018). This subject is particularly relevant in aquaculture (chapter 9).

Genetic markers are used to estimate the degree of admixture in feral, wild, and domestic populations. This technique was used for example with pigs in Greece (Koutsogiannouli et al. 2010) and Hawaii (Linderholm et al. 2016) based on melanocortin 1 receptor (*MC1R*) gene alleles. Changes in gene pools caused by domestication will likely impact selection differentials among animals recolonizing the wild (e.g., McPhee 2004; Björnerfeldt et al. 2006; Chen et al. 2018). In dogs (Cruz et al. 2008; Marsden et al. 2016) and horses (Schubert et al. 2014), a greater proportion of deleterious mutations than usual have been described as a cost of domestication (Gering et al. 2019). In contrast, the introgression of domesticated genes into the gene pool of feral chickens must have contributed to their adaptation to a new environment in Hawaii (Callaway 2016).

Feralization does not cause a return to the wild phenotype

Feralization would seem to be just domestication in reverse, and one could naively expect a return to the wild condition in different aspects of the biology of the populations in questions (e.g., morphology, behavior, physiology). However, feral animals never return to the wild state in all their features. An expectation of feral-to-wild conversion obviates that the starting point in the evolution of a population is important to understand its future. Furthermore, although habitats in which feral animals live can be similar to those of the ancestral wild environments, this is usually not the case. Environmental changes postdating domestication, anthropogenic disturbance, and dispersal beyond the native range can result in novel habitats for those populations (Gering et al. 2019).

Brain size has been reported to have decreased in domesticated mammals when compared to their wild living ancestors (chapter 7). Changes in brain size examined in feral populations of different species, including pigs (Kruska and Röhrs 1974), minks (Kruska and Sidorovich 2003), and cats (Derenne and Mougin 1976) supported the conclusion that feral mammals tend to increase in brain size but do not return to the brain size of their ancestors—the latter estimated based on the existing populations of those wild forms (Röhrs and Ebinger 1999; Kruska 2005). Many feral populations live in impoverished ecosystems with no predators and competitors. This may influence the evolution of brain size and structures, a subject worth studying comparatively.

The quantitative examination of brain composition in feral pigs revealed a differential pattern of change in brain regions in domestication and in evolution in general (Kruska 2005). In feral pigs, the cerebellum is about 11% smaller than in domestic ones; telencephalon and mesencephalon of ferals similar in volume size to those of domestics; and diencephalon and medulla oblongata are 6.5% and 10% larger in ferals, respectively, as estimated in comparisons of populations of these animals (Kruska and Röhrs 1974).

Wild boar and domestic pigs differ in their olfactory function and structure. A study of feral pigs reported that they have comparable cell density in the olfactory mucosa to wild forms, which is higher than in domestic pigs (Maselli et al. 2014). In contrast, they lack a full expression of some proteins that are important in olfactory function in wild boar, being in this sense more like domestic pigs. This is thus a case of structural but not biochemical change in a sensory system resulting from feralization.

A geometric morphometrics study of skull and mandibular shape in feral pigs in comparison with wild boar and domestic pigs found that numerous traits associated with domestication are present in feral specimens (Neaux et al. 2020). The shape changes induced by domestication are broadly maintained in feral populations. The

skull of ferals is significantly smaller than that of domesticated breeds. This difference in size was interpreted as perhaps resulting from founding events (Neaux et al.
2020).

The dingo is an emblematic animal in Australia, and for many years it has been
discussed whether it is a domesticated dog that became feral, or a wild canid
which has interacted with humans and interbred with domestic dogs over long
periods of time (Smith and Savolainen 2015; Koungoulos 2020). Some genomic
work has indeed proposed that dingoes originate from dogs in southern East
Asia, and that they diverged into a genetically distinct population in Australia
(Zhang et al. 2020b). Expanded methods and data will test these results. A study
of the endocranial volume in free-ranging dingoes showed that their brain/body
size clusters within the variation of domestic dogs, but dingoes have a larger brain
than most domestic dogs of the same body weight, as well as greater values of
encephalization than dogs (Smith et al. 2018). Thus, here the prediction of an
increase in brain size following feralization not quite reaching the level of the wild
form is fulfilled. But here it is important to consider that the condition of the
dingo's ancestor is unknown, so it could be the case that the dingo's brain size did
not decrease with domestication (Smith et al. 2018). It is also possible that compared to the dingo's domestic ancestor, the current dog breeds have an even more
reduced brain size, and that the dingo brain size has not changed with feralization. Perhaps our search for universals causes misinterpretations of data. Longitudinal, historical series of dingoes in which one could trace estimated changes in
brain size are lacking. Earlier work on singing dogs from New Guinea, presumably
related to the dingo, also reported brain size unlike wolves and within the values
of dogs (Schultz 1969). In terms of the shape of the semicircular canals, dingoes
exhibit the mean shape of variation recorded in wolves and dogs (Schweizer et al.
2017).

Works on brain size and composition in feral birds should provide another relevant subject for understanding the plasticity of this organ under different evolutionary parameters (Henriksen et al. 2018). Feral pigeons, because of their abundance and wide distribution in many environments, might offer a rich subject in
this regard. However, precisely the case of feral pigeons illustrates well the diversity of the environments in which domestics "go wild," given how many feral pigeons live in cities, having become habituated to human presence and dependent
on the habitats we provide (Haag-Wackernagel 1993).

Experimental studies of domestication

Domestication is often referred to as an experiment in evolution. But domestication is actually pretty poor as experiments go; there are too many variables involved with little control, and no records of how things started. Many real experiments have been conducted intending to understand domestication itself. The
numerous animal species kept in laboratories have gone through a process of taming and accommodation to life with humans that also gives clues about the domestication process (Arbuckle 2005). A large body of literature on experimental do-

Feralization in literature and film

The subject of feralization is rich in connections to major ideas on the wild and the domestic and the environment (Wilson et al. 2017), and as such it has inspired significant works of art, including well-known novels and films.

Two such novels by the American writer Jack London are *Call of the Wild*, published in 1903, and its less well-known companion novel in 1906, *White Fang*. The central character in *Call of the Wild* is a domestic dog named Buck, stolen from its home in California and sold to become a sled dog in Alaska. Buck becomes feral, learning to cope with the new, harsh environment, and eventually becomes a leader among other dogs. *White Fang* is the reverse story, in which a wild wolf-dog gradually becomes domesticated by a new owner. These two stories have been adapted to numerous films since as early as 1923 and 1925, respectively, including more recent *White Fang* films in 1991 and 2018.

In *L'Enfant sauvage*, from 1970, the nouvelle-vague French film director François Truffaut tells a story based on presumably true events of a child who spent the first 11 or 12 years of his life with no human contact and was then incorporated into society. From "feral to domestic" has fascinated humans, with much literature attempting to learn from these cases, to develop methods of teaching, for example, language, and others questioning the validity of a true "wildness" in the lives of the unfortunate children, who were reported as having many physical scars and other signs of harsh living.

These stories are about individuals, not populations, and as such, strictly speaking, concern rewilding and taming, respectively, rather than feralization and domestication. But they do illustrate some perceptions by Western society of these subjects.

mestication has focused on insects (Bull 2000), in particular the fruit fly *Drosophila* (Simões et al. 2009). These experiments allow real-time monitoring of trajectory changes. Aquaculture also offers case studies (Witten and Cancela 2018), as discussed in chapter 9. Intense cases of selection for particular traits constitute experiments, some documented quite extensively, as in the selection for milk production in cattle (VandeHaar and St-Pierre 2006), for racing speed in horses (Gaffney and Cunningham 1988), and for increased egg production in turkeys (Anthony et al. 1991). These examples have been important in their practical application for the food industry and entertainment but of limited value for understanding the first phases of domestication.

Controlled experiments have been conducted, however, aiming to understand the evolutionary effects of taming and captivity. Past and ongoing experiments concern rats and mice, ferrets, and chickens. The most influential experiment on domestication was conducted on the silver fox (Dugatkin and Trut 2017).

Domesticated birds and the origin of human language

Human language is so complex and apparently unique, it would seem impossible to explain its origins. Since Descartes (1596–1650), a fundamental scientific approach has been to divide a problem into parts to explain it, thus making the problem more tractable. This approach can be applied to language, as some fundamental aspects of human language are shared by other species of land vertebrates (Townsend et al. 2018). Comparative and experimental studies shed light on the origin and evolution of specific features of language. Experimental approaches using neuroinformatic tools on domestic birds provide major insights (Fee et al. 2004). Two domestic species are particularly relevant in this endeavour, the Bengalese finch and the canary.

The (domestic) Bengalese finch sings songs that are more complex in their temporal organizations than songs of the wild form, the white-backed munia (*Lonchura striata*), as shown by comparisons of syntactical and acoustical parameters of songs (Okanoya 2015; Figure 8.1). Acoustical features have been found to be strain-specific. Whereas the songs of munias are highly stereotyped, those of Bengalese finches are complex, with one song note followed by several possible song notes. It has been hypothesized that the risks of predation in the wild form constrained the complexity of songs; in domestic forms this was no longer an issue, songs could undergo changes. Relaxed evolution associated with domestication has led to more variation and fewer constraints or biases in the evolution and direction of change in songs. This process has informed ideas on human language evolution, as our language could also be operating under a similar model, one of "anything goes" (Deacon 2010). Here, identifying a potential to increase complexity in vocal communication is important to evaluate how a change in selection pressures can result in significant changes and language evolution (Thomas and Kirby 2018).

Fig 8.1. The domestic Bengalese finch, *Lonchura striata* (*right*), and its wild counterpart, Munia (*left*). Both are sexually monomorphic in appearance.

Variation in the domesticated breeds of canaries, which have a short generation time, is another optimal system for studying the evolution of vocal behavior, and in particular its genetic bases (Güttinger 1985; Mundinger 1995).
Using population genomics and pedigrees generated under laboratory conditions, it is possible to identify genomic regions underlying differences between
breeds and individuals, and the developmental processes and brain anatomical adaptations through which vocal learners acquire vocalizations (Condro
and White 2014). Bird singing is an acquired behavior learned from auditory
experiences early in life. By examining variation in songbirds within a species,
or between closely related species, one can identify genes and environmental
factors associated with predisposition of individuals for learning or for producing specific vocalizations (Petkov and Jarvis 2012).

The silver fox experiment

The experimental study of domestication of silver foxes in Siberia has been crucial in shaping current views and questions around the appearance of domestication
traits; it addresses issues of morphology, physiology and behavior. This long-term
study was initiated by D. K. Belyaev (1917–1985) and executed by Lyudmila Trut
(b. 1933) and colleagues at the Institute for Cytology and Genetics in Novosibirsk
(Russia), where rats and ferrets were also subjected to experimental domestication
(Trut 1999; Dugatkin and Trut 2017). In all cases strong selection was made for tame
and aggressive behavior in groups of animals derived from farmed or wild populations. Foxes were selected based on a measured reaction when a hand was placed in
their cages.

Belyaev hypothesized that the appearance of domestication traits can occur in
the absence of human intentionality or selection pressures focused on individual
traits. In fact, the silver fox experiment showed that, over a few generations, selection for tameness resulted in the acquisition of several other phenotypic traits.
After 10 generations, 18% of the foxes sought human contact and showed little
fear, and after 35 they were doglike in many of their morphological features (Trut
1999; Trut et al. 2009; Shipman 2011; Hare et al. 2005), including shortened
snouts, piebald coats, drooping ears, and upturned tails. The domestic foxes became people-oriented. Their adrenal glands made less corticosteroids—the "fight-
or-flight" hormones; basal hormone levels in domesticated foxes were only about
25% of that documented in wild foxes (Shipman 2011). Likewise, the neurochemis-
try of domesticated foxes greatly changed as well as several life history parameters.
Females reached sexual maturity and mated on average earlier in the reproductive
season, and some individuals even mated twice a year; furthermore, many gave
birth to one more pup than the untamed foxes (Belyaev and Trut 1975; Trut 1999).

Historical records, confirmed by genetic analyses (Statham et al. 2011), indicate
that the original breeding stock of foxes came from farm-bred individuals from
Prince Edward Island, Canada and included individuals expressing some, many, or
most of the traits of the domestication syndrome then reported in domesticated

foxes (Lord et al. 2020b). This fact has been discussed as one questioning the great influence of this experiment on ideas about the syndrome and its developmental bases (Lord et al. 2020b). In a way, the experiment had already started with farming (Statham et al. 2011; Zeder 2020), as captivity can indeed lead to phenotypic changes (including behavior and physiology) similar to those characteristic of domestication, whereby captivity involves taming and habituation to humans (O'Regan and Kitchener 2005). This experiment remains influential, in spite of these caveats, in discussions on developmental patterns associated with selection for tameness (Wilkins 2019).

Tamed and laboratory rat experiments

The studies of domestication from Novosibirsk included other species besides the silver fox, in particular a population of wild caught rats (*Rattus norvegicus*). These were selected over 60 generations for either reduced or enhanced aggression toward humans (Belyaev and Borodin 1982). The line of selected "tame" rats consisted of animals unafraid of humans, which tolerated handling and being picked up. The "aggressive" rats on the other hand attacked or fled from an approaching human hand.

Albert et al. (2009) studied the genetic bases of tameness in two lines of rats using populations they established in a laboratory in Leipzig by crossing them over 60 generations for increased tameness and increased aggression against humans followed by QTL (quantitative trait loci) analysis to identify loci affecting tameness. They measured 45 traits in 700 rats, including tameness and aggression, anxiety-related traits, organ weights, and levels of serum components. A set of genetic markers across the genome were used to identify genomic regions (QTL) that influence traits differing between the two lines, for example white spots, besides tameness. Using 201 genetic markers, Albert et al. (2009) confirmed the polygenic basis for tameness. They identified two significant QTL for tameness, which were part of an epistatic network of five loci. One of the QTLs affected adrenal weight and would be a good candidate for containing a neural crest cell gene (Wilkins et al. 2017a). An additional QTL influences the occurrence of white coat spots, with no significant effect on tameness.

Predating those in Novosibirsk and Leipzig are several other studies of *Rattus norvegicus* rats. One pioneering study (to my knowledge the first) aiming at re-enacting domestication, at least in the form of captivity, resulted in numerous publications and insights into the genetics of phenotypic traits. It was conducted at the Wistar Institute in Philadelphia by Helen Dean King, who in 1919 started rearing in captivity the progeny of wild rats captured in the vicinity of Philadelphia (Figure 8.2). She reported on changes observed in the first 10 generations of captive gray rats (King and Donaldson 1929) and subsequently after 26 generations (King 1939). These papers documented what organs changed in relative size/weight and in body weight and its growth rate. Captivity led to tameness, increased body size resulting from accelerated growth rate (20% heavier at the 25th generation); reduction in brain and thyroid gland weight, and mutations in color or structure of the hair. The genetics of the color mutations were the subject of several papers. King (1939) documented

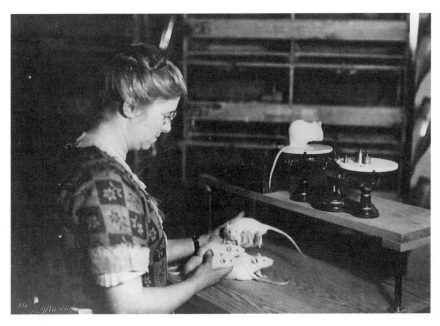

Fig 8.2. Helen Dean King (1869–1955). H. D. King led research in genetics, physiology and morphology on the effects of captivity on wild rats, and created the important biomedical laboratory model known as the Wistar rat, named after The Wistar Institute in Philadelphia, where she worked for more than four decades.

important changes in life history traits, with increased fertility and reproductive season in rats of the 25th generation. Subsequent studies compared maturation, fertility and some physiological and behavioral traits in captive wild and domestic rats (Castle 1947; Richter 1952; Clark and Price 1981). Rats are ubiquitous in biomedical research today, and this ubiquity originates from the rat captivity efforts at the Wistar institute by Helen Dean King and colleagues.

Longitudinal study of barn mice

The silver fox study of domestication concentrated on intense and directed selection regimes, important for figuring out the effects of selection for some traits (e.g., tameness) but not useful for reconstructing the conditions that approximate the initial phase of domestication, as for example interactions with humans in the commensal pathway of domestication. This initial phase was investigated in a study involving a population of wild house mice (*Mus musculus*) established in 2002 in an uninhabited barn in Illnau, Kanton Zurich, Switzerland, on the effects of interactions with humans. This longitudinal study was established with the introduction of 12 founder, wild-caught individuals (König and Lindholm 2012); the study population increased considerably in size, reaching 250–430 individuals at any given time (König et al. 2015). A standardized ad libitum feeding regime mirrored the case of commensal mice in farms, which typically have abundant food resources and a human-made environment (Pocock et al. 2004). The regular handling and monitoring of mice over 14 years in the course of other research projects resulted in habituated mice, unintentionally selected for tameness for approximately 20 generations (Manser et al. 2011). Geiger et al. (2018b) examined data based on this experimental population and quantified how, in the course of a decade, animals exhibited significantly decreased head length and increased occurrence of white patches of fur. These phenotypic changes fit well with the "domestication syndrome" in two of its most common—but not universal—features.

Behavioral selection in mice and in minks

Four studies discussed here serve as examples of experiments concerning behavioral aspects related to domestication.

As opposed to the silver fox experiment, which involved selection for reduced aggression toward people, a study in breeding mice examined the effects of decreased aggression toward other mice (Gariépy et al. 2001). The experiment examined whether selective breeding for differential aggression over several generations produces heterochronic changes in social development at different ages. Adult mice are usually aggressive toward strangers. In general, data showed that when the newly bred friendly mice encountered a strange mouse, instead of being aggressive, they shivered like a juvenile individual, a behavior characterized as heterochronic pattern of change.

Connor (1975) used a longitudinal approach and conducted a battery of nine behavioral tests to study the effect of captivity on wild house mice (*Mus musculus*). After ten generations, reportedly no behavioral differences were found between wild mice reared in laboratory conditions and wild mice maintained in a simulated naturalistic environment. Inbreeding did result in differences, though, as it partially reduced resistance to recapture by humans, and strongly reduced intermale aggression.

In 1988 a selection experiment was initiated at the Danish Institute of Agricultural Sciences with mink (*Neovison vison*) that resulted in two genetic lines. A "confident" line exhibited a high degree of approach, whereas the other showed avoidance (fearful line) toward humans (Malmkvist and Hansen 2002). The experiments showed the correlation of behavioral traits when compared with an unselected control line, since there were generalized fear responses of mink across several social and nonsocial situations in the different selection lines (Hansen 1996). This result was interpreted as behavioral selection that improved the welfare of farmed mink given the overall reduced level of fearfulness, including for example a reduction of chewing behavior (Malmkvist and Hansen 2002). More recent literature on minks' behavior and life history under domestication has explored diverse issues on the conditions of life and welfare (Díez-León and Mason 2016; Díez-León et al. 2016).

Insights can be gained in laboratory settings, even when selection experiments have not been conducted. In a quantitative and controlled study of behavior and endocrinology, Künzl and Sachser (1999) showed that cavies show reduced aggressiveness and increased social tolerance in contrast to a proxy of their wild counter-part (as represented by *Cavia aperea*). Domestic cavies were distinctly less attentive to their physical environment than the wild ones. The domestic cavies had lower basal activity and reactivity of the sympathetic–adrenomedullary SAM system, and lower reactivity of the pituitary–adrenocortical PAC system. The activity of the pituitary–gonadal system in males was higher in domestics than in wild forms.

Red junglefowl and White Leghorn chicken experiments

Per Jensen and colleagues in Linkoping are conducting a long-term experiment selecting for tameness in the red junglefowl, focusing on resulting genetic and be-havioral changes. They systematically aim at studying the genetic and epigenetic mechanisms behind those changes (Agnvall et al. 2018). Initially two captive popu-lations of red junglefowl differing significantly in their fear responses were main-tained for more than five generations at 70–80 individuals per population and generation to create an outbred population of the proxy for the wild, ancestral fowl, from which chickens originate (Agnvall et al. 2012). The details of the crossing, handling, and features of the populations and their changes are well recorded. All outbred red junglefowl were bred and maintained similarly, but with groups subject to directional selection for high or low fear in subsequent generations and an un-selected control population (Agnvall et al 2012, 2014; Bélteky et al. 2016). Crosses between red junglefowl and White Leghorn chicken are used to map the genetic loci affecting different traits. Jensen and colleagues also analyzed DNA-methylation differences between the two groups in brain and germ cells.

A study compared two groups of red junglefowl selected for high versus low fear of humans for five generations, examining changes in gene expression in the thala-mus/hypothalamus region of the brain, involved in fear and stress (Bélteky et al. 2016). After only five generations, the S5 generation grew larger and produced larger offspring than the outbred parental generation, and the two selected lines (high and low fear) differed in expression of 33 genes related to male reproduction (sperm) and immunological functions, with only a few known to be relevant in fear- and stress-related pathways. Thus, intentional selection for reduced fear of humans (increased tameness) in this study affected a range of apparently unrelated traits.

Agnvall et al. (2012) conducted an experiment on the reaction of red junglefowl in a standardized test for fear of humans. After just three generations of selection for high versus low fear of humans, the authors could establish that the variation in response had a significant genetic component and was genetically correlated with other behavioral traits related to fearfulness and to exploration.

Agnvall et al. (2018) studied red junglefowl selected for reduced fear of humans for eight generations and recorded that they exhibited a series of features contrast-ing with those of the opposite selection line showing high fear of humans. These features were: "Size: Increased hatch weight, Increased growth"; "Increased basal metabolic rate"; "Feed conversion: Increased growth per g food intake in females"; "Hormones: Increased blood serotonin in males"; "Boldness: Reduced latency to ap-proach a novel object"; "Social behavior: Increased social dominance"; "Organ size:

reduced brain size, reduced size of heart, liver, testes and spleen"; and "Plumage: Increased plumage quality" and "Possibly loss of pigmentation."

In another study of the putative role of tameness as a driver of domestication-related behavioral phenotypes, Katajamaa et al. (2018) studied the effects of selection in red junglefowl for divergent levels of fear of human for eight generations. They considered general activity, social behavior, and male courtship and found that differences between selection lines changed with age. Whereas adult low-fear individuals were more active, high-fear males showed more intense courtship behavior. Together, these studies demonstrate the role of tameness as a driver of domestication related phenotypes.

The experimental approach is showing the effects of selection for certain behavioral traits on physiological, life history, and morphological variables. Comparisons with already domesticated chickens are usually performed with White Leghorns as subjects. A potential expansion of this study, using the experimental results as guidance, would be a comparative study of behavior across chicken breeds and their genomes, which would reveal the whole extent of variation generated by selective breeding and the potential for segregation of traits, as has been quantified for integumental characters in this species (Núñez-León et al. 2019).

Developmental plasticity in pigs under experimental conditions

Experiments that evaluate the effects of anthropogenic environments in conditions that resemble the first phases of domestication could be particularly important for zooarchaeology, as they would serve to evaluate morphological markers of the domestication process.

A study tested whether changes in locomotor behavior due to mobility reduction could be quantified as bone shape modifications in wild boar and pigs (Harbers et al. 2020a). The shape of the calcaneus, the ankle bone, often well preserved archaeologically, was used as a phenotypic marker in captive-reared and wild-caught wild boars, and contrasted with changes induced in domestic pigs by selective breeding over the last 200 years. Mobility reduction was found to induce shape changes, with resulting shapes beyond the variation recorded in wild boars in their natural habitat. This variation recorded in wild boars under restricted mobility also differed from a much greater calcaneus shape change recorded in domestic pigs. It is a significant discovery that this experiment resulted in no bone size reduction in captive-reared wild boars—size is usually regarded as the variable most prone to change. The changes recorded in calcaneus shape for the animals in captivity differ from those induced by selective breeding—the latter had a much greater effect.

Other experiments studied the effects of captivity and domestication on humerus cortical thickness volume and topography in wild boar and pigs (Mainland et al. 2007; Harbers et al. 2020b). Patterns for each examined condition and population could be discerned. As just a portion of the humerus is required for study, fragmented specimens such as those encountered in the archaeological record can thus provide an assessment of the animal condition, independent from morphological or isotopic studies.

Other experimental studies of domestication and quantifications of rate change

Conducting experiments is an approach to understanding mechanisms associated with domestication. The silver fox experiment has been highly influential, but it should be emphasized that this is an experiment about selective breeding—intense selection for tameness in this case—and not the commensal situation hypothesized for the origin of many domesticated species. Long-term data on barn mice from a site in Switzerland presumably replicate well the processes and phenotypic patterns resulting from the domestication process (Geiger et al. 2018b), but the available data on these animals are more restricted. A promising way to conduct experimental research on domestication and its effects is by using fish models, as discussed in chapter 9.

Understanding mechanisms and results of selection in experiments can be useful in welfare applications. Long-term selection experiments have shown that it is possible to reduce fearfulness in farmed mink; farmers could add behavioral measurements to their normal selection criteria and thus improve the welfare of farmed mink (Díez-León and Mason 2016).

Other experiments that can provide major insights involve manipulations and documentation that help understand the developmental mechanisms leading to morphological change. Artificial selection in mice for relatively long limbs produced "longshanks"—these have been used to examine the mechanisms and consequences in postnatal limb development (Rolian 2020). Morphometric studies and quantification at the histological level of growth plates in the tibia showed that longshanks are produced not due to prolonged growth but instead through accelerated growth rates—there is an increase in the number of proliferating chondrocytes (Marchini and Rolian 2018). This selection had also a significant effect on trabecular bone morphology—longshank mice had significantly less trabecular bone at skeletal maturity, characterized by fewer, thinner trabeculae—suggesting that rapid linear bone growth may influence the risk of cancellous bone fragility (Farooq et al. 2017).

Experiments have been conducted over many years in poultry to determine which of three factors affect efficiency of genetic gain under artificial selection (variability exhaustion, nonadditive variability, and balance between fitness and the selected trait). One experiment (1938–1949) selected for long shanks (tarsometatarsus length) at maturity in females in a small population of White Leghorn chickens, showing that shank length increased considerably during the first half of the experiment, with no further gain afterward (Lerner and Dempster 1951). Parents with genotypes for longer shanks had a lower reproductive rate; the authors state that the data suggested adverse natural selection was at least in part responsible.

Studies of urban fauna

Industrialization in the nineteenth century brought with it the beginning of a trend that continued to this day, namely the migration to the cities from rural areas—this means that more and more people need to be fed by fewer and fewer people, thus the agricultural expansion based on selective breeding for desirable traits. Expansive agriculture and the demographic explosion of humans has led to habitat destruction. Humans produce a lot of waste, and cities are an intricate mesh of potential new habitats for diverse animals (and plants). The study of urban fauna has gained prominence (Szulkin et al. 2020). Diverse species in cities across the world have evolved and adapted to new living conditions in these human-shaped environments. There are obvious parallels in domestication, and one of them concerns the behavioral change to habituation, to tameness, in the case of mammals living commensaly with us. An example are the many populations of foxes living in many European cities. A study of urban and rural populations of red foxes (*Vulpes vulpes*) in the London area showed that urban individuals have wider and shorter snouts relative to rural individuals, as well as smaller braincases and reduced sexual dimorphism (Parsons et al. 2020). A study of longitudinal data in cranial capacity in two small rodent and one bat species that lived in urban environments showed greater capacity in the rodents and no changes in the bat (DePasquale et al. 2020).

Rates of evolution in the domestication process

Evolutionary transformations that are morphologically conspicuous have occurred in geological time. Those taking place in some features during selective breeding appear to operate at much faster speed (Gingerich 2019). It follows that domestic animals are examples of rapid evolution, as urban animals also are. Or are they? Rate of evolution can be calculated in different ways and over different time spans, and the method and parameters used surely affect the results. Over long periods of times, fluctuations in change may be averaged, thus much change is not captured. Concerning domestication, diverse interactions and their degree of intensity mean that change depends on the phase or aspect of domestication (e.g., unintended selection for tameness versus intensive selection). In the natural world, only exceptional situations reach the level of intense selection typical of selective breeding; hurricanes affecting island populations are perhaps an example (Donihue et al. 2018).

Gingerich (2019) analyzed the data provided by Lerner and Dempster (1951) for 14 generations that led to an increase in shank length of 13% in the selected line, in contrast to the random walk leading to almost no change in the control population. The median rates of change were 0.307 and 0.175 standard deviations per generation for the selected and control populations, respectively. Among experiments concerning egg production, there is Gowe's study of 30 years that started in 1950 and ran for 30 chicken generations (Gowe et al 1959; Gowe and Fairfull 1985), in which two selected strains increased egg production by 72% and 88%, whereas the control strain exhibited a lower gradual increase of 17%. Long-term data such as these for chicken are a great source for evaluating rates of phenotypic change and how these rates are affected by the conditions of existence.

In their study on habituated, commensal mice over approximately 20 generations, Geiger et al. (2018a) reported several rates of changes. These include increased proportion of adult mice with white patches of fur from 2.5% in 2010 to 5.4% in 2016, and significant decrease in head length and body size

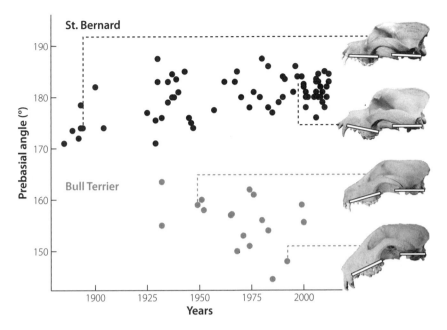

Fig 8.3. Historical changes in a skull variable in two domestic dog breeds. Change throughout decades can be discerned in both breeds, with dorsal bending in the St. Bernard and ventral bending in the Bull Terrier. A significant change in the prebasial angle—the angle between the hard palate and the cranial base of the skull, indicated with white bars—can be found in the St. Bernard. Skulls are scaled to the same neurocranial length.

in 13-day-old mice between 2007 and 2016. Further work could evaluate hypotheses to explain these changes, including genetic drift and inbreeding, and morphological skull alterations as adaptations to different food items.

In a metric study of skulls of wild and domestic pigs and domestic dogs, Geiger and Sánchez-Villagra (2018) showed that marked differences among breeds can arise in spite of apparently similar evolutionary rates between wild and domestic forms in at least some phases of their evolution. Similar rates for wild and domestic forms were also reported for at least some species of plants at the early phase of domestication (Purugganan and Fuller 2011). Examination without quantification of rates in the Berkshire breed of pigs is available (Herre 1938).

The studies above demonstrate that human intervention per se does not necessarily accelerate evolutionary rates of organisms (Hendry et al. 2008). The different levels of human-animal/plant interactions can affect evolutionary rates differently. There are marked differences among domestic groups and traits (Geiger and Sánchez-Villagra 2018), as shown for improved production in livestock (Bell 2015). For some domestic dog breeds, breeding standards, or their interpretation, have changed, bringing significant changes in skull shape and size at high rates (Figure 8.3). This is the case of skull shape changes in the St. Bernard (Drake and Klingenberg 2008; Nussbaumer 1982, 2000).

Selective breeding and animal welfare

The lives of animals are greatly affected by artificial selection, and thus questions arise about the obligations of humans to provide proper treatment and handling. Humans are causally responsible for the hardships faced by many domesticated forms. The distinction between wild and domestic is important here, as some philosophers have argued that different obligations arise. Humans may not harm wild animals but may choose to assist them, and they gain the obligation to assist them if responsible for the animals' plight. In the case of domesticated animals, our "contractual" and "causal" relation to them certainly creates an obligation (Palmer 2010).

Over decades, ethical issues have become more important as industrialization and mass-food production have intensified. Manipulations related to breeding can result in animals with organs ill-suited to healthy living, causing pain and suffering—as in some brachycephalic cats and dogs. Another example is the prevalence of keel bone fractures in laying hens. This bone serves as attachment for wing flight muscles, and its deficient constitution—surely resulting from trade-offs in breeds optimized for production—results in fractures that can impede proper locomotion important for the performance of natural species-specific behavior and cause pain. Thus, the skeletal biology but also the management conditions of animals with this condition or a propensity to it are investigated to increase animal welfare (MacLachlan et al. 2020). The goal is to optimize the design of laying hen feeders, as well as determine optimal bird density and spatial arrangement that considers production goals while maximizing welfare.

Another subject of intense investigation aimed at welfare is the design and size and kind of cage or habitat for laboratory or farmed animals. Here individual variables are examined and their effect tested in behavioral experiments (Díez-León et al. 2017). Research in this area of veterinary medicine is fundamental. One may wish that, parallel to it, more would be invested in anthropological and cultural evolution for insights on confronting consumer behavior and global and local economies that lead to the demands and consumption that generate the management challenges in the first place.

The welfare of domesticated animals in their relationship with humans is a multivariate, nonlinear matter. It is sometimes claimed that captivity per se poses welfare issues, but captivity can also help avoid physical stressors routinely experienced in the wild. Well-managed captive populations can be provided essential psychological outlets. It will be important to develop a paradigm of animal welfare that is not driven by some standard metric of physical well-being and is informed instead by the organismal biology of the species (Veasey 2017).

Summary: Variation in feralization and in rates of evolution, and the use of experiments in domestication studies

The feralization process, like the domestication one, involves few and general universals, with diverse mechanisms and diverse resulting patterns. Although feralization may seem like the domestication process in reverse, feral animals never return globally in their phenotype to the wild state. This is not surprising, as the founding population before feralization is not the same that gave rise to the wild population in the first place. Feralization does not imply any common environmental context, and feral animals can occur in many different kinds of places, including cities.

The use of the concept of feralization varies, both in terms of biology and in its connotation. Feral animals are seen at one extreme as reservoirs of genetic and phenotypic diversity worth preserving and at the other as invasive species and pests.

Population fitness is often reduced due to lower phenotypic variation in specialized breeds. Feral animals can hybridize with domestic and wild forms, which leads to gene pool changes that may be either beneficial or detrimental.

Using experimental (and control) specimens and data, robust interpretations of mechanisms associated with the domestication process can be drawn. Experiments usually concern one aspect of the domestication process (e.g., some kind of selective breeding for some module of the organism, or the commensal, unintentional interactions between specific populations of some animal species and of humans). The explanatory power of experiments is thus restricted, so their conclusions should not be generalized for domestication or for any species.

Experiments on wild boar and pigs have shown that phenotypic signatures can be detected after a single generation of captivity, associated with phenotypic plasticity.

Urban populations of diverse species live in a commensal relation to humans and can serve as a model for the initial phases of domestication, giving insights into phenotypic plasticity and genetic changes.

Human intervention per se does not necessarily accelerate evolutionary rates; domestic groups show marked differences, such as in livestock. These differences among current breeds can arise in spite of apparently similar evolutionary rates between wild and domestic forms, as shown by skulls of wild and domestic pigs, and of domestic dogs.

CHAPTER 9

Fish domestication

Until the mid-twentieth century, few fish species had been domesticated. The classic and "old" domesticates among fishes are carps and goldfishes and more recently salmonids. Many more are being managed and have become effectively domesticated—in the sense that their reproductive biology has been changed under human influence (Vigne 2011)—in the last few decades. Fish aquaculture currently involves more than 160 species (Duarte et al. 2007; Bostock et al. 2010), managed for diverse usage goals, including as food and ornament, or for recreation, bio-manipulation, conservation, and research (Lorenzen et al. 2012). The many species now domesticated belong to diverse groups within the evolutionary tree of bony fishes (Figure 9.1). The zebrafish, *Danio rerio*, is one of the model organisms of molecular and developmental biology; populations of this species have gone and are still going through a domestication process. Zebrafish offers a great model for studying developmental evolution and its relevance to variation in domesticated animals, including the trajectory of variation in development (Irmler et al. 2004), the role of tissue interactions in dental development (Oralová et al. 2020), and dietary effects on skeletal development (Cotti et al. 2020).

The domestication of the carp (*Cyprinus carpio*)

The common carp (*Cyprinus carpio*), the earliest truly domesticated fish, is one of the most extensively cultivated species. The natural distribution of common carp ranges from the Black, Caspian, and Aral sea drainages (Chistiakov and Voronova 2009) to an eastern population in China, Laos, and Vietnam. Perhaps it was independently domesticated twice, once in Asia, and separately in either Rome or medieval Europe.

In Asia, the carp became managed in association with rice agriculture, an ecologically sustainable and mutually beneficial relationship. As already practiced in the past, carp are kept in rice ponds, eating insects attracted to the rice, uprooting weed species, and naturally providing fertilizer (Xie et al. 2011). At the same time, the shade provided by the rice acts as a buffer against temperature extremes, and the rice absorbs fish waste (Lansing and Kremer 2011). Historical records document that carp have been raised in rice paddy fields since at least the first millennium B.C.

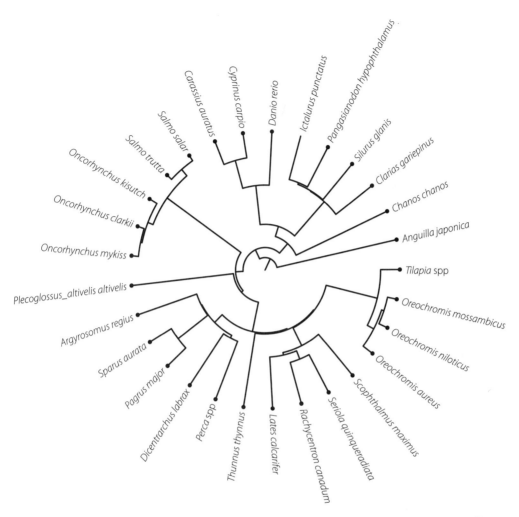

Fig 9.1. Evolutionary relationships among traditionally domesticated fish species. Examples include traditionally domesticated fish as well as more recently managed ones in the context of aquaculture, which have also experienced significant genetic and ecological modifications.

Since rice paddies date to the fifth millennium B.C., and archaeological evidence dates control of fish stocks for human use in Jiahu, China back to the sixth millennium B.C., carp origins may date to much earlier (Nakajima et al. 2019; Harland 2019). There are similarities in body length between archaeological specimens at the Jiahu site dating to 6000 B.C. and those of modern carp raised in a traditional Japanese rice and fish co-culture system (Nakajima et al. 2019).

Romans may have been the first to culture carp obtained from the Danube, and this tradition was continued in monasteries throughout the Middle Ages (Balon 1995). Given the long cultivation history of common carp in China, the ancestor of European domestic carp has been hypothesized to be Asian common carp, transported from Asia to Europe during ancient Greek and Roman periods (reviewed in Balon 1995). Studies of segments of mitochondrial DNA suggest different ancestors

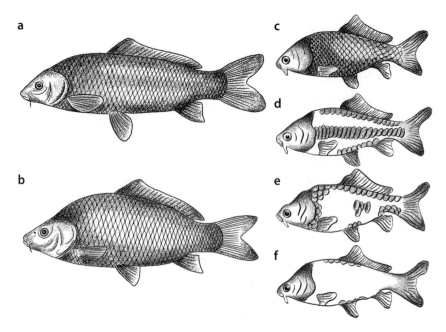

Fig 9.2. Examples of variations in body shape and squamation in common carp, *Cyprinus carpio*. (a) wild common carp and (b) its feral form. Scale patterns include from top to bottom: (c) scaly, (d) line, (e) mirror, and (f) leather.

for domestic carp in Europe: Russian scattered-scale mirror carp originated from the Asian subspecies *C. carpio haematopterus*, whereas German mirror carp was domesticated from the European subspecies *C. carpio carpio* (Zhou et al. 2003; Fig. 9.2).

A genomic study produced phylogenetic trees for carp fishes and concluded that the European and Asian carps formed distinct clades, but with exceptions and traces of exchange (Xu et al. 2014). The Asian strain Songpu, grouped into the European clade, apparently bred from mirror carp originally introduced from Europe in the 1950s. The results support the hypothesis that modern *C. carpio* evolved from the Caspian Sea ancestor and spread into Europe and the eastern mainland of Asia (Balon 1995).

Different strains of domesticated carp exist, and this species is also reported as feral (Fig. 9.2). The so-called "mirror" strains have a skin without scales. Some breeds have been the subject of intense selection for purely aesthetic criteria, including variable colors, depigmented spots or conversely richly colored ones, and fins of derived shape and length compared with the wild standard. These so-called "koi" carp are the subject of trade and can be luxury items (Guillaume 2010).

In China researchers developed a method of fish farming for the common carp (*Cyprinus carpio*) as well as for the grass carp (*Ctenopharyngodon idella*), the silver carp (*Hypophthalmichthys molitrix*), and the Indian carp (*Catla catla*). Each takes its food at a particular level of the trophic chain: the grass carp eats grass made available to it, the silver carp filters phytoplankton, and the Indian carp feeds on various invertebrates, while the common carp is omnivorous after the larval stage (Guillaume 2010).

Goldfish (*Carassius auratus*)

The goldfish (*Carassius auratus*) was the first ornamental domesticated fish (Dzwillo 1962) and is hugely popular worldwide to this day. The goldfish has also been used for physiology experiments (Ostrander 2000). The taxonomy and origins of goldfish are contested. The relation of the goldfish to the crucian carp *Carassius carassius* and to the common carp (*Cyprinus carpio*), with which it crosses, is complex. Hybridization, a long history of domestication, and numerous human translocations to many areas worldwide make the subject challenging (Hängling et al. 2005; Gao et al. 2012; Wang et al. 2014a; Podlesnykh et al. 2015). Crosses among these and other species are conducted to devise the genetic bases of some of the rich morphological diversification that has occurred during its domestication (Wang et al. 2018).

The wild goldfish is native to lateral waters of lakes and rivers of East and Central Asia, with the center of distribution in China (Gao et al. 2012). The goldfish has acquired global distribution in temperate and subtropical areas from escaped domesticated forms and deliberate releases. It has feral populations in all habitable continents, and in some localities it is sufficiently abundant to be sold as a food fish. The goldfish can reportedly live in wide ranges of temperature, water turbidity, and pH, and it can tolerate oxygen depletion and concentrations of heavy metals and insecticides (Balon 2004). This adaptability has surely been key in its widespread use and distribution worldwide.

The history of the creation and development of the goldfish as an ornamental fish in China is tied to Buddhist practices and sociocultural phenomena over hundreds of years (Chen 1956; Ota and Abe 2016). Many of the features of the peculiar varieties of goldfish, including the twin-tail, the "dragon eye," and the short body, were already in existence in the sixteenth century (Chen 1956). The goldfish was introduced to Europe and Japan from China in the sixteenth and seventeenth centuries already domesticated, including diverse peculiar, ornamental varieties (Balon 2004). There are many fancy varieties of domesticated goldfish living in aquaria and garden pools around the world: water bubble eye, blue fish, brown fish, brocade, pompon, fantail, veiltail, eggfish, telescope, calico, celestial, lionhead, tumbler, comet or meteor, and pearl scale (Balon 2004; Smartt 2001)—and their corresponding Asian names (Balon 2004; Ota and Abe 2016). The genetic bases of traits characterizing this great breed diversity are becoming known (Chen et al. 2020).

Aquaculture of salmonids

The first in vitro fertilizations of trout eggs with collected spermatozoa were conducted on common trout (*Salmo trutta*) as early as the eighteenth century. Larval rearing to the juvenile stage was not difficult because salmonids have the advantage of laying large eggs (several hundred milligrams each), and give birth to larvae that are easily fed. A species from the west coast of North America, the coho salmon (*Oncorhynchus kisutch*, formerly *Salmo gairdneri*), introduced in Europe at the end of the nineteenth century, has been intensively bred. Its nutrition, initially composed exclusively of slaughterhouse waste, has been gradually replaced by grain since the 1950s. The domestic form is in some cases raised in floating cages at sea from a certain stage of development. It is the most common domestic salmonid in many temperate countries in the Northern Hemisphere (Lien 2015).

The Atlantic salmon (*Salmo salar*) has been bred for production purposes only since 1960, including a large-scale national breeding program beginning in the 1970s in Norway. This program selects for increased growth and avoidance of early sexual maturation, as well as flesh color, fat content, and disease resistance (Gjøen and Bentsen 1997; Gjedrem 2010; Bolstad et al. 2017). The Atlantic salmon, like all salmonids, is a species that reproduces in freshwater and migrates to the sea after smoltification. It was necessary to study the hormonal bases of this phenomenon and control it in order to transfer the young salmon into floating cages at sea as soon they can live in salt water, without risk of massive mortality. The breeding has met with optimal conditions in the fjords of Scandinavia, especially in Norway, where salmon populations have grown at an exponential rate. Breeding has been extended to Scotland, Canada, Chile, and other countries with cold coastal waters. Salmon aquaculture, one of many examples of "marine aquaculture" includes production of the Atlantic as well as Pacific salmons (both *Oncorhynchus*), and salmon is the third largest aquaculture species worldwide (Guillaume 2010).

Examples of marine and freshwater aquaculture species

Aquaculture has benefited from research about the nutrition of fish larvae whose mass is in the order of a milligram and the farming of unicellular algae and roti-fers to feed them. This system is practiced in Japan for the seabream *Pagrus major*. The aquaculture of Japanese amberjack (*Seriola quiqueradiata*) has been enabled thanks to cage fattening of juveniles caught at sea (Guillaume 2010). There are now hatcheries and farms of the European sea bass (*Dicentrarchus labrax*), the "dorade" swordfish (*Chrysophrys aurata*), and the turbot (*Scophthalmus maximus*), especially in Mediterranean countries where thermal conditions are most favorable (Guillaume 2010). Among freshwater species raised in temperate climate, there is the American catfish (*Ictalurus punctatus*), the European or wels catfish (*Silurus glanis*), the Japanese ayu (*Plecoglossus altivelis*), and the perch from China, the so-called poisson mandarin (*Perca* sp.). The clarias or African sharptooth catfish (*Clarias gariepinus*) is an example among many other tropical species.

The various species belonging to the genus *Tilapia* and to related genera, all sold under the name of "tilapias," have had considerable success in aquaculture. The farming of the Nile tilapia, *Oreochromis niloticus*, dates back at least 3,500 years, with representations of farming preserved in bas-reliefs of Theban tombs in Egypt (Harache 2002). Although Nile tilapia is natively from Africa, more than 70% of its global production comes from Asia, with more than 1 million tons from China only (Teletchea 2016); production has increased exponentially since the late 1980s. Its speedy growth has contributed to this development, pattern analogous to that of chickens among domesticated terrestrial forms (Cressey 2009). Selection for growth and fillet yield in a group of Nile tilapia was shown to result in marked genetic im-provement of fillet weight after as few as six generations (Thodesen et al. 2011).

Another intensively cultured freshwater species is the Asian catfish called panga (*Pangasius hyophthalamus*). In the tropical marine area, the number of aquaculture species is at least as great as that of freshwater species. Among them are the bar tropical (*Lates calcarifer*), the milkfish (*Chanos chanos*), and the bluefin tuna (*Thunnus thynnus*), with the last sometimes raised in huge cages, so-called fattening farms, that can be about 200 m long. Another flagship species is the cobia (*Rachytcentron canadum*), a warm water fish that lives in temperatures above 20°C, is easy to grow in various conditions, including floating cages in the open sea. It grows very fast, to 2 m and 70 kg as an adult. Its level of production may become equivalent in tropical waters to that of salmon in freshwaters. The meagre (*Argyrosomus regius*) has recently been added to the range of Mediterranean aquaculture fish (Guillaume 2010).

The "doctor fish," *Garra rufa*, a freshwater cyprinid fish native from the Middle East, is widely used in foot spas for dead skin removal, and as therapy for patients with psoriasis (Ündar et al. 1990). It has been used in Turkey to cure skin diseases for more than 200 years.

Morphological changes associated with aquaculture and selection for ornaments

Body or head shape, size, proportions and features of the fins and the integument (e.g., color or size and distribution of scales), among other features, can be different in domestic fishes from their wild counterparts. Appearance traits are important in commercial fish, so much so that genomic findings in model fish are being used to find the genetic bases of traits and optimize the selection process (Colihueque and Araneda 2014).

There are several examples of body shape and scale pattern modifications in some domesticated fish, as exemplified by the common carp (Fig. 9.2). The Aischground and Galician strains of common carp are high-backed (Ankorion et al. 1992) and the variety Wuyuanensis, also called Purse red carp, has a broadly elliptical body and standard length to body height ratio of 2.3 (Zhang et al. 2013b). These changes were produced presumably to increase the aesthetics of body shape. The variety *haematopterus*, also called Amur wild carp, with spindle-shaped, elongated body and a standard length to body height ratio of 3.5, was developed to increase desirability for sport fishing (Zhang et al. 2013b). A variety of the rainbow trout (*Onchorhynchus mykiss*) from the Finnish national breeding program has been developed with a slender body, with low body height to length ratio, to produce a more visually appealing fish for the carcass market (Kause et al. 2003).

A study of coho salmon, *Oncorhynchus kisutch*, examined morphometric variation in captively reared and a population of wild individuals, in the state of Washington in the United States (Hard et al. 2000). The adults of the captive population had smaller heads and less hooked snouts, increased trunk depth, increase in dorsal and anal fin base lengths and in caudal peduncle size; shorter dorsal fins, and a forward shifting of pelvic fin placement. The captive individuals exhibited, in general, a reduction in body streamlining (Hard et al. 2000). The authors speculate that at least some of the differences were environmentally induced.

In terms of modifications in the integument, there are also many examples (Colihueque and Araneda 2014). Red strains of the tilapia *Oreochromis niloticus* (e.g., red Stirling, red Yumbo), lacking the signs of normal black pigmentation of

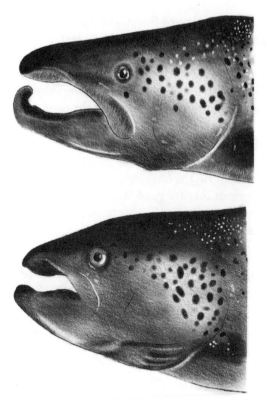

Fig 9.3. Heads of two male adults of the Atlantic salmon (*Salmo salar*), the one on top showing the typical kype of wild forms, the one below domesticated. This elongation and curvature of the lower jaw into a hooklike appendage is a secondary sexual characteristic, reduced in domesticated forms.

wild-type forms, were created to add market value (McAndrew et al. 1988; Moreira et al. 2005). Likewise to satisfy market demand, the blue-back variety of rainbow trout was created, with intense bluish back, whitish belly, and reduced number of dark spots on the back and below the lateral line (Colihueque et al. 2011).

Sexual dimorphism tends to decrease with domestication due to reduction in or lack of sexual selection. The relaxation of sexual selection in the domesticated Atlantic salmon, *Salmo salar*, has led to the reduction of the kype (Figure 9.3), a structure important for mate choice in the wild. The kype is an elongated and curved projection of the lower jaw, a hooklike appendage.

Variations of goldfish (*Carassius auratus*) are striking cases of morphological diversification led by domestication, in regard to both cranial and postcranial traits (Ota and Abe 2016). Skull differences between domestic and wild forms and among domestic varieties are notable (Bateson 1894; Dobkowitz 1962; Smartt 2001; Fig. 9.4) but have never been systematically described and compared. Some of the peculiarities in the many forms include, among others, changes in the neurocranium, a shortening and reduction of the hyopalatine arches, and a shortening of the lower jaw (Dobkowitz 1962), as well as sculpturing of the dorsal skull roof (Fig. 9.4).

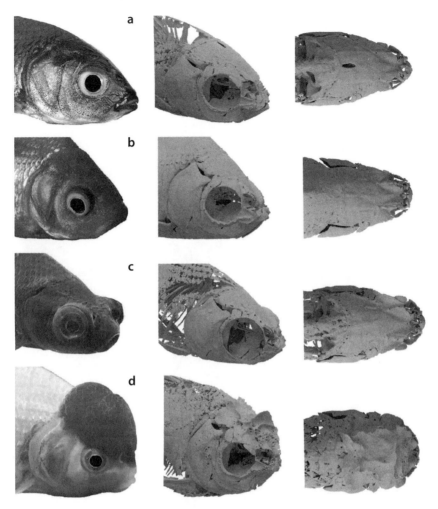

Fig 9.4. The head and high-resolution computer tomography images of the skull of four varieties of goldfish, *Carassius auratus*. (a) common goldfish in black variety. (b) Fantail goldfish. (c) black telescope goldfish. (d) Oranda goldfish. Scale 5 mm.

In the telescope goldfish, the eyes project outside the orbit to some extent, in some forms being located fully outside the rest of the head and attached to it by only a small peduncule (Bateson 1894). These and other forms are graphically documented in a 1780 treatise on fishes from China, starting with a quote by Voltaire exalting the creativity of the Chinese civilization; a testimony to the interest of Europeans in these animals (Billardon-Sauvigny 1780).

Some of the variations in goldfish skulls are so extreme that the ethical dimension becomes relevant. Breeding of some goldfish forms is prohibited in Switzerland (www.fischwissen.ch), and it is hoped that this kind of legislation becomes standard in other places as well and that public demand for these animals decreases.

In terms of modifications of the postcranial skeleton, a remarkable case is that of the tail of some goldfish strains established for ornamental purposes. Twin-tail goldfish have a bifurcated caudal axial skeleton. In spite of the extraordinary skeletal

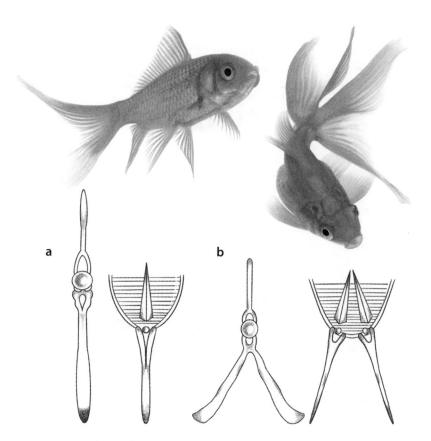

Fig 9.5. Caudal and anal fins of goldfish. Domestic goldfish, *Carassius auratus*, and cross-sectional skeletal anatomy in the caudal and anal fins of (a) normal and (b) twin-tailed goldfish. In each case, the penultimate vertebrae (*left*) and a transverse section through the anal fin region (*right*) are illustrated.

diversity of bony fishes, this feature is not found in any other species (Fig. 9.5). The basic architecture of caudal fins is the same and includes the fin ray and vertebral elements being located on the sagittal plane (Schultze and Arratia 1989). In the twin-tail goldfish, several caudal vertebral elements and their attaching fin-rays are duplicated to form a laterally bifurcated caudal axial skeleton (Watase 1887; Bateson 1894; Abe et al. 2014).

The embryology of the bifurcated median fins of two representative twin-tail ornamental goldfish strains compared with single-tail common goldfish revealed in the former a disrupted axial skeletal development at early larval stages and modified bilateral location of the pelvic fin (Li et al. 2019a). The ventral tissues of the twin-tail strain are enlarged and form the embryonic bifurcated fin fold (Abe et al. 2014). The developmental genetic bases of this bifurcated caudal skeleton are known; it arises from a stop-codon mutation in the chordin (*chdA*) gene, a duplicated gene that affects dorsal-ventral (DV) patterning during embryonic development (Abe et al. 2014).

Another remarkable feature of some goldfish variants is lack of a dorsal fin, for example, of the Osaka Ranchu and the Celestial goldfish. The Albino Ranchu goldfish exhibits complete ablation of the fin rays and endoskeletal elements of the dorsal fin (Kon et al. 2020). In early developmental stages during which the common goldfish shows an extended fin fold related to later fin development, the dorsal fin fold is lacking in Albino Ranchu embryos (Kon et al. 2020).

More subtle cases of postcranial variation than those of the goldfish are seen in the many kinds of spine deformities and fusion of vertebral bodies in the vertebral columns of farmed fishes. A study of vertebrae fusion in farmed Atlantic salmon established that this fusion can result from different processes: replacement of intervertebral cartilage by bone, shape alterations of vertebral body endplates, mineralization of the intervertebral cartilage, or transformation of intervertebral notochord tissue into cartilage (Witten et al. 2006). The continuous amalgamation of neighboring vertebrae into two vertebrae that have fused can lead to a severe malformation, causing spine shortening, with obvious implications for the animal's performance. This type of malformation is also well known in other fish species (Witten et al. 2006) and has been reported for humans presenting fused vertebrae and deformities in adjacent spine segments (Leivseth et al. 2005).

Reshaping and remodeling of two fused vertebral bodies into a functional single structure can contain progression of the problem, an example of developmental plasticity, as reported for Atlantic salmon. The progress of vertebral body fusion is stopped through remodeling of two or at most three fused vertebrae into one remodeled vertebra. Thus, deformed vertebrae do not necessarily lead to an aggravating spine malformation. These reshaped vertebrae can be identified on radiographs based on the presence of supernumerary neural and haemal arches (Figure 9.6).

The goldfish is a member of the most speciose clade of freshwater fishes, the Otophysi (Alfaro et al. 2009), a group diagnosed by the Weberian apparatus. This specialization of the vertebral column is a sound reception system consisting of a series of highly mobile ossicles linking the anterior chamber of the swim bladder to the inner ear, leading to hearing capabilities superior to those of other bony fishes (Bird and Hernandez 2007). An old description of the Weberian apparatus in one breed (Watson 1939) could be a starting reference from which to launch a study of variation in these structures, known to vary greatly among cypriniformes (Bird and Hernandez 2007).

The literature on fish morphological modification associated with domestication is related largely to aquaculture. The subject of a "domestication syndrome" in fishes has not been examined.

A study of farmed European sea bass (*Dicentrarchus labrax*) has argued that "early domesticates . . . , with no genetic differences with wild counterparts, contain epimutations in tissues with different embryonic origins" and that those epimutations coincide "with genes under positive selection in other domesticates" (Anastasiadi and Piferrer 2019, p. 2252). It is hypothesized, then, that the initial phase of domestication includes alterations in DNA methylation of developmental genes that affect the neural crest. As with testing for common patterns and mechanisms in the initial domestication in mammals and birds, there is a need for (1) comparative

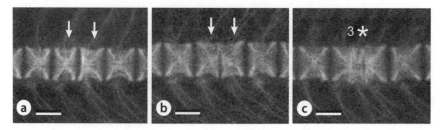

Fig 9.6. Vertebrae fusion in Atlantic salmon, *Salmo salar,* as documented by X-rays. White arrows (a, b) indicate the vertebrae prior to or in the process of fusion. The number 3 and the white star (c) indicate three fused vertebrae. Scale bars equal 500 µm."

studies of embryology, perhaps building on existing literature on domesticated forms (e.g., Nile tilapia *Oreochromis niloticus*: Fujimura and Okada 2007) and of general development (Ito et al. 2019); and (2) comparative examinations of the neural crest that not only address genomic changes in "neural crest genes" but also examine those changes in undomesticated species and consider the developmental aspect connecting the genotype with the phenotype (chapter 4).

Brain size changes

Several studies have shown how domestication affects behavioral traits in fishes, including social interactions, cognitive abilities, swimming capacities, and foraging (Pasquet 2018). This effect on behavior is due in great part to the environmental changes brought about by domestication, as both biotic and abiotic factors are known to affect fish behavior (Sommer-Trembo et al. 2017). Behavioral modifications are studied when considering welfare of fishes in farms. How brain size or anatomy is influenced by domestication, presumably in relation to behavior, has not been widely studied, but there is evidence that changes can be in some cases quite rapid. It is known for several fish species that a great degree of phenotypic plasticity exists in brain size (Kotrschal et al. 1998), but this plasticity does not characterize all species (Toli et al. 2017).

A comparative study of guppies (*Poecilia reticulata*) reported a large reduction in brain size in first generation, lab-reared female individuals compared with wild-caught ones (Burns et al. 2009). The laboratory guppies had a 19% smaller telencephalon and a 17% smaller optic tectum. Subsequent experiments manipulating the environment in lab-born guppies (e.g., varying in spatial complexity and size), however, did not show significant differences in phenotype (Burns et al. 2009).

In Chinook salmon (*Oncorhynchus tshawytscha*) from the same generation and genetic stock, hatchery-reared individuals had smaller brains, including both telencephalic lobes and olfactory bulbs, than wild-caught individuals (Kihslinger et al. 2006). The same study found no significant differences between conventional (i.e., unenriched) and enriched hatchery treatments. Another study, however, conducted on another salmon species had a contrasting result. A comparison of brain size between wild, juvenile steelhead salmon (*Oncorhynchus mykiss*) and fish from two hatchery treatments, one with small stones on the bottom and one without such stones, found that wild fish had larger total brain volumes than fish from both hatchery treatments (Kihslinger and Nevitt 2006). The same study also found that the cerebellum was larger in the wild and hatchery fish reared with a richer substrate of small stones, than in fish raised without stones.

Studying clonal rainbow trout lines (*Oncorhynchus mykiss*), it was found that total brain and olfactory volume are larger in domesticated lines than in wild-living ones (Campbell et al. 2015). A study of the same species also found smaller brains in hatchery than in wild reared stocks of rainbow trout (Marchetti and Nevitt 2003).

Variation in the observations from different studies is not unexpected, as surely the laboratory conditions are not all the same in all experiments, and each species has singular norms of reactions and responses in behavior and evolvability of the brain to abiotic and biotic factors.

A comparative study of the brains of 99 goldfish and 52 crucian carp found brain mass lower in the former, and weight of the vagal lobes prominently lower in the goldfish (Masai et al. 1982). The authors interpreted the result on the vagal lobe as indication of domestication-related changes in the primary gustatory centers associated with feeding behavior.

Life history and reproductive changes

Bony fishes are in general characterized by a higher rate of evolution and evolvability than other vertebrates (Rabosky et al. 2013), making them excellent subjects for intensive selection for specific life history or metabolic traits that are advantageous in managing conditions (Witten and Hall 2015). In a classic study involving translocation of natural populations to a site without predators, exceptional rates of evolution in life history and body size of guppies (*Poecilia reticulata*) were documented (Reznick et al. 1997).

Many examples of experiments concerning selection for traits in aquaculture have been published, as shown by studies on the Atlantic salmon (Table 9.1). What is particularly relevant about aquaculture research is the possibility of studying environmental effects and genetic changes simultaneously at a large scale, allowing investigation of epigenetics in a manner rarely possible in laboratory conditions (Harris et al. 2014). Among the variables usually considered in these studies are temperature, light regime, size, color, and shape of the tank, and composition, size, and shape of the food (Witten and Cancela 2018).

After just one generation of domestication in rainbow trout (*Oncorhynchus mykiss*), changes were reported in wound healing, immunity, and metabolism; these changes have been associated with genetic changes, and were interpreted as adaptations to highly crowded conditions (Christie et al. 2016). This is one of many examples of fast and significant changes produced by domestication. Wild rainbow trout reproduces only once a year. Applying selective breeding and modifying day length have enabled egg production virtually year-round (Fornshell 2002). This change has been accompanied by improvements in growth rate, disease resistance, and fecundity.

Table 9.1. Examples of studies of life history and metabolic changes driven by domestication in Atlantic salmon, *Salmo salar*. Domesticated Atlantic salmon and hybrids have altered growth patterns, life history traits, and phenology compared with wild salmon, as shown by common garden experiments (Skaala et al. 2019).

Population types	Subject	Reference
Wild and domestic	Effect of domestication on some life history traits, also with data on sea trout (*Salmo trutta*)	Petersson et al. 1996
Wild, feral, and hybrid	Genetic impact of escaped farmed salmon on survival, growth, migration, and parr maturity in a natural river	McGinnity et al. 1997
Wild, domestic, and hybrid	Comparison of size, maturation, and "quality traits" (e.g., fat content) after rearing in farming conditions	Glover et al. 2009
Wild, domestic, and hybrid	Quantification of trait differentiation between farmed and wild Atlantic salmon—hybridization data	Fraser et al. 2010
Wild, domestic, and hybrid	Comparison of survival, growth, and diet in a natural river environment	Skaala et al. 2012
Wild, domestic, and hybrid	Growth pattern under environmental stress	Solberg et al. 2013
Wild and domestic	Effects of domestication on parr maturity, growth, and vulnerability to predation	Debes & Hutchings 2014
Wild, domestic, and hybrid	Hatching time and alevin growth	Solberg et al. 2014
Wild and domestic	Plasticity of growth under different diets	Harvey et al. 2016
Wild and domestic	Transcriptomes of embryos, factors that influence heritability of gene expression.	Bicskei et al. 2016

Growth rates of many fishes can be greatly modified by growth hormone gene transgenesis, particularly if applied to wild strains as opposed to already modified, fast-growing domesticates. This change in growth rates has been shown in a comparison of gene expression in wild-type, domestic, and growth hormone (GH) transgenic strains of coho salmon (*Oncorhynchus kisutch*) (Devlin et al. 2009). Domestication and GH transgenesis apparently modify similar genetic pathways: nontransgenic strains of domesticated fish have higher muscle expression of GHs than wild populations, providing a possible mechanism for growth enhancement (Devlin et al. 2009).

Farming of monosex females has been developed for the rainbow trout based on production of sterile triploid individuals from pressure or thermal shock on eggs (Vandeputte et al. 2009). In this species, females mature later than males and have much better flesh quality. An advantage of this strategy is that if fish escape from rearing systems into the wild, they will not be able to reproduce (Fornshell 2002).

The increase in production of the Nile tilapia is tied to control of its reproduction in captivity. Monosex farming is conducted, including males only (Shelton and Rothbard 2006), which display a higher growth rate than females. Sex reversal is induced by hybridization, masculinizing hormones, or both combined (Lazard 2009).

Cultured fish are frequently released into natural environments, deliberately but also accidentally, where they may survive well. These releases may have negative impacts on wild populations, particularly where wild populations are already small and declining, and/or highly adapted to local conditions (Lorenzen et al. 2012).

Interbreeding between wild and domesticated animals occurs in several species of fishes. As a result of gene flow, genetic changes in life-history traits of wild populations occur, thereby influencing population dynamics and viability. Wild Atlantic salmon (*Salmo salar*) from 62 populations with high levels of domesticated ancestry (resulting from introgression) were reported to have altered age and size at maturation, including seven ancestral populations of breeding lines of the domesticated salmon (Bolstad et al. 2017). The abundance and continuity of escaped, domesticated fishes in the wild thus threatens wild populations by inducing genetic changes in fitness-related traits.

Fluctuating asymmetry

Fluctuating asymmetry has been used as a diagnostic tool for detecting the loss of genetic variation in cultured stocks of fish. An example is the study of some genetic loci and of five bilateral meristic traits in a hatchery stock of westslope cutthroat trout *Oncorhynchus clarkii lewisi*. This population of fish—characterized by reduced genetic variation at 35 isozyme loci compared with its wild counterpart and ancestor—had a high frequency of morphological deformities and an unusually high occurrence of asymmetry (Leary et al. 1985).

Fluctuating asymmetry has been reported as higher in interspecies hybrids (True and Haag 2001). Both meristic (number of gill rakers, pectoral and pelvic fin rays) and morphometric traits (length of head, opercular length, eye diameter, and length of maxilla) were studied in Atlantic salmon, *Salmo salar*, European trout, *Salmo trutta*, and hybrids between them. It was found that hybrids' fluctuating asymmetry is greater than that of wild populations in meristic traits but not in morphometric traits, in which hybrids show intermediate values (Wilkins et al. 1995).

Gene duplication and fish domestication

Gene duplication and divergence of function in regulatory regions of genes can result in viable morphological change, as has been studied in bony fishes (Santini et al. 2009). Gene duplication has resulted in zebrafish possessing two paralogs encoding the fibroblast growth factor receptor Fgfr1, leading to what has been interpreted as redundancy during embryogenesis, because only one of the paralogs is required for the formation of scales during juvenile development (Rohner et al. 2009). A study on the genetic basis of development of adult structures in zebrafish (Rohner et al. 2009) identified loss-of-function alleles changing the coding sequence of Fgfr1 that had been independently selected twice during the domestication of carp (Zhou et al. 2003). The *spiegeldanio* (*spd*) zebrafish mutation in the *fibroblast growth factor receptor 1* (*fgfr1*) gene leads to reduced scale formation in the adult.

The twin-tail morphology of some goldfish strains may have become possible by a previous genome duplication in the chordin genes (*chd*), via the subfunctionalized gene expression patterns of *chdA* and *chdB* (Abe et al. 2016). However, gene duplication alone does not explain the robustness of the developmental process that led to such a peculiar morphological novelty (Ota and Abe 2016). It is instructive to consider that a similar duplication of chordin genes occurred in the closely related common carp, but that no twin-tail morphology occurs in any strain of carp (Abe et al. 2016).

Duplication of the genome can occur through species hybridization. This can lead to allotetraploidization, leading to the possession of four sets of chromosomes derived from different species, as known for some amphibian species (Uno et al. 2013).

The genome of various common carp strains has been interpreted as showing that its genome duplication event is a case of allotetraploidization (Xu et al. 2014). It is claimed that allotetraploidization of the common carp and perhaps of other domesticated species may have been part of the process leading to or facilitating morphological diversification.

The subject of polyploidization in goldfish and the crucian carp *Carassius auratus* is not settled and requires further investigation. For the crucian carp diploid, triploid and tetraploid forms were reported (Qin et al. 2016).

The expression patterns of the subfunctionalized or neofunctionalized genes may have been modified, enabling morphological divergence. This has been hypothesized for goldfish based on analysis of transposable elements in its genome, assumed to be allotetraploid (Kon et al. 2020). Accumulated mutations in the asymmetrically evolved subgenomes may have led to the generation of diverse phenotypes such as dorsal fin loss, long-tail, telescope-eye, albinism, and twin-tail.

Summary: Variation in fish domestication

Fish domestication has increased in recent decades for diverse uses, such as food, as ornaments, and for research. There are currently more than 160 species of domesticated fish worldwide. The common carp, the goldfish, and salmonids are some of many examples of models of developmental evolution and evolvability.

Aquaculture research has revealed that rapid and diverse effects can happen in domesticated fish species, including gene-based changes in wound healing, immunity, and metabolism, life history changes, and changes in growth rate, disease resistance, and fecundity.

Aquaculture is associated with morphological changes, as exemplified by carp varieties showing either high-back or broadly elliptical or slender body shapes, probably a result of aesthetic selection. Integument selection has led to color diversity, as in the red strains of tilapias and the blue-back varieties of rainbow trout. Decreased sexual dimorphism after domestication can also result in morphological changes, such as a shorter kype in the domesticated Atlantic salmon. Spine deformities and fusion of vertebral bodies in the vertebral column of farmed fishes have been described, with cases of developmental plasticity via fused vertebrae remodeling.

Domesticated fish are frequently released into natural environments, with wild populations in many cases being negatively impacted when wild-domesticated interbreeding leading to genetic changes occurs .

Modern aquaculture will have to be improved to avoid further environmental degradation, loss of biodiversity, and damaging invasions typical of this trade. The rice–fish co-culture with carp and other species practiced in Asia is perhaps an example of sustainable farming.

The potential of domesticated fishes as research subjects of developmental evolution is immense, as they provide copious examples of regeneration, phenotypic plasticity, modularity, and life history change. Arguably the most remarkable morphological innovation brought up by domestication is the twin-tail of goldfishes, with this species also offering a rich and unexplored subject in its cranial evolution.

CHAPTER 10

Insect domestication

Humans have made many attempts to acclimatize insects to controlled environments, and to use insects to their advantage. Given the sheer diversity of insects and their potential for an ecological and ethical enrichment of human life, it is paradoxical that only one chapter of the book is devoted to them. This is a reflection of the relative lack of scientific literature on the domestication of insects compared with that for other animal groups. The future is likely to witness a major expansion of the already huge importance of insects as domestic, managed, and laboratory animals: silk moth, bees, and fruit flies, respectively, being major examples (Figure 10.1).

Different degrees of association and management have been defined to characterize the interactions of insects with humans (Lecocq 2018, p. 37): "Life cycle partially completed in man-controlled environments" (e.g., stingless bees); "Life cycle completed in man-controlled environments with significant gene flow from the wild" (e.g., *Dactylopius coccus, Kerria lacca*); "Life cycle completed in man-controlled environments without significant gene flow from the wild" (e.g., *Bombus terrestris, Acheta domesticus, Gromphadorhina portentosa*, whiteflies parasitoids); "Development of selective breeding programs or organisms engineering" (e.g., *Bombyx mori, Apis mellifera, Drosophila* spp.). Humans have thus developed different modes of interactions with insects, and some could lead to a domestication process.

In Japan, the human–wasp relationship has evolved rapidly due to local innovations in the past century, starting with wasp-keeping in hive boxes (earliest record from 1916) and a purpose-built house for multiple wasp hives (in 1990), followed by keeping new queens over the winter period in 1994 (Payne and Evans 2017). Current time-consuming practices by wasp collectors involve attracting worker wasps with pieces of meat or fish attached to a string with a marker for locating the nest, which is then dug out, transported and harvested for immediate consumption as cooked food, or preserved with soy sauce and mirin (Payne and Evans 2017).

Insects as food and uses for medicine

Insects, eaten regularly by more than 2 billion people, are a good source of protein, fat, vitamins, and minerals. They are stored and sold in many developing countries (Defoliart 1995; Mishra and Omkar 2017). Insects are considered cheaper and

Fig 10.1. Examples of domesticated insects. (a) silkmoth *Bombyx mori*, larva and moth. (b) fruit fly *Drosophila melanogaster*. (c) bumble bee *Bombus* sp. (d) honey bee *Apis mellifera*. (e) cochineal insect *Dactylopius coccus:* male left, female right. Not to scale.

more sustainable sources of protein for human consumption, or as parts of compound feeds for livestock, poultry, and aquaculture. Some species can biodegrade organic waste and transform it into high-quality, consumable biomass, and the most frequently reared edible insects emit lower levels of greenhouse gas emissions than conventional livestock (van Huis 2013).

Examples of commercially farmed insects for human consumption are the house cricket (*Acheta domesticus*), the palm weevil (*Rhynchophorus ferrugineus*), the giant water bug (*Lethocerus indicus*), and various species of water beetles (Coleoptera). In fact, more than 2,000 insect species are considered edible for humans, or used as animal feed (van Huis 2013).

The black soldier fly (*Hermetia illucens*) has received rapidly growing interest (van Huis 2020) for its ability to convert organic waste such as animal manure (Sheppard et al. 1994, 2002) or food waste (Nguyen et al. 2015) into high-quality protein and fat. Black soldier fly larvae can be used as animal feed for chickens or fish aquaculture (Sealey et al. 2011), or for human consumption, and for the production of biofuel (Li et al. 2011). Domestication of black soldier flies is in its early stages, but genetic manipulation has the potential to speed up the domestication process, for example with the creation of a wingless variant that may be easier to manage, or of enhanced feeding phenotypes (Zhan et al. 2020).

The giant waterbug *Lethocerus indicus* is an edible species common in several Southeast Asian cuisines. The taste of its flight muscles has been compared to sweet scallops and to shrimp. The larvae of the palm weevil *Rhynchophorus ferrugineus*, considered a delicacy in Vietnam, are usually eaten alive with fish sauce, or toasted or steamed with sticky rice and salad, or cooked with porridge.

Most of the drawings of insects in European Paleolithic art represent the larvae of warble flies (*Hypoderma*, maggots that live under the skin of reindeer), a delicacy among Arctic people today (Guthrie 2004).

Insects can also provide natural products for drugs that treat human diseases (Ratcliffe et al. 2011). Bees, wasps, flies, butterflies, moths, and cockroaches are commercially farmed for drug production. The following insect species have been both reared and bred (Price 2002a): giant roach (*Gomphadorhina* sp.), lac insect (*Laccifer* sp.), cochineal insect (*Dactylopius* sp.), parasitic wasp (*Aphytis* spp.), parasitic fly (*Eucelatoria* sp.), leaf beetles (*Chrysolina* sp.), dung beetle (*Onthophagus* spp.), carrion beetle (*Dermestes* sp.), and the fruit fly (*Drosophila* sp.). Efforts are under way to breed ghost moths (*Thitarodes* sp.) in captivity as hosts for *Ophiocordyceps sinensis,* a specialised entomopathogenic fungus highly prized in traditional Chinese medicine (Li et al. 2019b).

Cultural perception and insects as pets

Insects such as cicadas and crickets were kept as pets in ancient Greece and Rome, very likely in other places and times as well. Pet-keeping, though, is not the same as domestication, but it indicates a level of human-insect interaction that is rare in Western societies today. In contrast, some Asian cultures have close exposure to insects. Insects are intimately integrated in Japanese culture, particularly in children's education, including science class, summer festivals and books, toys, video games, manga cartoons, and animation films. Insects are kept as pets, in particular crickets, grasshoppers, and beetles. Live insects are available at beetle and butterfly houses (Kawahara 2007). Beetles are among the most popular insects in Japan, sold in pet stores and visited at the beetle "petting zoo" in Shizuoka. Beetles are bred and reared, including foreign imported species, they are sold with special food and even pheromone spray bottles for females that induce copulation behavior in males (Kawahara 2007).

Honey bee (*Apis mellifera*) — and other bees

Some authors do not consider bees domesticated, arguing that humans just obtain honey and other materials produced by bees, and that bees are not dependent on humans for their lives. Strictly seen, the reproduction of bees has generally not been affected; thus following some leading definitions of domestication (Vigne 2011), bees are not domestics. However, the management of bees has greatly influenced the extent of their reproduction, and their close association, the use of artificial insemination and managed hybridization argue for the inclusion of these wonderful animals in this book.

About 87% of angiosperm plants, including more than 70% of crop species, are pollinated by bees (Ollerton et al. 2011) that collect nectar from flowers and bring it to their hives, where it is used to make honey they feed from (Renauld 2016).

Honey bees live in permanent colonies with one reproductive female (the queen), many sterile female workers, and some males (drones) during the reproductive season. The nest is made from wax secreted from abdominal glands to form vertical hexagonal combs where brood are reared and food is stored for the dearth period caused by tropical drought or heavy rainfall and by cold in the temperate zones. Pollen collection is connected with brood rearing (Crane 1984). These two products, honey and beeswax, are used by humans and are the reason beekeepers have developed tools to take care of beehives and perform honey and wax extraction.

Humans interact with several honeybee species of the genus *Apis*. Examples include *A. florea* mostly in tropical Asia extending to southern Iran and Oman, where it has been domesticated but with a low yield, and *A. dorsata* with wild nests harvested in India (Crane 1984). *Apis mellifera*, the western honey bee or European honey bee, is the most common domesticated honeybee worldwide. Management of *Apis mellifera* has led to widespread admixture between different, previously isolated geographic populations, and to domestic honey bees with genomes more variable than those of their wild progenitors (Harpur et al. 2012). Artificial intelligence approaches (automated identification) and geometric morphometrics analysis, both of forewings, have been used to differentiate Africanized from European honeybees (Francoy et al 2008).

Early hominids probably used the Oldowan tool kit to open beehives, a hypothesis based on convergent evidence of honey and larvae consumption from nonhuman primates, historical and contemporary foraging populations, and artistic representations from Upper Paleolithic rock art (Crittenden 2011). This liquid honey harvesting may have provided energy dense food to early hominin foragers to supplement meat and plant food; some have even speculated that this activity may have favored the evolution of larger hominin brains and reduced size molars (Crittenden 2011).

Ancient Egyptian bee iconography dated to the Old Kingdom (approximately 2,400 BC) has been reported, and honey hunting by Stone Age people is suggested by rock art in a prehistoric Holocene context (Crane 1999). The timing of initial association of bees with humans has been estimated based on archaeological records containing beeswax lipid residues (e.g., pottery vessels of Neolithic Old World farmers), as these are very specific and provide a "chemical fingerprint" (Roffet-Salque 2015). Based on beeswax lipid analysis from pottery vessels across Neolithic Europe, the Middle East, and North Africa, bee product exploitation seems to have occurred continuously and extensively from the seventh millennium BCE, including in Neolithic farming communities and in some places for 8,000 years or more, therefore dating to the beginnings of agriculture or even earlier (Roffet-Salque et al. 2015; Figure 10.2). Beeswax was often detected in later archaeological periods in lipid extracts from unglazed pottery vessels (Mayyas et al. 2012), probably a residue of cooking honey or from vessels used for processing wax combs (Heron et al. 1994; Copley et al. 2005), as fuel in lamps, and in larger vessels used as protobeehives in Roman Greece (second century BCE to fourth century CE) (Evershed et al. 2003) and applied to waterproof pottery vessels (Salque et al. 2013).

Today, the honey bee *Apis mellifera* is of great economic importance and plays a critical role in agriculture across the globe by acting as a pollinator. Managed honeybee hives are used by farmers to ensure crop pollination, in particular in cases where wild bees do not visit agricultural fields (Klein et al. 2007). Some other bee species have also been managed for this purpose, including colonies of bumble bees (*Bombus* spp.) bred in captivity for greenhouse pollination (Velthuis and Van Doorn 2006), colonies of stingless bees (Meliponini), important pollinators in tropical countries

Fig 10.2. Early honey hunting. One of the earliest documenta-
tions of honey hunting is this depiction of humans collecting
honey from a wild honey bee nest in mesolithic rock art at Cue-
vas de la Araña in Valencia, Spain (Crane, 1999).

(Cortopassi-Laurino et al. 2006, Lecocq 2018), and some solitary bees, like the alfalfa leaf cutter bee (*Megachile rotundata*) in North America (Richards 1987), and
the red mason bee (*Osmia bicornis*) in Europe (Gruber et al. 2011). Arguably, bumble bees can be considered "more domesticated" than honey bees, as their reproduction is completely controlled in captivity in the case of production of colonies for
greenhouse pollination.

Man-made hives for bee colonies allow apiarists to collect honey and other bee
products by controlling colony reproduction and mating. There is a long history of
artificial insemination of the queen (Laidlaw 1944), which was reported as early as
1885, consisting of semen injection into the queen's vaginal opening (McLain 1885).

Environmental change, diseases, parasites, and pesticides affect honey bee
populations, a matter of study and concern (vanEngelsdorp and Meixner 2010).

One of many examples of a parasite causing significant mortality in honey bees is
Varroa destructor, a parasitic mite of honey bees originating from Asia and having
spread globally in the European honey bee (*Apis mellifera*) population through transport of honeybee hives and transmitting viral diseases, including Deformed Wing
Virus (Wilfert et al. 2016). It can be controlled with chemical treatments, as shown in
Europe and North America; however, use of this strategy prevents natural selection
for resistance (van Alphen and Fernhout 2020). Heritable traits in *A. mellifera* that
contribute to resistance against this parasite can be selected naturally upon *Varroa*
invasion, as shown by South African honeybee species living in large panmictic
populations (van Alphen and Fernhout 2020).

The role of the specific social bee microbiome in defense of managed bees against
their parasites has been studied (Koch and Schmid-Hempel 2011, Kwong and
Moran 2016). Decades of captive rearing of bumble bees may have come at the cost
of a loss of some bacterial taxa that confer protection against the gut parasite *Crithidia bombi* in wild bumble bees (Mockler et al. 2018), potentially rendering managed bumble bees more susceptible to infections. At the same time, *Snodgrassella
alvi*, a common gut bacterium of honey bees, has recently been genetically engineered to induce RNA interference (RNAi) responses in bees against *Varroa* mites,
to suppress the mites and the associated Deformed Wing Virus (Leonard et al.
2020).

Silkworm *Bombyx mori*

The best example of insect domestication is the silkworm, caterpillar of the moth *Bombyx mori* and main source of silk, which depends on humans to maintain tightly controlled conditions where it can reproduce indoors (Soumyaet al. 2017). Annual global silk production at one point reached 200,000 metric tons, an important industry that employs millions of people in rural and semirural areas, including reportedly 8 million in India (Chauhan and Tayal 2017). Furthermore, *B. mori* is an edible insect, a health food, a pet, and used in research because of its short life cycle and laboratory culture adaptation (van Huis et al. 2013; Nwibo et al. 2015).

Based on archaeological and molecular analyses, silkworm domestication has been timed to about 7,500 ybp from the Chinese extant wild silk moth *B. mandarina*, starting likely with a single event in a directed pathway and with exclusive Chinese silk production for millennia in spite of silk spreading to Eurasia (reviewed in Lecocq 2018). Although silk demand increased considerably in the eighteenth and nineteenth centuries, it declined afterward because of the emerging cotton industry and silkworm disease breakouts (Chauhan and Tayal 2017). The expansion of silkworm raising for silk or sericulture started only 2,000 years ago, initially to Korea and Japan and later via the Silk Road to Central Asia and Europe (Xiang et al. 2018).

Silkworm production is strictly human-controlled and scheduled, with caterpillars fed mulberry leaves and let climb on artificial support provided, where they spin their cocoons, which are collected (Chauhan and Tayal 2017). *B. mori* are either killed before metamorphosis or left alive to generate adults for breeding (Chauhan and Tayal 2017). In contrast to wild moth species such as *B. mandarina*, *B. mori* heavy adults cannot fly (Chauhan and Tayal 2017). Different breeding programs end environments/temperatures have led to more than 1,000 inbred strains of domesticated silkworms (Zanatta et al. 2009; Xiang et al. 2018).

Olfactory functions in adult antennae detect plant odorants in silkworm moths; however, domestication was shown to result in reduction in the number of olfactory sensilla in *B. mori* and affected a primary processing center of the brain (Bisch-Knaden et al. 2014).

Tarsal gustatory senses required for oviposition site choice have also been studied in both *B. mori* and its wild counterpart *B. mandarina*, both lacking mouthparts. The study of morphological features and response patterns of tarsal sensilla showed differences between the two species, suggested as a result of domestication (Takai et al 2018). This study showed that although both wild and domesticated female individuals had more tarsal sensilla than males, suggesting a role in females in oviposition site recognition, sensilla density was lower in *B. mori* females, and they showed lower or no sensitivity to water-soluble extract of mulberry leaf compared with *B. mandarina* females (Fig. 10.3).

Dactylopius coccus

Carmine has been used as a source of red for dyes, lake pigments, cosmetics, and food/pharmaceutical colorants (Cardon 2010). The red dye is produced from the scales of the cochineal insect *Dactylopius coccus,* a Hemiptera in (sub)tropical

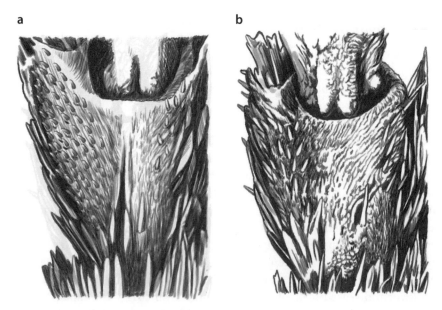

Fig 10.3. A morphological difference between wild and domesticated female individuals of the silkmoth. Tarsal sensilla in the fifth tarsomere of (a) wild and (b) domestic silkmoth, *Bombyx mandarina* and *B. mori*, respectively.

South and Central America, where it has been used to produce carmine since pre-Columbian times (Rodríguez et al. 2001). Another Hemiptera, *Porphyrophora* is also a source of red dye (Lecocq 2018). Cochineal farming is estimated to have begun in the tenth century CE in Mexico, with the earliest known cochineal-dyed textiles dating to the twelfth century (Van Dam et al. 2015).

Carmine was exported during the Spanish colonial period, and was later introduced in Australia, the Canary Islands, South Africa, and South Asia (Rodríguez et al. 2001). Its production fell sharply in the mid-nineteenth century when artificial red dyes became available. Although the cochineal trade declined considerably in the twentieth century, production restarted in the 1970s after carcinogenic and hazardous properties of synthesized dyes were revealed (Lecocq 2018).

The species *D. coccus* is an example of dependency on humans for propagation and protection (reviewed in Lecocq 2018). Wingless females are parasites of *Opuntia* cactus plants and are thus produced on cactus farms (Rodríguez et al. 2001) where they mate with winged males and produce nymphs which reach new host plants by wind. The infested cacti contain cochineals, some of which are harvested after 90 days (during this period humans control predators) and some are left for another reproduction cycle.

The laboratory fruit fly and mosquitoes

The laboratory fruit fly *Drosophila melanogaster* has been a model experimental organism used in genetics research since the early 1900s, and also in developmental as well as neurobiological research; its genome was the first major complex organism sequenced (Adams et al. 2000). As a model organism reared in the laboratory, it offers several advantages compared with mice, for example, lower maintenance cost and shorter life cycle. Using its different developmental stages (embryo, larva, pupa, and adult), research has focused on genetics and molecular biology underlying developmental processes. The adult fly brain is complex and regulates different behaviors, including sleep (Kempf et al. 2019). Vision has also been extensively studied using the fruit fly model, and it is also the subject of drug discovery for human disease studies (Pandey and Nichols 2011).

As a laboratory-model organism reared for prolonged periods, the fruit fly shows more docile, domesticated behaviors than wild *D. melanogaster*. There are unique laboratory-specific genome variants, which when analyzed suggest a rapid evolution of domesticated behavior via unintended early human selection on a broad pool of neurogenetic targets (Stanley and Kulathinal 2016).

The subject of phenotypic reversal to an ancestral condition given a changed selection environment has been addressed in many studies with feral mammals (chapter 9). It was also studied in *Drosophila melanogaster* by selecting over many generations in laboratory conditions for different reproductive and life history traits (Teotónio and Rose 2000), the latter so important in domestication (Vigne 2011). The four populations had the following regimes: selected for late-life fecundity, for resistance to starvation, for intermediate generation time, and for flies that developed into adults especially quickly. The flies from these four lines were returned to the original standard conditions of the laboratory. The monitored phenotypic traits in question were: male and female resistance to starvation; male and female development time; early fecundity at high and low population densities; and female lipid content and female dry body weight (Bull 2000). They found that the effects of captivity after about 100 generations in a particular environment were partially reversed in 20 generations, when the fruit flies were returned to their original environment. Most traits did revert, while some did not.

The experiments showed that reversals can occur, but neither for all traits nor at the same rate. Another case for the lack of universality of a pattern.

The highly influential twentieth-century geneticist T. Dobzhansky (1965) published a survey of *Drosophila* species recognized at the time and discussed what he called the wild versus domestic ones. He actually referred to the adaptability of some species to niches constructed by humans and how their geographic ranges had dramatically changed in some cases.

The interaction between mosquitoes, such as *Aedes albopictus*, and humans has been discussed as an example of the different kinds of animal–human interactions related to domestication (Dupé 2018). The dependence of a parasite on its host can make the latter create a management strategy (Budiansky 1992), such as sterilization. Mosquitoes can be used in strategies against mosquitoes. Laboratory-reared colonies of different *Anopheles* mosquito species connected to malaria have been used in diverse experiments. These include, for example, controlled human malaria infection to test malaria vaccine candidates (Arévalo-Herrera et al. 2016), to evaluate mosquitoes' gut-microbiota immunity to malaria parasites (Sharma et al. 2020), and to measure insecticidal effect on a mosquito malaria vector (Hancock et al. 2018).

Diseases and zoonoses

In the *Descent of Man*, Darwin (1871, p. 181) stated, "Humboldt saw in South America a parrot which was the sole living creature that could speak the language of a lost tribe." This tragic anecdote exemplifies the close connection between pets and domestic animals with human culture and human fate. The case of South America is particularly illustrative. After thousands of years of isolation from human populations in other regions of the planet, South Americans encountered Europeans who brought with them diseases to which they had no acquired immunity that decimated them (Crosby 1986). Hunter-gatherer groups in other parts of the world exposed to farmers faced a similar fate. In some areas of Europe and Asia, epidemic infectious diseases of social domestic animals spread to crowded farming populations, such as smallpox and measles. Farmers had evolved over many generations or acquired some resistance, enabling them to replace societies of hunter-gatherers they encountered when migrating (Diamond and Bellwood 2003). Domestication has historically brought new interactions between humans and animals that provide an expanded conduit for infection in both (Bartosiewicz and Gál 2013).

Zoonoses, defined as those diseases caused by a pathogen transmitted from a vertebrate animal to a human, are often asymptomatic and nonlethal in the natural reservoir host, but can be severe and potentially lethal in humans or other "spillover" hosts (Quammen 2012). These diseases come from a wide variety of animal species that act as reservoirs, both domesticated and wild, due in part to recent accelerated livestock production with expansion and intensification of farming (Recht et al. 2020).

Here I refer to a few of the many examples of zoonoses. Pigs are susceptible to infection by both mammalian and avian influenza viruses and can serve as "mixing vessels" in interspecies viral transmission, whereby novel strains are produced that can then infect humans (Zhang et al. 2020a). *Mycobacterium bovis*, the bovine tuberculosis agent found in cattle and other mammals including deer, can cause human (zoonotic) tuberculosis, which is transmitted mostly by consumption of unpasteurized milk and dairy products or during slaughter or hunting by wound contact. The incidence of this disease has decreased due to implementation of cattle control measures and routine pasteurization of cow's milk (reviewed in Recht et al. 2020). Chagas disease has also been tied to domesticated animals. Hunting dogs in some indigenous populations in Nicaragua were shown to have antibodies of *Trypanosoma cruzi* (the parasite that causes Chagas disease) and therefore at potential zoonotic risk for Chagas disease in these communities (Roegner et al. 2019). Highly infected populations of the insect vector of Chagas are found in pens of domestic cavies in Peru, raised as a human food source in the region (Levy et al. 2015).

Many important transmittable human diseases are animal diseases that have jumped species barriers, an example being bird flu (Peiris et al. 2004), which has also been introduced into other mammalian species, including swine, equines, dogs and cats (Kaplan and Webby 2013). It has been hypothesized that there is a causal correlation between the number of parasites and pathogens shared by humans and how long animals have been domesticated (Morand et al. 2014).

Myxomatosis, a disease caused by *Myxoma virus*, is an example of what occurs when a virus jumps from an adapted species to a naive host. The natural hosts are certain species of cottontail rabbit of the genus *Sylvilagus*, from the Americas. Whereas the myxoma virus causes only a mild disease in these species, it is the source of a usually fatal disease in European rabbits, *Oryctolagus cuniculus*. The virus was intentionally introduced in France, Chile, and Australia in the 1950s to control European rabbit populations, decimating populations but also creating a natural experiment in virulence evolution (Fenner and Fantini 1999). The newly evolved attenuated strains allowed rabbits to survive longer and came to dominate as they were more readily transmitted. Comparisons of the sequences of the protein-coding regions in the genome of rabbits collected before and after the pandemic showed that there is a polygenic basis of resistance (Alves et al. 2019). Furthermore, there is a strong pattern of parallel evolution of the same alleles in rabbit populations in Australia, France, and the United Kingdom.

The dispersal capacity of obligate ectoparasites including ticks, fleas, and lice that depend on their animal hosts for transport, and gut pathogens such as *Escherichia coli*, may have been affected by past extinction events (Doughty et al. 2020; Fig. 10.4). The decline in the Late Pleistocene/early Holocene of megafauna (animals greater than 44.5 kg or 98 lb body weight) leading to a decrease in dispersal of seeds and nutrients, may have also caused a reduction in the movement of ectoparasites and fecal microbes to ~15% of pre-extinction levels, with the largest declines occurring in the Americas and Eurasia. Following the different Neolithic transitions across the planet, humans and their domestic animals picked up the now diverged pathogens, leading to new dynamics and contacts and with that to new infectious diseases (Doughty et al. 2020).

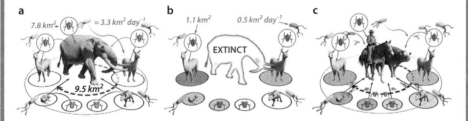

Fig 10.4. The effect of Pleistocene Megafauna Extinctions in pathogen dispersal and of domestication of large mammals. Illustrated are animal assemblages as hypothesized by Doughty et al. (2020). (a) South American Late Pleistocene assemblage with animal home ranges and tick-borne and fecal pathogens they host. Pathogen interactions were favored by the large home ranges of megafauna. (b) Early to middle Holocene assemblage is shown without the extinct megafauna, increasing immuno-naivety for all species. Diverging shades signify hypothetical speciation due to less megafauna-mediated interaction between pathogens. (c) Late Holocene assemblage including humans and their domestic animals picking up the now diverged pathogens which could cause infectious diseases in people and domestic animals. Numbers in (a) and (b) indicate the mean global dispersal distance for ectoparasites and fecal pathogens. Megafauna decline reduced pathogen dispersal, which may have increased emergent infectious diseases.

Summary: Variation in insect domestication

The many kinds of interactions of humans with insects span a broad range of intensity, from life cycles partially occurring in a human-controlled environment, to selective breeding programs that include genetic engineering. In different regions of the world the same species can be subject to various degrees of interactions. There is regional variation also in the species involved in practices that are similar in kind, as in managing of crop pollinators using bees, including not only the Western honey bee in Europe and adjacent areas but also bumble bees, stingless bees, and some solitary bees in other areas of the world. The silkworm moth is the classic example, and the oldest one, of intense domestication, one of great historical importance and biological consequences, resulting in more than 1,000 inbred strains.

Insect model organisms include the laboratory fruit fly, used in foundational work on genetics, also the first major complex organism whose genome was sequenced. Experiments with domesticated fruit flies involving life history and reproductive variables have shown variation in the reversibility of evolution—different rates of evolution or stasis in spite of changed selection environment. As in feralization in other animal groups, work on an insect species showed how some traits revert to an ancestral condition fully or partially, while others do not, and how this can occur at different rates. This variation in reversibility is another case of lack of universality and of variation associated with domestication and evolution.

Insects have been and are currently used by humans as important sources of medicine, as pets, and growingly so, as a food source. Insects are a more sustainable protein source than livestock in some areas and can become an alternative to livestock, perhaps helping to circumvent ecological and ethical issues raised by standard and current practices of animal production. Introduction of insects as a food source in Western society areas of the world may offer local options. Should consumption of insects transform into a large industry, it remains to be seen what new environmental or ethical challenges it will bring about.

Epilogue

A Culturesphere of domesticated and not domesticated animals

The human cultural web-of-life has been properly called the ethnosphere, the myriad of ways of being, languages, and myths (Davis 2001). The nature of humans' relation to animals is one dimension of the ethnosphere, and here human variation has found another realm of expression. Domesticated animals in their tremendous variety are the result of evolution resulting from some of those relations.

This book has explored the morphological responses or consequences of the domestication process across species and populations, and some of the mechanisms behind those patterns. The behavioral dimension of the response of animals to the environment, or simply their interaction, is barely treated here, although this aspect is fundamental to understanding the response of organism to their environment (Diogo 2017). Here, responses that concern knowledge and skills acquired through any mechanism of social learning in local populations—culture—can be relevant. A fresh and open look at the evidence gained from field and experimental studies of animal populations has shown that culture is not a unique human trait (Whitehead et al. 2019). Evidence comes not just from chimpanzees, whales, and parrots, but also from fruit flies (Danchin et al. 2018) and goats (Landau and Provenza 2020), among other species.

Comparative studies across species could reveal the evolution of culture and its many manifestations, rates of change, species repertoires, mechanisms of social transmission, and the details of gene-culture coevolution. As there is a biosphere and an ethnosphere, the biological web-of-life, there is also a culturesphere, the myriad of animal cultures. That manifestation of biodiversity is one aspect of domesticated and of wild animals we are only beginning to comprehend. Charting geographic behavioral variation among populations can be a starting point in studying the culturesphere. Intense studies of socially learned skills are needed for a better comprehension of diversity in animal cultures and their ecological context (Schuppli and van Schaik 2019). A great future for understanding of the culturesphere may reside in the analytical investigation of natural history observations of citizen scientists in their barns, and in data from studies in research centers of hundreds of populations of domesticated animals. This was suggested as a possibility in response to early reports of cultural transmission in cetaceans (Pryor 2001), and has become more evident in view of conceptual and empirical observations accumulated since then (Guillo and Claidière 2020).

Acknowledgments

When studying the evolution of life history patterns and development of extinct mammals from islands, I was struck by certain facts in the distribution of traits and their relation to selection regimes resembling those of domesticated forms. Developmental patterns of the animals I studied seemed to play a role. These facts seemed to me to merit thorough consideration from an organismal perspective of the domestication process—that process of such importance in human history and in the history of this planet.

I learned much from my dear Zurich colleagues in domestication matters, Madeleine Geiger, Marcus Clauss, and Lilo Meier. They greatly inspired this work and helped directly or indirectly with most of the chapters. For enlighting discussion or specific ideas or advice over the last years, I thank Allowen Evin, Greger Larson, Dieter Kruska, Richard A. Schneider, Carel van Schaik, and Adam Wilkins.

Exceptional colleagues were generous with valuable insights, information, and editorial help with many of the chapters: Miguel Carneiro (chapters 1–3), Allowen Evin (1–3), Michael Matschiner (1–6, 9), Gerardo Antonio Cordero (4–6), Daniel Núñez-León (3–4), Ana Balcarcel (7), Eckhard Witten (9), and Hauke Koch (10).

László Bartosiewicz, Lennart Olsson, Lisandro Milocco, Isaac Salazar-Ciudad, Ingmar Werneburg, Laura AB Wilson, Valentina Segura, David Flores, Mietje Germonpré, Caroline Weckerle, Nico Michiels, Torsten Scheyer, Hiroshi Nagashima, Evelyn Hüppi, Gerald Mayr, Kinya Ota, Matthias Starck, Robert Asher, Philip Gingerich, Chiara Barbieri, Javier Fernández, and Carolin Sommer-Trembo provided useful inputs on diverse topics.

In addition to those mentioned above I thank those former and current students and lab associates whose work informed much of what I present here: Christian Kolb, Evelyn Hüppi, Madlen Stange, Kristoff Veitschegger, Gabriel Aguirre-Fernández, Carlos Manuel Herrera, Sinéad Lynch, Laura Heck, Anita Schweizer, Jorge Carrillo-Briceño, Georgios Georgalis, and Anne-Claire Fabre.

The speakers at the 2017 Ringvorlesung at the University of Zurich, organized together with Marcus Clauss, informed many of the topics treated here, as did those participating in the symposium at the 2016 Euro Evo Devo conference in Uppsala, hosted together with Leif Andersson. I thank Pablo Teta and the Museo de Ciencias Naturales in Buenos Aires for access to the *vaca ñata* and to other specimens. I thank those providing permissions and proper figure versions, including Martin Fischer, Matthias Starck, and Eckhard Witten.

Gabriel Aguirre-Fernández generously solved many problems and creatively generated figures or coordinated the production of many of them, graciously performed by Diego A-F and Vera Primavera. I also thank Jaime Chirinos for his excellent artwork.

I thank Evelyn Hüppi, Jorge Carrillo-Briceño, Alexandra Wegmann, and Gabriel Aguirre-Fernández for translations, fact-checking, and technical support, and Sabine Schenk for much help in locating references. Alexandra Wegmann effectively coordinated the reference system, a gargantuan task. I enjoyed the effective and gracious administrative and IT support of Heike Götzmann and Heini Walter, respectively, as well as the camaraderie of the museum curator, Christian Klug. Coaching by Christian Bartsch helped me face challenges and Humbug at the workplace.

Judith Recht was a gracious and effective proofreader and scientific editor for the entire text, and she contributed much to summarizing literature on diseases as well as species accounts.

Alison A. Kalett at Princeton University Press kindly welcomed this project, and she and Whitney Rauenhorst coordinated many of the tasks needed to bring it to completion. Lucinda Treadwell thoroughly and constructively improved the manuscript at the copy-editing stage, and Ali Parrington guided the many steps of the production process.

I thank the Swiss National Fund SNF and the University of Zurich (SWF and Lehrkredit grants; Einrichtungskredit; URPP Evolution in Action symposium; MNF Faculty of Science) for financial support on much of my own research presented here.

The writing took place in the company of my dear partner, Aline Ramires, and while also developing and starting another wonderful and growing project.

References

Abbott R, et al. 2013. Hybridization and speciation. J Evol Biol 26:229–246.

Abe G, Lee SH, Chang M, et al. 2014. The origin of the bifurcated axial skeletal system in the twin-tail goldfish. Nat Commun 5:3360.

Abe G, Lee SH, Li J, et al. 2016. Open and closed evolutionary paths for drastic morphological changes, involving serial gene duplication, sub-functionalization, and selection. Sci Rep 6:26838.

Abraham K. 1901. Beiträge zur Entwicklungsgeschichte des Wellensittichs (*Melopsittacus undulatus*). Anat Hefte 17:590–669.

Abramyan J, Richman JM. 2018. Craniofacial development: discoveries made in the chicken embryo. Int J Dev Biol 62:97–107.

Acharya RM. 1982. Sheep and goat breeds of India. FAO Animal Health and Production Paper no. 30.

Acker DA, et al. 2001. Morphologic stages of the equine embryo proper on days 17 to 40 after ovulation. Am J Vet Res 62:1358–1364.

Ackermans NL. 2020. The history of mesowear: a review. PeerJ 8:e8519.

Ackermans NL, et al. 2019. Root growth compensates for molar wear in adult goats (*Capra aegagrus hircus*). J Exp Zool A 331:139–148.

Adams MD, et al. 2000. The genome sequence of *Drosophila melanogaster*. Science 287:2185–2195.

Adams JR, Leonard JA, Waits LP. 2003. Widespread occurrence of a domestic dog mitochondrial DNA haplotype in southeastern US coyotes. Mol Ecol 12:541–546.

Aeschbach M, Carrillo JD, Sánchez-Villagra MR. 2016. On the growth of the largest living rodent: postnatal skull and dental shape changes in capybara species (*Hydrochoerus* spp.). Mamm Biol 81:558–570.

Agassiz JLR. 1860. Review of On the Origin of Species. Am J Sci Arts 2:142–154.

Agnvall B, et al. 2012. Heritability and genetic correlations of fear-related behaviour in red junglefowl—possible implications for early domestication. PLoS ONE 7:e35162.

Agnvall B, et al. 2014. Red junglefowl (*Gallus gallus*) selected for low fear of humans are larger, more dominant and produce larger offspring. Animal 8:1498–1505.

Agnvall B, et al. 2015. Is domestication driven by reduced fear of humans? Boldness, metabolism and serotonin levels in divergently selected red junglefowl (*Gallus gallus*). Biol Lett 11:20150509.

Agnvall B, Bélteky J, Jensen P. 2017. Brain size is reduced by selection for tameness in red junglefowl—correlated effects in vital organs. Sci Rep 7:3306.

Agnvall B, et al. 2018. Is evolution of domestication driven by tameness? A selective review with focus on chickens. Appl Anim Behav Sci 205:227–233.

Ahola M, et al. 2018. On the scent of an animal skin: new evidence on Corded Ware mortuary practices in Northern Europe. Antiquity 92:118–131.

Ainsworth SJ, Stanley RL, Evans DJR. 2010. Developmental stages of the Japanese quail. J Anat 216:3–15.

Ajayi SS. 1975. Observations on the biology, domestication and reproductive performance of the African giant rat *Cricetomys gambianus* Waterhouse in Nigeria. Mammalia 39:343–364.

Albarella U, Dobney K, Rowley-Conwy P. 2009. Size and shape of the Eurasian wild boar (*Sus scrofa*), with a view to the reconstruction of its Holocene history. Environ Archaeol 14:103–136.

Alberch P. 1982a. The generative and regulatory roles of development in evolution. In: Environmental Adaptation and Evolution: A Theoretical and Empirical Approach (Mossakowski D, Roth G eds). G. Fischer-Verlag, 19–36.

Alberch P. 1982b. Developmental constraints in evolutionary processes. In: Evolution and Development (Bonner JT ed). Springer Verlag, 313–332.

Alberch P. 1985. Developmental constraints: why St. Bernards often have an extra digit and poodles never do. Am Nat 126:430–433.

Alberch P. 1989. The logic of monsters: evidence for internal constraint in development and evolution. Geobios 22:21–57.

Alberch P, Gale EA. 1983. Size dependence during the development of the amphibian foot. Colchicine-induced digital loss and reduction. Development 76:177–197.

Alberch P, et al. 1979. Size and shape in ontogeny and phylogeny. Paleobiology 5:296–317.

Albert FW, et al. 2009. Genetic architecture of tameness in a rat model of animal domestication. Genetics 182:541–554.

Albert FW, et al. 2012. A comparison of brain gene expression levels in domesticated and wild animals. PLoS Genet 8:e1002962.

Alberto FJ, et al. 2018. Convergent genomic signatures of domestication in sheep and goats. Nat Commun 9:813.

Albertsdóttir E, et al. 2011. Genetic analysis of 'breeding field test status' in Icelandic horses. J Anim Breed Genet 128:124–132.

Alfaro ME. et al. 2009. Nine exceptional radiations plus high turnover explain species diversity in jawed vertebrates. PNAS 106:!3410–13414.

Almathen F, et al. 2016. Ancient and modern DNA reveal dynamics of domestication and cross-continental dispersal of the dromedary. PNAS 113:6707–6712.

Almlöf J. 1961. On achondroplasia in the dog. Zentralbl Veterinaermed 8:43–56.

Al-Nasser A, et al. 2007. Overview of chicken taxonomy and domestication. Worlds Poult Sci J 63:285–300.

Al-Sagair O, ElMougy S. 2002. Post-natal development in the linear and tric morphometrics of the Camelidae skull. Anat Histol Embryol 31:232–236.

Alves JM, et al. 2015. Levels and patterns of genetic diversity and population structure in domestic rabbits. PLoS ONE 10:e0144687.

Alves JM, et al. 2019. Parallel adaptation of rabbit populations to myxoma virus. Science 363:1319–1326.

Ameen C, et al. 2017. A landmark-based approach for assessing the reliability of mandibular tooth crowding as a marker of dog domestication. J Archaeol Sci 85:41–50.

Ammann P, Kump E, Egloff N. 2012. Nutztierkompass. Wissenswertes über die ProSpecieRara-Rassen. ProSpecieRara, Binkfert Duck.

Anastasiadi D, Piferrer F. 2019. Epimutations in developmental genes underlie the onset of domestication in farmed European sea bass. Mol Biol Evol 36:2252–2264.

Ancel A, Liess S, Girard H. 1995. Embryonic development of the domestic guinea fowl (*Numida meleagris*). J Zool 235:621–634.

Anderson PW. 1972. More is different. Science 177:393–396.

Andersson L. 2010. Studying phenotypic evolution in domestic animals: a walk in the footsteps of Charles Darwin. In: Proceedings, Cold Spring Harbor Symposia on Quantitative Biology. Cold Spring Harbor Laboratory Press, 319–325.

Andersson L. 2016. Domestic animals as models for biomedical research. Ups J Med Sci 121:1–11.

Andersson L, Georges M. 2004. Domestic-animal genomics: deciphering the genetics of complex traits. Nat Rev Genet 5:202–212.

Andersson L, et al. 2012. Mutations in DMRT3 affect locomotion in horses and spinal circuit function in mice. Nature 488:642–646.

Ankorion Y, Moav R, Wohlfarth G. 1992. Bidirectional mass selection for body shape in common carp. Genet Sel Evol 24:43–52.

Anthony N, Emmerson D, Nestor K. 1991. Genetics of growth and reproduction in the turkey. 12. Results of long-term selection for increased 180-day egg production. Poult Sci 70:1314–1322.

Antipa G. 1909. Fauna Ichtiologica României. Inst. de Arte Grafice "Carol Göbl".

Antonelli A, et al. 2010. Absence of mammals and the evolution of New Zealand grasses. Proc R Soc B 278:695–701.

APA (American Poultry Association). 2015. The American Standard of Perfection. Andesite Press.

Araújo ICSd, et al. 2019. Macroscopic embryonic development of Guinea fowl compared to other domestic bird species. Rev Bras Zootec 48:e20190056.

Arbuckle BS. 2005. Experimental animal domestication and its application to the study of animal exploitation in prehistory. In: Proceedings of the 19th Conference of the International Council of Archaeozoology, August 2002. Durham, 18–33.

Arévalo-Herrera M, et al. 2016. Protective efficacy of *Plasmodium vivax* radiation-attenuated sporozoites in Colombian volunteers: a randomized controlled trial. PLoS Negl Trop Dis 10:e0005070.

Arisawa K, et al. 2006. Skeletal analysis and characterization of gene expression related to pattern formation in developing limbs of Japanese Silkie fowl. J Poult Sci 43:126–134.

Arnaout B, et al. 2021. Development of the chicken skull: a complement to the external staging table of Hamburger and Hamilton. Anat Rec.

Arnold P, Amson E, Fischer MS. 2017. Differential scaling patterns of vertebrae and the evolution of neck length in mammals. Evolution 71:1587–1599.

Arthur W. 2011. Evolution: A Developmental Approach. Wiley-Blackwell.

Ash A, et al. 2016. Regional differences in health, diet and weaning patterns amongst the first Neolithic farmers of central Europe. Sci Rep 6:29458.

Asher RJ, Lehmann T. 2008. Dental eruption in afrotherian mammals. BMC Biol 6:14.

Asher RJ, et al. 2011. Variability and constraint in the mammalian vertebral column. J Evol Biol 24:1080–1090.

Asher RJ, et al. 2017. Dental eruption and growth in Hyracoidea (Mammalia, Afrotheria). J Vert Paleontol 37:e1317638.

Asma ST. 2001. Stuffed Animals and Pickled Heads: The Culture and Evolution of Natural History Museums. Oxford University Press.

Axelsson E, et al. 2013. The genomic signature of dog domestication reveals adaptation to a starch-rich diet. Nature 495:360–364.

Baker C, Manwell C. 1981. "Fiercely feral": on the survival of domesticates without care from man. Z Tierzuecht Zuechtungsbiol 98:241–257.

Balasse M, et al. 2016. Wild, domestic and feral? Investigating the status of suids in the Romanian Gumelnița (5th mil. cal BC) with biogeochemistry and geometric morphometrics. J Anthropol Archaeol 42:27–36.

Balasse M, et al. 2018. Wild game or farm animal? Tracking human-pig relationships in ancient times through stable isotope analysis. In: Hybrid Communities: Biosocial Approaches to Domestication and other Trans-Species Relationships, (Stépanoff C, Vigne J-D eds). Routledge, 81–96.

Balcarcel AM, et al. 2021a. Singular patterns of skull shape and brain size change in the domestication of South American camelids. J Mammal 102:220–235.

Balcarcel AM, et al. 2021b. Intensive human contact correlates with smaller brains: differential brain size reduction in cattle types. Proc R Soc Lond B, 288:20210813.

Balcarcel et al. 2021c. The mammalian brain under domestication: discovering patterns after a century of old and new analyses. J Exp Zool B, in press.

Ballard WW. 1981. Morphogenetic movements and fate maps of vertebrates. Am Zool 21:391–399.

Ballari SA, Barrios-García MN. 2014. A review of wild boar *Sus scrofa* diet and factors affecting food selection in native and introduced ranges. Mammal Rev 44:124–134.

Balon EK. 1995. Origin and domestication of the wild carp, *Cyprinus carpio*: from Roman gourmets to the swimming flowers. Aquaculture 129:3–48.

Balon EK. 2004. About the oldest domesticates among fishes. J Fish Biol 65:1–27.

Baranowska Körberg I, et al. 2014. A simple repeat polymorphism in the MITF-M promoter is a key regulator of white spotting in dogs. PLoS ONE 9:e104363.

Baranowski P. 2017. Craniometric characteristics and cranial indices of Polish heath sheep rams. Int J Morphol 35:133–140.

Bar-On YM, Phillips R, Milo R. 2018. The biomass distribution on Earth. PNAS 115:6506–6511.

Barrickman NL, et al. 2008. Life history costs and benefits of encephalization: a comparative test using data from long-term studies of primates in the wild. J Hum Evol 54:568–590.

Barriga EH, et al. 2015. Animal models for studying neural crest development: is the mouse different? Development 142:1555–1560.

Barrios N, et al. 2016. Estudio descriptivo del color de manto y señas del perro Ovejero Magallánico, Chile. Arch Zootec 65:99–101.

Barrios N, et al. 2019. The Patagonian Sheepdog: Historical Perspective on a Herding Dog in Chile. Diversity 11:245.

Bartels T, et al. 2001. Intrakraniale Fettkörper bei Hausenten (*Anas platyrhynchos* f. dom.). Tierarztl Prax 29:384–390.

Bartley MM. 1992. Darwin and domestication: studies on inheritance. J Hist Biol 25:307–333.

Barton L, et al. 2009. Agricultural origins and the isotopic identity of domestication in northern China. PNAS 106:5523–5528.

Bartosiewicz L, Gál E. 2013. Shuffling Nags, Lame Ducks: The Archaeology of Animal Disease. Oxbow Books.

Bartosiewicz L. 2015. The ghost in the corridor . . . Some remarks on "Animal Secondary Products," edited by Haskel J. Greenfield. Germania 93:233–245.

Bartosiewicz L, Bonsall C. 2018. 'Herd' mentality. In: Social Dimensions of Food in the Prehistoric Balkans (Ivanova M, et al. eds). Oxbow Books.

Bastide P, et al. 2018. Phylogenetic comparative methods on phylogenetic networks with reticulations. Syst Biol 67:800–820.

Bateson W. 1894. Materials for the Study of Variation: Treated with Special Regard to Discontinuity in the Origin of Species. Macmillan and Company.

Bateson P. 1978. Sexual imprinting and optimal outbreeding. Nature 273:659–660.

Bateson P. 2002. William Bateson: a biologist ahead of his time. J Genet 81:49–58.

Beaudoin S, Barbet P, Bargy F. 2003. Developmental stages in the rabbit embryo: guidelines to choose an appropriate experimental model. Fetal Diagn Ther 18:422–427.

Behringer V, et al. 2014. Age-related changes in thyroid hormone levels of bonobos and chimpanzees indicate heterochrony in development. J Hum Evol 66:83–88.

Bell G. 2015. The Evolution of Life. Oxford University Press.

Bélteky J, et al. 2016. Domestication and tameness: brain gene expression in red junglefowl selected for less fear of humans suggests effects on reproduction and immunology. R Soc Open Sci 3:160033.

Bélteky J, et al. 2018. Epigenetics and early domestication: differences in hypothalamic DNA methylation between red junglefowl divergently selected for high or low fear of humans. Genet Sel Evol 50:13.

Belyaev DK, Borodin PM. 1982. The influence of stress on variation and its role in evolution. Biol Zentralbl 101:705–714.

Belyaev DK, Trut LN. 1975. Some genetic and endocrine effects of selection for domestication in silver foxes. In: The Wild Canids: Their Systematics, Behavioral Ecology, and Evolution. Van Nostrand Reinhold, 416–426.

Benecke N. 1994a. Archäozoologische Studien zur Entwicklung der Haustierhaltung in Mitteleuropa und Südskandinavien von den Anfängen bis zum ausgehenden Mittelalter. Akademie Verlag.

Benecke N. 1994b. Der Mensch und seine Haustiere. Die Geschichte einer jahrtausendealten Beziehung. Theiss.

Benítez-Burraco A, Kempe V. 2018. The emergence of modern languages: has human self-domestication optimized language transmission? Front Psychol 9:551.

Benítez-Burraco A, Lattanzi W, Murphy E. 2016. Language impairments in ASD resulting from a failed domestication of the human brain. Front Neurosci 10:373.

Benítez-Burraco A, Theofanopoulou C, Boeckx C. 2018. Globularization and domestication. Topoi 37:265–278.

Benito BM, et al. 2017. The ecological niche and distribution of Neanderthals during the Last Interglacial. J Biogeogr 44:51–61.

Benjamini Y, Hochberg Y. 1995. Controlling the false discovery rate: a practical and powerful approach to multiple testing. J R Stat Soc Series B 57.

Bennett DK. 1980. Stripes do not a zebra make, Part I: A cladistic analysis of *Equus*. Syst Biol 29:272–287.

Bennett PM, Harvey PH. 1985. Relative brain size and ecology in birds. J Zool 207:151–169.

Bennett CE, et al. 2018. The broiler chicken as a signal of a human reconfigured biosphere. R Soc Open Sci 5:180325.

Berge S. 1948. Genetical researches on the number of vertebrae in the pig. J Anim Sci 7:233–238.

Bergström A, et al. 2020. Origins and genetic legacy of prehistoric dogs. Science 370:557–564.

Bertin A, et al. 2018. Incubation temperature affects the expression of young precocial birds' fear-related behaviours and neuroendocrine correlates. Sci Rep 8:1857.

Betteridge KJ, et al. 1982. Development of horse embryos up to twenty two days after ovulation: observations on fresh specimens. J Anat 135:191.

Bicskei B, et al. 2016. Comparing the transcriptomes of embryos from domesticated and wild Atlantic salmon (*Salmo salar* L.) stocks and examining factors that influence heritability of gene expression. Genet Sel Evol 48:20.

Billardon-Sauvigny E. 1780. Histoire naturelle des dorades de la Chine. Billardon de Sauvignon.

Billet G, et al. 2017. The hidden anatomy of paranasal sinuses reveals biogeographically distinct morphotypes in the nine-banded armadillo (*Dasypus novemcinctus*). PeerJ 5:e3593.

Bininda-Emonds ORP, Jeffery JE, Richardson MK. 2003. Inverting the hourglass: quantitative evidence against the phylotypic stage in vertebrate development. Proc R Soc Lond B 270:341–346.

Bird NC, Hernandez LP. 2007. Morphological variation in the Weberian apparatus of Cypriniformes. J Morphol 268:739–757.

Bird DJ, et al. 2018. Olfaction written in bone: cribriform plate size parallels olfactory receptor gene repertoires in Mammalia. Proc R Soc Lond B 285:20180100.

Birkhead T. 2003. The Red Canary: The Story of the First Genetically Engineered Animal. Orion.

Bisch-Knaden S, et al. 2014. Anatomical and functional analysis of domestication effects on the olfactory system of the silkmoth *Bombyx mori*. Proc R Soc Lond B 281:20132582.

Bischoff TLW. 1842. Entwicklungsgeschichte des Kaninchen-Eies. Vieweg und Sohn.

Bischoff TLW. 1845. Entwicklungsgeschichte des Hunde-Eies. Vieweg und Sohn.

Björnerfeldt S, Webster MT, Vilà C. 2006. Relaxation of selective constraint on dog mitochondrial DNA following domestication. Genome Res 16:990–994.

Blandford P. 1987. Biology of the polecat *Mustela putorius*: a literature review. Mammal Rev 17:155–198.

Blumberg MS. 2009. Freaks of Nature: What Anomalies Tell Us about Development and Evolution. Oxford University Press.

Bocherens H, et al. 2015. Reconstruction of the Gravettian food-web at Předmostí I using multi-isotopic tracking (^{13}C, ^{15}N, ^{34}S) of bone collagen. Quat Int 359–360:211–228.

Bochno R, Brzozowski W, Murawska D. 2005. Age-related changes in the distribution of lean, fat with skin and bones in duck carcasses. Br Poult Sci 46:199–203.

Bochno R, Murawska D, Brzostowska U. 2006. Age-related changes in the distribution of lean fat with skin and bones in goose carcasses. Poult Sci 85:1987–1991.

Boer EF, et al. 2019. Pigeon foot feathering reveals conserved limb identity networks. Dev Biol 454:128–144.

Boessneck J, von den Driesch A. 1978. The significance of measuring animal bones from archaeological sites. Approaches to Faunal Analysis in the Middle East 2:25–39.

Bohlken H. 1961. Haustiere und Zoologische Systematik. Z Tierzuecht Zuechtungsbiol 76:107–113.

Böhmer C. 2017. Correlation between Hox code and vertebral morphology in the mouse: towards a universal model for Synapsida. Zool Lett 3:1–11.

Böhmer C, Böhmer E. 2017. Shape variation in the craniomandibular system and prevalence of dental problems in domestic rabbits: a case study in evolutionary veterinary science. Vet Sci 4:5.

Bohren B, Siegel P. 1975. Light effects during incubation on lines of White Leghorns selected for fast and slow hatching. Poult Sci 54:1372–1374.

Boitani L, Ciucci P, Ortolani A. 2007. Behaviour and social ecology of free-ranging dogs. In: The Behavioural Biology of Dogs (Jensen P ed). CAB International, 147–165.

Bolstad GH, et al. 2017. Gene flow from domesticated escapes alters the life history of wild Atlantic salmon. Nat Ecol Evol 1:0124.

Bonnet R. 1897. Beiträge zur Embryologie des Hundes. Anat Hefte 9:419–512.

Bonnet R. 1901. Beiträge zur Embryologie des Hundes. Erste Fortsetzung. Anat Hefte 16:231–332.

Bonnet R. 1902. Beiträge zur Embryologie des Hundes. Zweite Fortsetzung. Anat Hefte 64/65: 323–499.

Bonser RHC, Casinos A. 2003. Regional variation in cortical bone properties from broiler fowl—a first look. Br Poult Sci 44:350–354.

Bonsmann A, et al. 2016. Anatomical atlas of the quail's ear (*Coturnix coturnix*). Anat Histol Embryol 45:399–404.

Borchers N, Reinsch N, Kalm E. 2004. The number of ribs and vertebrae in a Piétrain cross: variation, heritability and effects on performance traits. J Anim Breed Genet 121:392–403.

Bostock J, et al. 2010. Aquaculture: global status and trends. Philos Trans R Soc B 365:2897–2912.

Botelho JF, et al. 2014. The developmental origin of zygodactyl feet and its possible loss in the evolution of Passeriformes. Proc R Soc Lond B 281:20140765.

Botelho JF, et al. 2015a. Skeletal plasticity in response to embryonic muscular activity underlies the development and evolution of the perching digit of birds. Sci Rep 5:9840.

Botelho JF, Smith-Paredes D, Vargas AO. 2015b. Altriciality and the evolution of toe orientation in birds. Evol Biol 42:502–510.

Botigué LR, et al. 2017. Ancient European dog genomes reveal continuity since the Early Neolithic. Nat Commun 8:16082.

Bottema S. 1989. Some observations on modern domestication processes. In: The Walking Larder: Patterns of Domestication, Pastoralism, and Predation (Clutton-Brock J ed). Routledge, 31–45.

Bowler PJ. 1992. The Eclipse of Darwinism: Anti-Darwinian Evolution Theories in the Decades around 1900. Johns Hopkins University Press.

Bowles S. 2011. Cultivation of cereals by the first farmers was not more productive than foraging. PNAS 108:4760–4765.

Boyer CC. 1953. Chronology of development for the golden hamster. J Morphol 92:1–37.

Boyko AR, et al. 2010. A simple genetic architecture underlies morphological variation in dogs. PLoS Biol 8:e1000451.

Bradford GE, Spearow JL, Hanrahan JP. 1991. Genetic variation and improvement in reproduction. In: Reproduction in Domestic Animals, 4th ed. (Cupps P, ed). Elsevier, 605–636.

Bradley BJ, Lawler RR. 2011. Linking genotypes, phenotypes, and fitness in wild primate populations. Evol Anthropol 20:104–119.

Brander P, Perentes E. 1995. Intracranial lipoma in a laboratory rat. Vet Pathol 32:65–67.

Brianza SZM, et al. 2006. Cross-sectional geometrical properties of distal radius and ulna in large, medium and toy breed dogs. J Biomech 39:302–311.

Bridault A, et al. 2000. Wild boar: age at death estimates. The relevance of new modern data for archaeological skeletal material. 1. Presentation on the corpus. Dental and epiphyseal fusion ages. Anthropozoologica 31:11–18.

Bright JA, et al. 2016. The shapes of bird beaks are highly controlled by nondietary factors. PNAS 113:5352–5357.

Bristol F. 2000. Manual of equine reproduction. Can Vet J 41:332.

Brocal J, et al. 2018. C7 vertebra homeotic transformation in domestic dogs—are Pug dogs breaking mammalian evolutionary constraints? J Anat 233:255–265.

Brock DA, et al. 2011. Primitive agriculture in a social amoeba. Nature 469:393.

Brody S. 1945. Bioenergetics and Growth with Special Reference to the Efficiency Complex in Domestic Animals. First published: Reinhold. Reprinted: Hafner, 1964.

Bromham L. 2009. Why do species vary in their rate of molecular evolution? Biol Lett 5:401–404.

Bronchain OJ, et al. 2017. Implication of thyroid hormone signaling in neural crest cells migration: evidence from thyroid hormone receptor beta knockdown and NH3 antagonist studies. Mol Cell Endocrinol 439:233–246.

Bronson RT. 1979. Brain weight-body weight scaling in breeds of dogs and cats. Brain Behav Evol 16:227–236.

Brookman P, et al. 2010. Helios: Eadweard Muybridge in a Time of Change. Steidl.

Brooks SA, et al. 2010. Morphological variation in the horse: defining complex traits of body size and shape. Anim Genet 41:159–165.

Brothwell D. 1979. Roman evidence of a crested form of domestic fowl, as indicated by a skull showing associated cerebral hernia. J Archaeol Sci 6:291–293.

Brown WH, Pearce L. 1945. Hereditary achondroplasia in the rabbit: I. Physical appearance and general features. J Exp Med 82:241–260.

Browne J. 1995. Charles Darwin: Voyaging. Volume 1. Alfred Knopf.

Brüne M. 2007. On human self-domestication, psychiatry, and eugenics. Philos Ethics Humanit Med 2:21.

Bryden MM, Evans HE, Binns W. 1972a. Embryology of the sheep. I. Extraembryonic membranes and the development of body form. J Morphol 138:169–185

Bryden MM, Evans HE, Binns W. 1972b. Embryology of the sheep. II. The alimentary tract and associated glands. J Morphol 138:187–205.

Bryden MM, Evans H, Binns W. 1973. Embryology of the sheep. III. The respiratory system, mesenteries and celom in the fourteen to thirty-four day embryo. Anat Rec 175:725–735.

Buchholtz EA, et al. 2012. Fixed cervical count and the origin of the mammalian diaphragm. Evol Dev 14:399–411.

Buckingham KJ, et al. 2013. Multiple mutant T alleles cause haploinsufficiency of Brachyury and short tails in Manx cats. Mamm Genome 24:400–408.

Budiansky S. 1992. The Covenant of the Wild: Why Animals Chose Domestication. William Morrow & Company.

Buffon GLL, Daubenton LJM. 1755. Histoire Naturelle, générale et particulière, avec la description du Cabinet du Roi. Imprimerie Royale.

Bull J. 2000. Déjà vu. Nature 408:416–417.

Bull G, Payne S. 1982. Tooth eruption and epiphysial fusion in pigs and wild boar. In: Ageing and sexing animal bones from archaeological sites, (Wilson B, Grigson C, Payne S eds), 55–71.

Bunbury N, et al. 2018. Late stage dynamics of a successful feral goat eradication from the UNESCO World Heritage site of Aldabra Atoll, Seychelles. Biol Invasions 20:1735–1747.

Burger PA. 2016. The history of Old World camelids in the light of molecular genetics. Trop Anim Health Prod 48:905–913.

Burger IH, Johnson JV. 1991. Dogs large and small: the allometry of energy requirements within a single species. J Nutr 121:S18-S21.

Burk RF, et al. 2006. A combined deficiency of vitamins E and C causes severe central nervous system damage in guinea pigs. J Nutr 136:1576–1581.

Burns JG, Saravanan A, Helen Rodd F. 2009. Rearing environment affects the brain size of guppies: lab-reared guppies have smaller brains than wild-caught guppies. Ethology 115:122–133.

Burt DW. 2007. Emergence of the chicken as a model organism: implications for agriculture and biology. Poult Sci 86:1460–1471.

Byrom A, et al. 2015. Feral ferrets (*Mustela furo*) as hosts and sentinels of tuberculosis in New Zealand. N Z Vet J 63:sup1:42–53.

Cadieu E, et al. 2009. Coat variation in the domestic dog is governed by variants in three genes. Science 326:150–153.

Caldwell PJ, Snart AE. 1974. A photographic index for aging mallard embryos. J Wildl Manage 38:298–301.

Callaway E. 2016. When chickens go wild. Nature 529:270–274.

Cambra-Moo O, et al. 2015. Multidisciplinary characterization of the long-bone cortex growth patterns through sheep's ontogeny. J Struct Biol 191:1–9.

Campbell KH, et al. 1996. Sheep cloned by nuclear transfer from a cultured cell line. Nature 380:64–66.

Campbell JM, et al. 2015. Aggressive behavior, brain size and domestication in clonal rainbow trout lines. Behav Genet 45:245–254.

Cardini A, Polly PD. 2013. Larger mammals have longer faces because of size-related constraints on skull form. Nat Commun 4:2458.

Cardon D. 2010. Natural Dyes—Sources, Tradition, Technology and Science. Archetype Books.

Cardozo A. 1975. Origen y filogenia de los camélidos sudamericanos. Academia Nacional de Ciencias de Bolivia.

Carlborg Ö, et al. 2003. A global search reveals epistatic interaction between QTL for early growth in the chicken. Genome Res 13:413–421.

Carlisle A, et al. 2017. Testing hypotheses of developmental constraints on mammalian brain partition evolution, using marsupials. Sci Rep 7:4241.

Carlson KJ, Marchi D. 2014. Introduction: towards refining the concept of mobility. In: Reconstructing Mobility: Environmental, Behavioral, and Morphological Determinants (Carlson KJ, Marchi D eds). Springer, 1–11.

Carneiro M, et al. 2011. The genetic structure of domestic rabbits. Mol Biol Evol 28:1801–1816.

Carneiro M, et al. 2014. Rabbit genome analysis reveals a polygenic basis for phenotypic change during domestication. Science 345:1074–1079.

Carneiro M, et al. 2017. Dwarfism and altered craniofacial development in rabbits is caused by a 12.1 kb deletion at the HMGA2 locus. Genetics 205:955–965.

Carrera I, et al. 2009. Use of magnetic resonance imaging for morphometric analysis of the caudal cranial fossa in Cavalier King Charles Spaniels. Am J Vet Res 70:340–345.

Carrier D. 2002. Functional tradeoffs in specialization for fighting versus running. In: Topics in Functional and Ecological Vertebrate Morphology (Aerts P, et al. eds). Shaker Publishing, 237–255.

Carril J, Tambussi CP. 2017. Skeletogenesis of *Myiopsitta monachus* (Psittaciformes) and sequence heterochronies in Aves. Evol Dev 19:17–28.

Carroll SB. 2005. Endless Forms Most Beautiful: The New Science of Evo-Devo. WW Norton & Company.

Casinos A, et al. 1986. On the allometry of long bones in dogs (*Canis familiaris*). J Morphol 190:73–79.

Castle W. 1906. The origin of a polydactylous race of guinea-pigs. Carnegie Institution of Washington.

Castle W. 1947. The domestication of the rat. PNAS 33:109.

Ceballos G, et al. 2015. Accelerated modern human–induced species losses: entering the sixth mass extinction. Sci Adv 1:e1400253.

Cesarini L, et al. 2009. Cognitive profile of disorders associated with dysregulation of the RAS/MAPK signaling cascade. Am J Med Genet A 149A:140–146.

Chamoli U, Wroe S. 2011. Allometry in the distribution of material properties and geometry of the felid skull: why larger species may need to change and how they may achieve it. J Theor Biol 283:217–226.

Chang G, et al. 2009. Behavior differentiation between wild Japanese quail, domestic quail, and their first filial generation. Poult Sci 88:1137–1142.

Chang J, et al. 2019. An R package and online resource for macroevolutionary studies using the ray-finned fish tree of life. Methods Ecol Evol 10:1118–1124.

Charlesworth B. 2009. Effective population size and patterns of molecular evolution and variation. Nat Rev Genet 10:195–205.

Charvet CJ, Striedter GF. 2011. Developmental modes and developmental mechanisms can channel brain evolution. Front Neuroanat 5:4.

Chauhan TPS, Tayal MK. 2017. Mulberry sericulture. In: Industrial Entomology (Omkar ed). Springer, 197–263.

Chawla S, Reece J. 2002. Timing of oestrus and reproductive behaviour in Indian street dogs. Vet Rec 150:450–451.

Chen SC. 1956. A history of the domestication and the factors of the varietal formation of the common goldfish, *Carassius auratus*. Sci Sin 5:287–321.

Chen S-Y, et al. 2005. Mitochondrial diversity and phylogeographic structure of Chinese domestic goats. Mol Phylogenet Evol 37:804–814.

Chen J, et al. 2018. Population size may shape the accumulation of functional mutations following domestication. BMC Evol Biol 18:4.

Chen D, et al. 2020. The evolutionary origin and domestication history of goldfish (*Carassius auratus*). PNAS 117: 29775–29785.

Cherubin RC, et al. 2019. Feral horse impacts on threatened plants and animals in sub-alpine and montane environments in Victoria, Australia. Ecol Manag Restor 20:47–56.

Chessa B, et al. 2009. Revealing the history of sheep domestication using retrovirus integrations. Science 324:532–536.

Childe VG. 1928. The most ancient East: the Oriental prelude to European prehistory. Kegan Paul, Trench, Trubner & Co.

Childe VG. 1936. Man Makes Himself. Watts & Co.

Chinsamy-Turan A. 2005. The Microstructure of Dinosaur Bone: Deciphering Biology with Fine Scale Techniques. Johns Hopkins University Press.

Chistiakov DA, Voronova NV. 2009. Genetic evolution and diversity of common carp *Cyprinus carpio* L. Cent Eur J Biol 4:304–312.

Christie MR, et al. 2016. A single generation of domestication heritably alters the expression of hundreds of genes. Nat Commun 7:10676.

Cieri RL, et al. 2014. Craniofacial feminization, social tolerance, and the origins of behavioral modernity. Curr Anthropol 55:419–433.

Cieslak M, et al. 2011. Colours of domestication. Biol Rev 86:885–899.

Clark BR, Price EO. 1981. Sexual maturation and fecundity of wild and domestic Norway rats (*Rattus norvegicus*). J Reprod Fertil 63:215–220.

Clauss M, et al. 2014. Low scaling of a life history variable: analysing eutherian gestation periods with and without phylogeny-informed statistics. Mamm Biol 79:9–16.

Clayton G. 1972. Effects of selection on reproduction in avian species. J Reprod Fertil Suppl. 15:1–21.

Clayton G, Robertson A. 1957. An experimental check on quantitative genetical theory II. The long-term effects of selection. J Genet 55:152–170.

Clayton G. 1984a. Muscovy duck. In: Evolution of Domesticated Animals (Mason IL ed). Longman, 340–344.

Clayton G. 1984b. Common duck. In: Evolution of Domesticated Animals (Mason IL ed). Longman, 334–339.

Clutton-Brock J. 1987. A Natural History of Domesticated Mammals. Cambridge University Press.

Clutton-Brock J. 1999. A Natural History of Domesticated Mammals. 2nd ed. Cambridge University Press.

Clutton-Brock J. 2012. Animals as Domesticates: A World View Through History. Michigan State University Press.

Cnotka J. 2012. Haubenenten im Focus morphometrischer und ethologischer Untersuchungen. Eine Studie mit Aspekten zur Neuroanatomie, Domestikation und zum Tierschutz. Akademiker Verlag.

Cole RJ, et al. 2012. Vegetation recovery 16 years after feral pig removal from a wet Hawaiian forest. Biotropica 44:463–471.

Colihueque N, et al. 2011. Skin color characterization in rainbow trout by use of computer-based image analysis. N Am J Aquac 73:249–258.

Colihueque N, Araneda C. 2014. Appearance traits in fish farming: progress from classical genetics to genomics, providing insight into current and potential genetic improvement. Front Genet 5:251.

Colli L, et al. 2018a. Genome-wide SNP profiling of worldwide goat populations reveals strong partitioning of diversity and highlights post-domestication migration routes. Genet Sel Evol 50:58.

Colli L, et al. 2018b. New insights on water buffalo genomic diversity and post-domestication migration routes from medium density SNP chip data. Front Genet 2:53.

Colten R, et al. 2017. Were Hutia Domesticated in the Caribbean? In: Proceedings of the 81st Annual Meeting of the Society for American Archaeology. Vancouver, British Columbia.

Coltman DW, et al. 2003. Undesirable evolutionary consequences of trophy hunting. Nature 426:655–658.

Concannon PW. 2000. Canine pregnancy: predicting parturition and timing events of gestation. In: Recent Advances in Small Animal Reproduction (Concannon PW, et al. eds). International Veterinary Information Service (www.ivis.org).

Condro MC, White SA. 2014. Recent advances in the genetics of vocal learning. Comp Cogn Behav Rev 9:75–98.

Connor JL. 1975. Genetic mechanisms controlling the domestication of a wild house mouse population (*Mus musculus* L.). J Comp Physiol Psychol 89:118–130.

Conway Morris S. 2003. Life's Solution: Inevitable Humans in a Lonely Universe. Cambridge University Press.

Cooper KL, et al. 2014. Patterning and post-patterning modes of evolutionary digit loss in mammals. Nature 511:41–45.

Copley MS, et al. 2005. Dairying in antiquity. III. Evidence from absorbed lipid residues dating to the British Neolithic. J Archaeol Sci 32:523–546.

Coppinger L, Coppinger RP. 1982. Livestock-guarding dogs that wear sheep's clothing. Smithsonian 13:64–73.

Coppinger R, et al. 1987. Degree of behavioral neoteny differentiates canid polymorphs. Ethology 75:89–108.

Corbett S, et al. 2018. The transition to modernity and chronic disease: mismatch and natural selection. Nat Rev Genet 19:419–430.

Cordero GA, Berns CM. 2016. A test of Darwin's 'lop-eared' rabbit hypothesis. J Evol Biol 29:2102–2110.

Cordero GA, Sánchez-Villagra MR, Werneburg I. 2020. An irregular hourglass pattern describes the tempo of phenotypic development in placental mammal evolution. Biol Lett 16:20200087.

Cornevin C. 1895. Traité de Zootechnie Spéciale: les oiseaux de Basse-cour. J.-B. Baillière et fils.

Cortopassi-Laurino M, et al. 2006. Global meliponiculture: challenges and opportunities. Apidologie 37:275–292.

Costford SR, et al. 2007. Gain-of-function R225W mutation in human AMPKγ3 causing increased glycogen and decreased triglyceride in skeletal muscle. PLoS ONE 2:e903.

Côté SD, Festa-Bianchet M. 2001. Life-history correlates of horn asymmetry in mountain goats. J Mammal 82:389–400.

Cotti S, et al. 2020. More bone with less minerals? The effects of dietary phosphorus on the post-cranial skeleton in zebrafish. Int J Mol Sci 21:5429.

Crane E. 1984. Honeybees. In: Evolution of Domesticated Animals (Mason IL ed). Longman, 403–415.

Crane E. 1999. The World History of Beekeeping and Honey Hunting. Routledge.

Craven BA, et al. 2007. Reconstruction and morphometric analysis of the nasal airway of the dog (*Canis familiaris*) and implications regarding olfactory airflow. Anat Rec 290:1325–1340.

Crawford RD. 1984. Turkey. In: Evolution of Domesticated Animals (Mason IL ed). Longman.

Crawford RD. 1990. Poultry Breeding and Genetics. Elsevier.

Crawford S, et al. 2010. Naturally occurring R225W mutation of the gene encoding AMP-activated protein kinase (AMPK)γ3 results in increased oxidative capacity and glucose uptake in human primary myotubes. Diabetologia 53:1986–1997.

Cressey D. 2009. Aquaculture: future fish. Nature 458:398–400.

Crispo M, et al. 2015. Efficient generation of myostatin knock-out sheep using CRISPR/Cas9 technology and microinjection into zygotes. PLoS ONE 10:e0136690.

Crittenden AN. 2011. The importance of honey consumption in human evolution. Food Foodways 19:257–273.

Crockford SJ. 2002. Animal domestication and heterochronic speciation. In: Human Evolution through Developmental Change (Minugh-Purvis N, McNamara KJ eds). Johns Hopkins University Press, 122–153.

Crockford SJ. 2005. Native dog types in North America before arrival of European dogs. In: Proceedings of the World Small Animal Veterinary Association World Congress. Mexico City, Mexico.

Crockford SJ, Kuzmin YV. 2012. Comments on Germonpré et al., Journal of Archaeological Science 36, 2009 "Fossil dogs and wolves from Palaeolithic sites in Belgium, the Ukraine and Russia: osteometry, ancient DNA and stable isotopes", and Germonpré, Lázkičková-Galetová, and Sablin, Journal of Archaeological Science 39, 2012 "Palaeolithic dog skulls at the Gravettian Předmostí site, the Czech Republic". J Archaeol Sci 39:2797–2801.

Crosby AW. 1986. Ecological Imperialism. Cambridge University Press.

Cruz F, Vilà C, Webster MT. 2008. The legacy of domestication: accumulation of deleterious mutations in the dog genome. Mol Biol Evol 25:2331–2336.

Cryan JF, O'Mahony SM. 2011. The microbiome-gut-brain axis: from bowel to behavior. Neurogastroenterol Motil 23:187–192.

Cucchi T, et al. 2011. Early Neolithic pig domestication at Jiahu, Henan Province, China: clues from molar shape analyses using geometric morphometric approaches. J Archaeol Sci 38:11–22.

Cucchi T, et al. 2016. Social complexification and pig (Sus scrofa) husbandry in ancient China: a combined geometric morphometric and isotopic approach. PLoS ONE 11:e0158523.

Cucchi T, et al. 2017. Detecting taxonomic and phylogenetic signals in equid cheek teeth: towards new palaeontological and archaeological proxies. R Soc Open Sci 4:160997.

Cummings EG, et al. 1981. Epithelial-directed mesenchyme differentiation in vitro: model of murine odontoblast differentiation mediated by quail epithelia. Differentiation 20:1–9.

Curth S, Fischer MS, Kupczik K. 2017. Patterns of integration in the canine skull: an inside view into the relationship of the skull modules of domestic dogs and wolves. Zoology 125:1–9.

da Silva Nunes Barreto R, et al. 2016. Organogenesis of the musculoskeletal system in horse embryos and early fetuses. Anat Rec 299:722–729.

Dahr E. 1941. Über die Variation der Hirnschale bei wilden und zahmen Caniden. Ark Zool 33A:1–56.

Daly KG, et al. 2018. Ancient goat genomes reveal mosaic domestication in the Fertile Crescent. Science 361:85–88.

Danchin E, et al. 2018. Cultural flies: conformist social learning in fruitflies predicts long-lasting mate-choice traditions. Science 362:1025–1030.

Darwin C. 1845. Journal of Researches into the Geology and Natural History of the Various Countries Visited by H.M.S. Beagle round the World, under the Command of Capt. Fitz Roy, R.N. 2nd ed. John Murray.

Darwin C. 1859. On the Origin of Species by Means of Natural Selection. John Murray.

Darwin C. 1868. The Variation of Animals and Plants under Domestication. Popular ed. John Murray.

Darwin C. 1871. The Descent of Man. John Murray.

Darwin F. 1887. The Life and Letters of Charles Darwin, revised 2nd ed.

Davies TG, et al. 2017. Open data and digital morphology. Proc R Soc Lond B 284:20170194.

Davis SJM. 1981. The effects of temperature change and domestication on the body size of Late Pleistocene to Holocene mammals of Israel. Paleobiology 7:101–114.

Davis SJM. 1982. The taming of the few. New Sci 95:697–700.

Davis SJM. 1987. The Archaeology of Animals. Yale University Press.

Davis W. 2001. Light at the Edge of the World: A Journey Through the Realm of Vanishing Cultures. Douglas & McIntyre.

Davison A, et al. 1999. Hybridization and the phylogenetic relationship between polecats and domestic ferrets in Britain. Biol Conserv 87:155–161.

de Beer GR. 1930. Embryos and Evolution. Clarendon Press.

de Beer GR. 1937. The Development of the Vertebrate Skull. Clarendon Press.

de Beer GR. 1954. Embryos and Ancestors. Clarendon Press.

De Boom HPA. 1965. Anomalous animals. S Afr J Sci 61:159–171.

De Marinis AM, Asprea A. 2006. Hair identification key of wild and domestic ungulates from southern Europe. Wildl Biol 12:305–320.

De Queiroz K. 1998. The general lineage concept of species, species criteria, and the process of speciation: a conceptual unification and terminological recommendations. In: Endless Forms: Species and Speciation (Howard DJ, Berlocher SH eds). Oxford University Press, 57–75.

de Roode JC, Lefèvre T, Hunter MD. 2013. Self-medication in animals. Science 340:150–151.

de Saint-Exupéry A. 1943. The Little Prince. Reynal & Hitchcock.

Deacon TW. 2010. A role for relaxed selection in the evolution of the language capacity. PNAS 107:9000–9006.

Debes PV, Hutchings JA. 2014. Effects of domestication on parr maturity, growth, and vulnerability to predation in Atlantic salmon. Can J Fish Aquat Sci 71:1371–1384.

Dechambre E. 1949. La theorie de la foetalisation et la formation des races de chiens et de porcs. Mammalia 13:129–137.

Decker JE, et al. 2014. Worldwide patterns of ancestry, divergence, and admixture in domesticated cattle. PLoS Genet 10:e1004254.

DeFoliart GR. 1995. Edible insects as minilivestock. Biodivers Conserv 4:306–321.

Dehnel A. 1949. Studies on the genus *Sorex* L. Ann Univ Mariae Curie-Sklodowska Sect C Biol 4:17–102.

Delany ME. 2004. Genetic variants for chick biology research: from breeds to mutants. Mech Dev 121:1169–1177.

DePasquale C, et al. 2020. Selection for increased cranial capacity in small mammals during a century of urbanization. J Mammal 101:1706–1710.

Der Sarkissian C, et al. 2015. Evolutionary genomics and conservation of the endangered Przewalski's horse. Curr Biol 25:2577–2583.

Derenne P, Mougin JL. 1976. Données craniométriques sur le lapin et le chat haret de l'Île aux Cochons, Archipel Crozet (46°06′S, 50°14′E). Mammalia 40:495–516.

Derrickson EM. 1992. Comparative reproductive strategies of altricial and precocial eutherian mammals. Funct Ecol 6:57–65.

Desbonnet L, et al. 2014. Microbiota is essential for social development in the mouse. Mol Psychiatry 19:146–148.

Descola P. 2013. Beyond Nature and Culture. University of Chicago Press.

Devlin RH, et al. 2009. Domestication and growth hormone transgenesis cause similar changes in gene expression in coho salmon (*Oncorhynchus kisutch*). PNAS 106:3047–3052.

d'Huy J. 2015. Mythology and ancient origins of the domestic dog. Zenodo. doi: 10.5281/zenodo.31043

Diamond J. 1997. Guns, Germs, and Steel. WW Norton & Co.

Diamond J. 2002. Evolution, consequences and future of plant and animal domestication. Nature 418:700–707.

Diamond J, Bellwood P. 2003. Farmers and their languages: the first expansions. Science 300:597–603.

Diaz Heijtz R, et al. 2011. Normal gut microbiota modulates brain development and behavior. PNAS 108:3047–3052.

Diez-Leon M, Mason G. 2016. Effects of environmental enrichment on aspects of maternal behavior, and infant growth and mortality, in American mink (*Neovison vison*). Zoo Biol 35:19–28.

Díez-León M, et al. 2016. Environmentally enriching American mink (*Neovison vison*) increases lymphoid organ weight and skeletal symmetry, and reveals differences between two sub-types of stereotypic behaviour. Appl Anim Behav Sci 177:59–69.

Díez-León M, Quinton M, Mason G. 2017. How tall should a mink cage be? Using animals' preferences for different ceiling heights to improve cage design. Appl Anim Behav Sci 192:24–34.

Dimitriadis JN. 1937. Das Skyrospony. Z Züchtung B, Tierzüchtung Züchtungsbiol einschließlich Tierernährung 37:343–385.

Diogo R. 2017. Evolution Driven by Organismal Behavior: A Unifying View of Life, Function, Form, Mismatches and Trends. Springer.

Diogo R, Guinard G, Diaz Jr RE. 2017. Dinosaurs, chameleons, humans and Evo-Devo-Path: linking Étienne Geoffroy's teratology, Waddington's homeorhesis, Alberch's logic of "monsters", and Goldschmidt hopeful "monsters". J Exp Zool B 328:207–229.

Diogo R, et al. 2019. Musculoskeletal study of cebocephalic and cyclopic lamb heads illuminates links between normal and abnormal development, evolution and human pathologies. Sci Rep 9:1–12.

do Prado Paim T, et al. 2019. New World goat populations are a genetically diverse reservoir for future use. Sci Rep 9:1476.

Dobkowitz L. 1962. Vom Wandel der Karausche (Giebel) zum Goldfisch: Bemerkungen zu einer Demonstration. Z Tierzuecht Zuechtungsbiol 77:234–237.

Dobney K, Larson G. 2006. Genetics and animal domestication: new windows on an elusive process. J Zool 269:261–271.

Dobzhansky T. 1951. Genetics and the Origin of Species. 3rd ed. Columbia University Press.

Dobzhansky T. 1965. "Wild" and "domestic" species of *Drosophila*. In: The Genetics of Colonizing Species (Baker HG, Stebbins GL eds). Academic Press, 533–546.

Doherty TS, et al. 2016. Invasive predators and global biodiversity loss. PNAS 113:11261–11265.

Domazet-Lošo T, Tautz D. 2010. A phylogenetically based transcriptome age index mirrors ontogenetic divergence patterns. Nature 468:815–818.

Domyan ET, et al. 2016. Molecular shifts in limb identity underlie development of feathered feet in two domestic avian species. eLife 5:e12115.

Doncaster L. 1906. On the inheritance of coat colour in rats. Proc Camb Philos Soc 13:215–227.

Donihue CM, et al. 2018. Hurricane-induced selection on the morphology of an island lizard. Nature 560:88–91.

Donkin R. 1989. The Muscovy Duck, *Cairina moschata domestica*. A A Balkema.

Dorshorst B, Okimoto R, Ashwell C. 2010. Genomic regions associated with dermal hyperpigmentation, polydactyly and other morphological traits in the Silkie chicken. J Hered 101:339–350.

Doughty CE, et al. 2020. Megafauna decline have reduced pathogen dispersal which may have increased emergent infectious diseases. Ecography 43:1107–1117.

Dounias E. 2018. Cooperating with the wild: past and present auxiliary animals assisting humans in their foraging activities. In: Hybrid Communities: Biosocial Approaches to Domestication and other Trans-Species Relationships (Stépanoff C, Vigne J-D eds). Routledge.

Dove WF. 1935. The physiology of horn growth: a study of the morphogenesis, the interaction of tissues, and the evolutionary processes of a Mendelian recessive character by means of transplantation of tissues. J Exp Zool 69:347–405.

Dove WF. 1936. Artificial production of the fabulous unicorn. Sci Monthly 42(5):431–436.

Downs KM, Davies T. 1993. Staging of gastrulating mouse embryos by morphological landmarks in the dissecting microscope. Development 118:1255–1266.

Drachman DB, Sokoloff L. 1966. The role of movement in embryonic joint development. Dev Biol 14:401–420.

Drake AG. 2011. Dispelling dog dogma: an investigation of heterochrony in dogs using 3D geometric morphometric analysis of skull shape. Evol Dev 13:204–213.

Drake AG, Klingenberg CP. 2008. The pace of morphological change: historical transformation of skull shape in St Bernard dogs. Proc R Soc Lond B 275:71–76.

Drake AG, Klingenberg CP. 2010. Large-scale diversification of skull shape in domestic dogs: disparity and modularity. Am Nat 175:289–301.

Drake AG, Coquerelle M, Colombeau G. 2015. 3D morphometric analysis of fossil canid skulls contradicts the suggested domestication of dogs during the late Paleolithic. Sci Rep 5:8299.

Drake AG, et al. 2017. Three-dimensional geometric morphometric analysis of fossil canid mandibles and skulls. Sci Rep 7:9508.

Drescher HE. 1975. Der Einfluß der Domestikation auf die Körper-und Organproportionierung des Nerzes. Z Tierzuecht Zuechtungsbiol 92:272–281.

Drew IM, Perkins Jr D, Daly P. 1971. Prehistoric domestication of animals: effects on bone structure. Science 171:280–282.

Dries B, et al. 2019. Variation of actin filament length in dogs. J Anat 234:694–699.

Driscoll CA, et al. 2009. The taming of the cat. Sci Am 300:68–75.

Drögemüller C, et al. 2008. A mutation in hairless dogs implicates FOXI3 in ectodermal development. Science 321:1462–1462.

Duarte CM, Marbá N, Holmer M. 2007. Rapid domestication of marine species. Science 316:382–383.

Duboule D. 1994. Temporal colinearity and the phylotypic progression: a basis for the stability of a vertebrate Bauplan and the evolution of morphologies through heterochrony. Development 1994:135–142.

Duerst JU. 1926. Das Horn der "Cavicornia": Seine Entstehungsursache, seine Entwicklung, Gestaltung und Einwirkung auf den Schädel der horntragenden Wiederkäuer: Eine Monographie der Hörner. Denkschr allg Schweiz Ges gesamten Nat 63.

Duerst JU. 1931. Grundlagen der Rinderzucht. Springer-Verlag.

Dugatkin LA, Trut L. 2017. How to Tame a Fox (and Build a Dog): Visionary Scientists and a Siberian Tale of Jump-Started Evolution. University of Chicago Press.

Dunn IC, et al. 2011. The chicken polydactyly (Po) locus causes allelic imbalance and ectopic expression of Shh during limb development. Dev Dyn 240:1163–1172.

Dunnum JL, Salazar-Bravo J. 2010. Molecular systematics, taxonomy and biogeography of the genus Cavia (Rodentia: Caviidae). J Zool Syst Evol Res 48:376–388.

Dupé S. 2018. From parasite to reared insect: humans and mosquitoes in Réunion Island. In: Hybrid Communities: Biosocial Approaches to Domestication and Other Trans-species Relationships (Stépanoff C, Vigne J-D eds). Routledge, 289–301.

Dutta P, et al. 2020. Whole genome analysis of water buffalo and global cattle breeds highlights convergent signatures of domestication. Nat Commun 11:4739.

Dyson-Hudson R, Smith EA. 1978. Human territoriality: an ecological reassessment. Am Anthropol 80:21–41.

Dzwillo M. 1962. Domestikation bei Fischen. Z Tierzuecht Zuechtungsbiol 77:172–185.

Ealba EL, et al. 2015. Neural crest-mediated bone resorption is a determinant of species-specific jaw length. Dev Biol 408:151–163.

Eames BF, Schneider RA. 2005. Quail-duck chimeras reveal spatiotemporal plasticity in molecular and histogenic programs of cranial feather development. Development 132:1499–1509.

Eames BF, Schneider RA. 2008. The genesis of cartilage size and shape during development and evolution. Development 135:3947–3958.

Ebinger P. 1972. Vergleichend-quantitative Untersuchungen an Wild-und Laborratten. Z Tierzuecht Zuechtungsbiol 89:34–57.

Ebinger P. 1974. A cytoarchitectonic volumetric comparison of brains in wild and domestic sheep. Z Anat Entwicklungs 144:267–302.

Ebinger P. 1980. Zur Hirn-Körpergewichtsbeziehung bei Wölfen und Haushunden sowie Haushundrassen. Z Saugetierkd 45:148–153.

Ebinger P. 1996. Domestikationsbedingte Änderungen von Hirn und Verhalten beim Hausgeflügel. Acta Biol Benrodis Suppl. 3:37–52.

Ebinger P, Löhmer R. 1984. Comparative quantitative investigations on brains of rock doves, domestic and urban pigeons (*Columba l. livia*). J Zool Syst Evol Res 22:136–145.

Ebinger P, Löhmer R. 1985. Zur Hirn-Körpergewichtsbeziehung bei Stock- und Hausenten. Zool Anz 214:285–290.

Ebinger P, Löhmer R. 1987. A volumetric comparison of brains between greylag geese (*Anser anser* L.) and domestic geese. J Hirnforsch 28:291–299.

Ebinger P, Röhrs M. 1995. Volumetric analysis of brain structures, especially of the visual system in wild and domestic turkeys (*Meleagris gallopavo*). J Hirnforsch 36:219–228.

Ebinger P, Röhrs M, Pohlenz J. 1989. Veränderungen von Hirn-und Augengrößen bei wilden und domestizierten Truthühnern (*Meleagris gallopavo* L., 1758). J Zool Syst Evol Res 27:142–148.

Eda M. 2020. The osteological microevolution of red junglefowl and domestic fowl under the domestication process. Quat Int, doi: https://doi.org/10.1016/j.quaint.2020.10.001.

Ekarius C. 2007. Storey's Illustrated Guide to Poultry Breeds. Storey Publishing.

Ekarius C. 2008. Storey's Illustrated Breed Guide to Sheep, Goats, Cattle and Pigs: 163 Breeds from Common to Rare. Storey Publishing.

Ekdale EG. 2016. Form and function of the mammalian inner ear. J Anat 228:324–337.

Eleroğlu H, et al. 2018. Analysis of growth curves of guinea fowl (*Numida meleagris*) fed diets containing dry oregano (*Origanum vulgare* L.) in an organic system. J Agric Nat Resour 45: 99–108.

Ellis JL, et al. 2009. Cranial dimensions and forces of biting in the domestic dog. J Anat 214: 362–373.

Enlow DH, Brown SO. 1956. A comparative histological study of fossil and recent bone tissues. Part I. Tex J Sci 9:405–443.

Epstein H. 1969. Domestic Animals of China. Commonwealth Agricultural Bureaux.

Epstein H. 1971. The Origin of the Domestic Animals of Africa. Africana Publishing Corporation.

Epstein H. 1977. Domestic Animals of Nepal. Holmes & Meier.

Erikson P. 2000. The social significance of pet-keeping among Amazonian Indians. In: Companion Animals and Us Exploring the Relationships between People and Pets (Podberscek AL, Paul ES, Serpell JA eds). Cambridge University Press, 7–26.

Eriksson J, et al. 2008. Identification of the yellow skin gene reveals a hybrid origin of the domestic chicken. PLoS Genet 4:e1000010.

Ervynck A, et al. 2001. Born Free? New Evidence for the Status of *Sus scrofa* at Neolithic Çayönü Tepesi (Southeastern Anatolia, Turkey). Paléorient 27:47–73.

Ervynck A, et al. 2007. Paradise Lost: an investigation into the transition from forest dwelling pigs to farm animals in medieval Flanders, Belgium. In: Pigs and Humans: 10,000 Years of Interaction (Albarella U, et al. eds). Cambridge University Press, 171–193.

Esteve-Altava B, et al. 2015. Anatomical networks reveal the musculoskeletal modularity of the human head. Sci Rep 5:8298.

Esteve-Altava B, Rasskin-Gutman D. 2015. Evo-Devo insights from pathological networks: Exploring craniosynostosis as a developmental mechanism for modularity and complexity in the human skull. J Anthropol Sci 93:1–15.

Evans HE, De Lahunta A. 2012. Miller's Anatomy of the Dog. Saunders.

Evans K, McGreevy P. 2006. Conformation of the equine skull: a morphometric study. Anat Histol Embryol 35:221–227.

Evans HE, Sack WO. 1973. Prenatal development of domestic and laboratory mammals: growth curves, external features and selected references. Anat Histol Embryol 2:11–45.

Evans AR, et al. 2012. The maximum rate of mammal evolution. PNAS 109:4187–4190.

Evershed RP, et al. 2003. New chemical evidence for the use of combed ware pottery vessels as beehives in ancient Greece. J Archaeol Sci 30:1–12.

Evershed RP, et al. 2008. Earliest date for milk use in the Near East and southeastern Europe linked to cattle herding. Nature 455:528–531.

Evin A. 2020. Des hommes, des plantes et des animaux: domestication et évolution morphologique [Habilitation]. Université Montpellier.

Evin A, et al. 2013. The long and winding road: identifying pig domestication through molar size and shape. J Archaeol Sci 40.

Evin A, et al. 2015a. Phenotype and animal domestication: a study of dental variation between domestic, wild, captive, hybrid and insular *Sus scrofa*. BMC Evol Biol 15:6.

Evin A, et al. 2015b. Unravelling the complexity of domestication: a case study using morphometrics and ancient DNA analyses of archaeological pigs from Romania. Philos Trans R Soc B 370.

Evin A, et al. 2016. The use of close-range photogrammetry in zooarchaeology: creating accurate 3D models of wolf crania to study dog domestication. J Archaeol Sci Rep 9:87–93.

Evin A, et al. 2017. A test for paedomorphism in domestic pig cranial morphology. Biol Lett 13:20170321.

Evin A, et al. 2020. Building three-dimensional models before destructive sampling of bioarchaeological remains: a comment to Pálsdóttir et al. (2019). R Soc Open Sci 7:192034.

Ewart JC. 1897. Critical Period in the Development of the Horse. Adam and Charles Black.

Ewart JC. 1917. VII. Studies on the development of the horse. I. The development during the third week. Earth Env Sci Trans R Soc 51:287–329.

Fang M, et al. 2009. Contrasting mode of evolution at a coat color locus in wild and domestic pigs. PLoS Genet 5:e1000341.

Farooq S, et al. 2017. Cortical and trabecular morphology is altered in the limb bones of mice artificially selected for faster skeletal growth. Sci Rep 7:10527.

Farrell MJ, Davies TJ. 2019. Disease mortality in domesticated animals is predicted by host evolutionary relationships. PNAS 116:7911–7915.

Fausto C. 2007. Feasting on people: eating animals and humans in Amazonia. Curr Anthropol 48:497–530.

Faux C, Field DJ. 2017. Distinct developmental pathways underlie independent losses of flight in ratites. Biol Lett 13:20170234.

Fawcett E. 1918. The primordial cranium of *Erinaceus europaeus*. J Anat 52:211–250.

Feder JL, Egan SP, Nosil P. 2012. The genomics of speciation-with-gene-flow. Trends Genet 28:342–350.

Fee MS, Kozhevnikov AA, RHR Hahnloser RHR. 2004. Neural mechanisms of vocal sequence generation in the songbird. Ann NY Acad Sci 1016:153–170.

Feiner N, et al. 2018. Developmental plasticity in reptiles: insights from temperature-dependent gene expression in wall lizard embryos. J Exp Zool A Ecol Integr Physiol 329:351–361.

Felice RN, Goswami A. 2018. Developmental origins of mosaic evolution in the avian cranium. PNAS 115:555–560.

Félix M-A, Wagner A. 2008. Robustness and evolution: concepts, insights and challenges from a developmental model system. Heredity 100:132–140.

Fenner F, Fantini B. 1999. Biological Control of Vertebrate Pests: The History of Myxomatosis—an Experiment in Evolution. CABI Publishing.

Fernandes IR, et al. 2017. Macroscopic and microscopic analysis of 2 embryos and 1 foetus derived from a sheep (*Ovis aries*) without breed. Braz J Vet Res Anim Sci 54:168–177.

Fiddyment S, et al. 2019. So you want to do biocodicology? A field guide to the biological analysis of parchment. Herit Sci 7:35.

Fiems LO. 2012. Double muscling in cattle: genes, husbandry, carcasses and meat. Animals 2:472–506.

Figueirido B. 2018. Phenotypic disparity of the elbow joint in domestic dogs and wild carnivores. Evolution 72:1600–1613.

Finarelli JA. 2006. Estimation of endocranial volume through the use of external skull measures in the Carnivora (Mammalia). J Mammal 87:1027–1036.

Finarelli JA. 2011. Estimating endocranial volume from the outside of the skull in Artiodactyla. J Mammal 92:200–212.

Finlay BL, Hersman MN, Darlington RB. 1998. Patterns of vertebrate neurogenesis and the paths of vertebrate evolution. Brain Behav Evol 52:232–242.

Fiorello CV, German RZ. 1997. Heterochrony within species: craniofacial growth in giant, standard, and dwarf rabbits. Evolution 51:250–261.

Firth EC. 2006. The response of bone, articular cartilage and tendon to exercise in the horse. J Anat 208:513–526.

Fischer C-J. 1973. Vergleichende quantitative Untersuchungen an Wildkaninchen und Hauskaninchen. Inaugural-Dissertation Tierärztliche Hochschule Hannover.

Fischer MS, Lilje KE. 2014. Dogs in Motion. 2nd ed. VDH Service GmbH.

Fischer MS, Lehmann SV, Andrada E. 2018. Three-dimensional kinematics of canine hind limbs: in vivo, biplanar, high-frequency fluoroscopic analysis of four breeds during walking and trotting. Sci Rep 8:16982.

Fish JL, et al. 2014. Multiple developmental mechanisms regulate species-specific jaw size. Development 141:674–684.

Fisher AD, et al. 2001. Handling and stress response traits in cattle: identification of putative genetic markers. In: Proceedings of the 35th International Congress of the ISAE. Davis, USA (Garner JP, Mench JA, Heekin SP eds). Center for Animal Welfare, University of California, Davis, 100.

Fisher RA. 1934. Crest and hernia in fowls due to a single gene without dominance. Science 80:288–289.

Fitak R, et al. 2020. Genomic signatures of domestication in Old World camels. Commun Biol 3, 316.

Flad RK, Yuan J, Li S. 2007. Zooarcheological evidence for animal domestication in northwest China. Dev Quat Sci 9:167–203.

Flores DA, Giannini N, Abdala F. 2018. Evolution of post-weaning skull ontogeny in New World opossums (Didelphidae). Org Divers Evol 18:367–382.

Flux JEC, Fullagar PJ. 1992. World distribution of the rabbit Oryctolagus funiculus on islands. Mammal Rev 22:151–205.

Foley CW, Lasley JF, Osweiler GD. 1979. Abnormalities of companion animals: analysis of heritability. Iowa State University Press.

Fondon JW, Garner HR. 2004. Molecular origins of rapid and continuous morphological evolution. PNAS 101:18058–18063.

Fornshell G. 2002. Rainbow trout—challenges and solutions. Rev Fish Sci 10:545–557.

Fox MW. 1978. The Dog. Its Domestication and Behavior. Garland STPM Press.

Frahm HD, Rehkämper G. 1998. Allometric comparison of the brain and brain structures in the white crested polish chicken with uncrested domestic chicken breeds. Brain Behav Evol 52:292–307.

Frahm HD, Rehkämper G, Werner CW. 2001. Brain alterations in crested versus non-crested breeds of domestic ducks (Anas platyrhynchos f. d.). Poult Sci 80:1249–1257.

Frahm H, Stephan H. 1976. Vergleichende Volumenmessungen an Hirnen von Wild- und Gefangenschaftstieren des Spitzhörnchens (Tupaia). J Hirnforsch 17:449–462.

Francis RC. 2015. Domesticated: Evolution in a Man-Made World. WW Norton & Company.

Francoy TM, et al. 2008. Identification of Africanized honey bees through wing morphometrics: two fast and efficient procedures. Apidologie 39:488–494.

Frantz LA, et al. 2015. Evidence of long-term gene flow and selection during domestication from analyses of Eurasian wild and domestic pig genomes. Nat Genet 47:1141–1148.

Frantz LA, et al. 2016. Genomic and archaeological evidence suggest a dual origin of domestic dogs. Science 352:1228–1231.

Frantz LA, et al. 2019. Ancient pigs reveal a near-complete genomic turnover following their introduction to Europe. PNAS 116:17231–17238.

Frantz LA, Larson G. 2019. A genetic perspective on the domestication continuum. In: Hybrid Communities: Biosocial Approaches to Domestication and Other Trans-species Relationships (Stépanoff C, Vigne J-D eds). Routledge, 306.

Frantz LA, et al. 2020. Animal domestication in the era of ancient genomics. Nat Rev Genet:1–12.

Fraser DJ, et al. 2010. Potential for domesticated–wild interbreeding to induce maladaptive phenology across multiple populations of wild Atlantic salmon (*Salmo salar*). Can J Fish Aquat Sci 67:1768–1775.

Freiberger K, et al. 2020. Secondary palate development in the dog (*Canis lupus familiaris*). Cleft Palate Craniofac J:1055665620943771.

Frémondeau D, et al. 2015. Standardized pork production at the Celtic village of Levroux Les Arènes (France, 2nd c. BC): evidence from kill-off patterns and birth seasonality inferred from enamel δ18O analysis. J Archaeol Sci Rep 2:215–226.

Fromm E. 1964. Die Seele des Menschen. Ihre Fähigkeit zum Guten und zum Bösen. Deutsche Verlags-Anstalt.

Froy H, et al. 2021. Heritable variation in telomere length predicts mortality in Soay sheep. PNAS 118:e2020563118.

Fuenzalida Á. 2006. El perro Ovejero Magallánico; testimonio de reconstrucción histórica [Tesis Licenciatura en Historia]. Pontificia Universidad Católica. Chile.

Fujimura K, Okada N. 2007. Development of the embryo, larva and early juvenile of Nile tilapia *Oreochromis niloticus* (Pisces: Cichlidae). Dev Growth Differ 49:301–324.

Fuller DQ. 2006. Agricultural origins and frontiers in South Asia: a working synthesis. J World Prehist 20:1–86.

Fuller DQ. 2010. An emerging paradigm shift in the origins of agriculture. Gen Anthropol 17:1–12.

Gaffney B, Cunningham E. 1988. Estimation of genetic trend in racing performance of thoroughbred horses. Nature 332:722–724.

Gaillard J-M, et al. 1997. Variation in growth form and precocity at birth in eutherian mammals. Proc R Soc Lond B 264:859–868.

Gál E, et al. 2010. Evidence of the crested form of domestic hen (*Gallus gallus* f. *domestica*) from three post-medieval sites in Hungary. J Archaeol Sci 37:1065–1072.

Galis F. 1999. Why do almost all mammals have seven cervical vertebrae? Developmental constraints, Hox genes, and cancer. J Exp Zool 285:19–26.

Galis F, van Alphen JJM, Metz JAJ. 2001. Why five fingers? Evolutionary constraints on digit numbers. Trends Ecol Evol 16:637–646.

Galis F, et al. 2007. Do large dogs die young? J Exp Zool B 308:119–126.

Galton F. 1865. The first steps towards the domestication of animals. Transactions of the Ethnological Society of London 3:122–138.

Gao Y, et al. 2012. Quaternary palaeoenvironmental oscillations drove the evolution of the Eurasian *Carassius auratus* complex (Cypriniformes, Cyprinidae). J Biogeogr 39:2264–2278.

García-Borrón JC, Sánchez-Laorden BL, Jiménez-Cervantes C. 2005. Melanocortin-1 receptor structure and functional regulation. Pigm Cell Res 18:393–410.

Gariépy J-L, Bauer DJ, Cairns RB. 2001. Selective breeding for differential aggression in mice provides evidence for heterochrony in social behaviours. Anim Behav 61:933–947.

Gaspar J. 1930. Analyse der Erbfaktoren des Schädels bei einer Paarung von Ceylon-Nackthund × Dackel. Jenaische Zeitschrift für Naturwissenschaft 65:245–274.

Gaunitz C, et al. 2018. Ancient genomes revisit the ancestry of domestic and Przewalski's horses. Science 360:111–114.

Gaupp E. 1906. Die Entwicklung des Kopfskelettes. In: Handbuch der vergleichenden und experimentellen Entwickelungslehre der Wirbeltiere (Hertwig O ed). Fischer Verlag.

Gautier A. 1993. "What's in a name?" A short history of the Latin and other labels proposed for domestic animals. In: Skeletons in her cupboard: Festschrift for Juliet Clutton-Brock (Clason AT ed). Oxbow Books, 91–98.

Gazda MA, et al. 2020. A genetic mechanism for sexual dichromatism in birds. Science 368:1270–1274.

Geiger M, Asher RJ. 2019. Schultz's rule in domesticated mammals. Mamm Biol 98:36–42.

Geiger M, Haussman S. 2016. Cranial suture closure in domestic dog breeds and its relationships to skull morphology. Anat Rec 299:412–420.

Geiger M, et al. 2016. Unaltered sequence of dental, skeletal, and sexual maturity in domestic dogs compared to the wolf. Zool Lett 2:16.

Geiger M, Sánchez-Villagra MR. 2018. Similar rates of morphological evolution in domesticated and wild pigs and dogs. Front Zool 15:23.

Geiger M, et al. 2017. Neomorphosis and heterochrony of skull shape in dog domestication. Sci Rep 7:13443.

Geiger M, et al. 2018a. The influences of domestication and island evolution on dental growth in sheep. J Mamm Evol:1–16.

Geiger M, et al. 2018b. A longitudinal study of phenotypic changes in early domestication of house mice. R Soc Open Sci 5:172099.

Geiger M, et al. 2021. Exceptional changes in skeletal anatomy under domestication: the case of brachycephaly. Integr Org Biol, doi: https://doi.org/10.1093/iob/obab023.

Gentry A, Clutton-Brock J, Groves CP. 2004. The naming of wild animal species and their domestic derivatives. J Archaeol Sci 31:645–651.

Geoffroy Saint-Hilaire I. 1854. Domestication et Naturalisation des Animaux Utiles. 13th ed. Dusacq.

Gerbault P, et al. 2014. Storytelling and story testing in domestication. PNAS 111:6159–6164.

Gering E, et al. 2019. Maladaptation in feral and domesticated animals. Evol Appl 12:1274–1286.

Germain D, Laurin M. 2009. Evolution of ossification sequences in salamanders and urodele origins assessed through event-pairing and new methods. Evol Dev 11:170–190.

Germonpré M, et al. 2009. Fossil dogs and wolves from Palaeolithic sites in Belgium, the Ukraine and Russia: osteometry, ancient DNA and stable isotopes. J Archaeol Sci 36:473–490.

Germonpré M, et al. 2015. Palaeolithic dogs and Pleistocene wolves revisited: a reply to Morey (2014). J Archaeol Sci 54:210–216.

Germonpré M, et al. 2018. Self-domestication or human control? The Upper Palaeolithic domestication of the wolf. In: Hybrid Communities: Biosocial Approaches to Domestication and Other Trans-species Relationships (Stépanoff C, Vigne J-D eds). Routledge, 306.

Gifford-Gonzalez D, Hanotte O. 2011. Domesticating animals in Africa: implications of genetic and archaeological findings. J World Prehist 24:1–23.

Gilbert AS. 1989. Microscopic bone structure in wild and domestic animals: a reappraisal. In: Early Animal Domestication and Its Cultural Context (Crabtree PJ, Ryan K, Campana D eds). University of Pennsylvania Museum of Archaeology, 47–86.

Gilbert SF, Epel D. 2009. Ecological Developmental Biology: Integrating Epigenetics, Medicine, and Evolution. Sinauer.

Gilbert MTP, Shapiro MD. 2014. Pigeons: domestication. In: Encyclopedia of Global Archaeology (Smith C ed). Springer New York, 5944–5948.

Gingerich PD. 2019. Rates of Evolution: A Quantitative Synthesis. Cambridge University Press.

Girdland Flink L, et al. 2014. Establishing the validity of domestication genes using DNA from ancient chickens. PNAS 111:6184–6189.

Giuffra EJMH, et al. 2000. The origin of the domestic pig: independent domestication and subsequent introgression. Genetics 154:1785–1791.

Gjedrem T. 2010. The first family-based breeding program in aquaculture. Rev Aquac 2:2–15.

Gjøen H, Bentsen H. 1997. Past, present, and future of genetic improvement in salmon aquaculture. ICES J Mar Sci 54:1009–1014.

Glaus A. 1932. Untersuchungen über den Schneidezahnwechsel beim schweizerischen Braunvieh zum Zwecke der Altersbestimmung [Doctoral dissertation]. University of Zurich.

Glover KA, et al. 2009. A comparison of farmed, wild and hybrid Atlantic salmon (*Salmo salar* L.) reared under farming conditions. Aquaculture 286:203–210.

Godfrey LR, et al. 2001. Teeth, brains, and primate life histories. Am J Phys Anthropol 114: 192–214.

Godfrey LR, et al. 2005. Schultz's unruly rule: dental developmental sequences and schedules in small-bodied, folivorous lemurs. Folia Primatol 76:77–99.

Goepfert N. 2010. The llama and the deer: dietary and symbolic dualism in the central Andes. Anthropozoologica 45:25–45.

Goldschmidt R. 1940. The Material Basis of Evolution. Yale University Press.

Goliomytis M, Panopoulou E, Rogdakis E. 2003. Growth curves for body weight and major component parts, feed consumption, and mortality of male broiler chickens raised to maturity. Poult Sci 82:1061–1068.

Gombrich EH. 2002. The Preference for the Primitive: Episodes in the History of Western Taste and Art. Phaidon.

González-Porter GP, et al. 2011. Patterns of genetic diversity in the critically endangered Central American river turtle: human influence since the Mayan age? Conserv Genet 12:1229–1242.

Goswami A. 2006. Cranial modularity shifts during mammalian evolution. Am Nat 168:270–280.

Goswami A, Prochel J. 2007. Ontogenetic morphology and allometry of the cranium in the common European mole (*Talpa europaea*). J Mammal 88:667–677.

Goswami A, Weisbecker V, Sánchez-Villagra MR. 2009. Developmental modularity and the marsupial-placental dichotomy. J Exp Zool B 312B:186–195.

Goswami A, et al. 2012. Shape, variance and integration during craniogenesis: contrasting marsupial and placental mammals. J Evol Biol 25:862–872.

Goswami A, et al. 2014. The macroevolutionary consequences of phenotypic integration: from development to deep time. Philos Trans R Soc B 369.

Goswami A, et al. 2016. Do developmental constraints and high integration limit the evolution of the marsupial oral apparatus? Integr Comp Biol 56:404–415.

Gottlieb G. 1972. Zing-Yang Kuo: radical scientific philosopher and innovative experimentalist (1898–1970). J Comp Physiol Psychol 80:1–10.

Gould SJ. 1981. The Mismeasure of Man. Norton and Company.

Gowe RS, et al. 1959. Environment and poultry breeding problems: 4. The value of a random-bred control strain in a selection study. Poult Sci 38:443–461.

Gowe RS, Fairfull RW. 1985. The direct response to long-term selection for multiple traits in egg stocks and changes in genetic parameters with selection. In: Poultry Genetics and Breeding (Hill WG, Manson JM, Hewitt D eds). British Poultry Sci, 125–146.

Gracy KN, et al. 1990. The accumulation of oxidized isoforms of chicken triosephosphate isomerase during aging and development. Mech Ageing Dev 56:179–186.

Gradé JT, Tabuti JRS, Van Damme P. 2009. Four footed pharmacists: indications of self-medicating livestock in Karamoja, Uganda. Econ Bot 63:29–42.

Grant PR, et al. 2003. Inbreeding and interbreeding in Darwin's finches. Evolution 57:2911–2916.

Gray AP. 1958. Bird Hybrids. A Check-List with Bibliography. Farnham Royal: Commonwealth Agricultural Bureaux.

Green RM, et al. 2017. Developmental nonlinearity drives phenotypic robustness. Nat Commun 8:1970.

Greer AE. 1991. Limb reduction in squamates: identification of the lineages and discussion of the trends. J Herpetol 25:166–173.

Gremillion KJ, Barton L, Piperno DR. 2014. Particularism and the retreat from theory in the archaeology of agricultural origins. PNAS 111:6171–6177.

Grobet L, et al. 1998. Molecular definition of an allelic series of mutations disrupting the myostatin function and causing double-muscling in cattle. Mamm Genome 9:210–213.

Groenen MAM, et al. 2012. Analyses of pig genomes provide insight into porcine demography and evolution. Nature 491:393–398.

Gruber B, et al. 2011. On managing the red mason bee (*Osmia bicornis*) in apple orchards. Apidologie 42:564.

Guagnin M, Perri AR, Petraglia MD. 2018. Pre-Neolithic evidence for dog-assisted hunting strategies in Arabia. J Anthropol Archaeol 49:225–236.

Guillaume J. 2010. Ils ont domestiqué plantes et animaux. Editions Quae.

Guillo D, Claidière N. 2020. Do guide dogs have culture? The case of indirect social learning. Humanit Soc Sci Commun 7:31.

Gündemir O, et al. 2020. A study on morphology and morphometric parameters on skull of the *Bardhoka autochthonous* sheep breed in Kosovo. Anat Histol Embryol 49:365–371.

Gunnarsson U, et al. 2011. The Dark brown plumage color in chickens is caused by an 8.3-kb deletion upstream of *SOX10*. Pigment Cell Melanoma Res 24:268–274.

Günzel-Apel A-R, Bostedt H eds. 2016. Reproductive Medicine and Neonatology in Dogs and Cats. Schattauer.

Gupta M, et al. 2017. Feldman et al. do protest too much, we think. J Genet 96:509–511.

Gurevitch J, Padilla DK. 2004. Are invasive species a major cause of extinctions? Trends Ecol Evol 19:470–474.

Gustafson KD, et al. 2018. Founder events, isolation, and inbreeding: Intercontinental genetic structure of the domestic ferret. Evol Appl 11:694–704.

Guthrie RD. 2004. The Nature of Paleolithic Art. University of Chicago Press.

Güttinger HR. 1985. Consequences of domestication on the song structures in the canary. Behaviour 94:254–278.

Haag WG. 1948. An osteometric analysis of some aboriginal dogs. Univ Ky Rep Anthropol 7:107–264.

Haag-Wackernagel D. 1993. Street pigeons in Basel. Nature 361:200.

Haag-Wackernagel D, Moch H. 2004. Health hazards posed by feral pigeons. J Infect 48:307–313.

Haase E. 2000. Comparison of reproductive biological parameters in male wolves and domestic dogs. Z Saugetierkd 65:257–270.

Haeckel E. 1891. Anthropogenie oder Entwickelungsgeschichte des Menschen. Keimesgeschichte des Menschen. Engelmann.

Habermehl K-H. 1975. Die Altersbestimmung bei Haus- und Labortieren. 2. vollständig neubearbeitete Auflage ed. Paul Parey.

Hall BK. 2009. The Neural Crest and Neural Crest Cells in Vertebrate Development and Evolution. Springer Science & Business Media.

Hall BK. 2015. Bones and Cartilage. Developmental and Evolutionary Skeletal Biology. 2nd ed. Academic Press/Elsevier.

Hall BK, Herring S. 1990. Paralysis and growth of the musculoskeletal system in the embryonic chick. J Morphol 206:45–56.

Hall BK, Miyake T. 2004. Divide, accumulate, differentiate: cell condensation in skeletal development revisited. Int J Dev Biol 39:881–893.

Hall J, et al. 2014. Evolution of a developmental mechanism: species-specific regulation of the cell cycle and the timing of events during craniofacial osteogenesis. Dev Biol 385:380–395.

Hallgrímsson B, et al. 2003. Embryological origins of developmental stability: size, shape and fluctuating asymmetry in prenatal random bred mice. J Exp Zool Part B 296:40–57.

Hallgrímsson B, et al. 2009. Deciphering the palimpsest: studying the relationship between morphological integration and phenotypic covariation. Evol Biol 36:355–376.

Hallgrímsson B, et al. 2012. The generation of variation and the developmental basis for evolutionary novelty. J Exp Zool Part B 318:501–517.

Hamburger V, Hamilton HL. 1951. A series of normal stages in the development of the chick embryo. J Morphol 88:49–92.

Hammer K. 1984. The domestication syndrome. Die Kulturpflanze 32:11–34.

Hammond J. 1947. Animal breeding in relation to nutrition and environmental conditions. Biol Rev 22:195–213.

Hammond J. 1962. Some changes in the form of sheep and pigs under domestication. Z Tierzuecht Zuechtungsbiol 77:156–158.

Hammond CL, Simbi BH, Stickland NC. 2007. In ovo temperature manipulation influences embryonic motility and growth of limb tissues in the chick (*Gallus gallus*). J Exp Biol 210:2667–2675.

Hancock PA, et al. 2018. Associated patterns of insecticide resistance in field populations of malaria vectors across Africa. PNAS 115:5938–5943.

Hanot P, et al. 2017. Identifying domestic horses, donkeys and hybrids from archaeological deposits: a 3D morphological investigation on skeletons. J Archaeol Sci 78:88–98.

Hanot P, Bochaton C. 2018. New osteological criteria for the identification of domestic horses, donkeys and their hybrids in archaeological contexts. J Archaeol Sci 94:12–20.

Hanotte O, et al. 2002. African pastoralism: genetic imprints of origins and migrations. Science 296:336–339.

Hansell J. 1998. The Pigeon in History: Or, the Dove's Tale. Millstream Books.

Hansen SW. 1996. Selection for behavioural traits in farm mink. Appl Anim Behav Sci 49:137–148.

Hansen Wheat C, et al. 2019. Behavioural correlations of the domestication syndrome are decoupled in modern dog breeds. Nat Commun 10:2422.

Hansen Wheat C, van der Bijl W, Wheat CW. 2020. Morphology does not covary with predicted behavioral correlations of the domestication syndrome in dogs. Evol Lett 4:189–199.

Harache Y. 2002. Development and diversification issues in aquaculture. A historical and dynamic view of fish culture diversification. Cah Options Mediterr 59:15–23.

Harari NH. 2011. Sapiens: A Brief History of Humankind. Harper.

Harbers H, et al. 2020a. The mark of captivity: plastic responses in the ankle bone of a wild ungulate (Sus scrofa). R Soc Open Sci 7:192039.

Harbers H, et al. 2020b. Investigating the impact of captivity and domestication on limb bone cortical morphology: an experimental approach using a wild boar model. Sci Rep 10:1–13.

Hard JJ, et al. 2000. Evidence for morphometric differentiation of wild and captively reared adult coho salmon: a geometric analysis. Environ Biol Fishes 58:61–73.

Hare B. 2017. Survival of the friendliest: Homo sapiens evolved via selection for prosociality. Annu Rev Psychol 68:155–186.

Hare B, et al. 2005. Social cognitive evolution in captive foxes is a correlated by-product of experimental domestication. Curr Biol 15:226–230.

Harland J. 2019. The origins of aquaculture. Nat Ecol Evol 3:1378–1379.

Harpur BA, et al. 2012. Management increases genetic diversity of honey bees via admixture. Mol Ecol 21:4414–4421.

Harris DR. 1972. The origins of agriculture in the tropics: Ecological analysis affords new insights into agricultural origins and suggests a fresh evaluation of the limited archaeological evidence. Am Sci 60:180–193.

Harris MP, et al. 2008. Zebrafish eda and edar mutants reveal conserved and ancestral roles of ectodysplasin signaling in vertebrates. PLoS Genet 4:e1000206.

Harris M, et al. 2014. Fish is fish: the use of experimental model species to reveal causes of skeletal diversity in evolution and disease. J Appl Ichthyol 30:616–629.

Harrison RG. 1935. Heteroplastic Grafting in Embryology. Williams & Wilkins Company.

Harrison LB, Larsson HCE. 2008. Estimating evolution of temporal sequence changes: a practical approach to inferring ancestral developmental sequences and sequence heterochrony. Syst Biol 57:378–387.

Harvati K, Weaver TD. 2006. Human cranial anatomy and the differential preservation of population history and climate signatures. Anat. Rec. Part A Discov. Mol. Cell. Evol. Biol. An Off. Publ. Am. Assoc. Anat. 288:1225–1233.

Harvey CE. 1985. Veterinary Dentistry. Saunders.

Harvey AC, et al. 2016. Plasticity in growth of farmed and wild Atlantic salmon: is the increased growth rate of farmed salmon caused by evolutionary adaptations to the commercial diet? BMC Evol Biol 16:264.

Hata H, Kato M. 2006. A novel obligate cultivation mutualism between damselfish and Polysiphonia algae. Biol Lett 2:593–596.

Hatt J-M, et al. 2019. The rumen washes off abrasives before heavy-duty chewing in ruminants. Mamm Biol 97:104–111.

Hausmann UF. 1840. Ueber die Zeugung und Entstehung des wahren weiblichen Eies bei den Säugethieren und Menschen. Helwing.

Haussler KK, Stover SM, Willits NH. 1997. Developmental variation in lumbosacropelvic anatomy of thoroughbred racehorses. Am J Vet Res 58:1083–1091.

Hautier L, et al. 2010. Skeletal development in sloths and the evolution of mammalian vertebral patterning. PNAS 107:18903–18908.

Hawkes K, O'Connell J. 1992. On optimal foraging models and subsistence transitions. Curr Anthropol 33:63–66.

Hängling et al. 2005. A molecular approach to detect hybridisation between crucian carp (*Carassius carassius*) and non-indigenous carp species (*Carassius* spp. and *Cyprinus carpio*). Freshwater Biology 50:403–417.

Hecht EE, et al. 2019. Significant neuroanatomical variation among domestic dog breeds. J Neurosci 39:7748–7758.

Heck L, Clauss M, Sánchez-Villagra MR. 2017. Gestation length variation in domesticated horses and its relation to breed and body size diversity. Mamm Biol 84:44–51.

Heck L, Clauss M, Sánchez-Villagra MR. 2018a. Do domesticated mammals selected for intensive production have less variable gestation periods? Mamm Biol 88:151–155.

Heck L, et al. 2018b. Shape variation and modularity of skull and teeth in domesticated horses and wild equids. Front Zool 15:14.

Heck L, Sánchez-Villagra MR, Stange M. 2019. Why the long face? Comparative shape analysis of miniature, pony, and other horse skulls reveals changes in ontogenetic growth. PeerJ 7:e7678.

Heikkinen ME, et al. 2020. Long-term reciprocal gene flow in wild and domestic geese reveals complex domestication history. G3 (Bethesda) 10:3061–3070.

Helskog K. 2011. Reindeer corrals 4700–4200 BC: myth or reality? Quat Int 238:25–34.

Hemmer H. 1978. Innerartliche Unterschiede der relativen Hirngröße und ihr Wandel vom Wildtier zum Haustier. Saeugetierkd Mitt 26:312–317.

Hemmer H. 1990. Domestication: The Decline of Environmental Appreciation. Cambridge University Press.

Henderson E. 2007. Platyrrhine dental eruption sequences. Am J Phys Anthropol 134:226–239.

Hendricks BL. 1995. International Encyclopedia of Horse Breeds. 1st ed. University of Oklahoma Press.

Hendrikse JL, Parsons TE, Hallgrímsson B. 2007. Evolvability as the proper focus of evolutionary developmental biology. Evol Dev 9:393–401.

Hendry AP, Farrugia TJ, Kinnison MT. 2008. Human influences on rates of phenotypic change in wild animal populations. Mol Ecol 17:20–29.

Henneberg B. 1908. Beiträge zur Entwickelung der Ohrmuschel. Anat Hefte 36:107–188.

Henriksen R, et al. 2016. The domesticated brain: genetics of brain mass and brain structure in an avian species. Sci Rep 6:34031.

Henriksen R, Gering E, Wright D. 2018. Feralisation—the understudied counterpoint to domestication. In: Origin and Evolution of Biodiversity (Pontarotti P ed). Springer, 3–22.

Henry M, et al. 1995. Mating pattern and chromosome analysis of a mule and her offspring. Biol Reprod 52:273–279.

Henry P. 2016. On domestication and feralization: re-rethinking invasive species from a genetic perspective. J Brief Ideas: zenodo.45084.

Henton E. 2013. Herding and settlement identity in the central Anatolian Neolithic: herding decisions and organisation in Çatalhöyük, elucidated through oxygen isotopes and microwear in sheep teeth. In: Proceedings of Tenth International Symposium on the Archaeozoology of South-Western Asia and Adjacent Areas, 69–99.

Hernádi A, et al. 2012. Man's underground best friend: domestic ferrets, unlike the wild forms, show evidence of dog-like social-cognitive skills. PLoS ONE 7:e43267.

Hernández F. 1651. Rerum medicarum Nouae Hispaniae thesaurus, seu Plantarum animalium mineralium Mexicanorum historia. Ex typographeio Vitalis Mascard.

Heron C, et al. 1994. The chemistry of Neolithic beeswax. Naturwissenschaften 81:266–269.

Herre W. 1938. Zum Wandel des Rassebildes der Haustiere: Studien am Schädel des Berkshireschweines. Sonderdruck aus Kühn-Archiv: 203–228.

Herre W. 1953. Studien am Skelett des Mittelohres wilder und domestizierter Formen der Gattung *Lama* Frisch. Acta Anat 19:271–289.

Herre W, Röhrs M. 1973. Haustiere—zoologisch gesehen. Gustav Fischer.

Herre W, Röhrs M. 1990. Haustiere—zoologisch gesehen. 2nd ed. Springer-Verlag.

Herschel JFW. 1831. A Preliminary Discourse on the Study of Natural Philosophy. Longmans.

Hetherington AJ, et al. 2015. Do cladistic and morphometric data capture common patterns of morphological disparity? Palaeontology 58:393–399.

Heuser CH, Streeter GL. 1929. Early stages in the development of pig embryos, from the period of initial cleavage to the time of the appearance of limb-buds. Contrib Embryol 20:1–29.

Higgs ES, Jarman MR. 1969. The origins of agriculture: a reconsideration. Antiquity 43:31–41.

Hill WG. 2014. Applications of population genetics to animal breeding, from Wright, Fisher and Lush to genomic prediction. Genetics 196:1–16.

Hilzheimer M. 1912. Geschichte unserer Haustiere. T. Thomas.

Hilzheimer M. 1926. Natürliche Rassengeschichte der Haussäugetiere. De Gruyter.

Hipsley CA, Sherratt E. 2019. Psychology, not technology, is our biggest challenge to open digital morphology data. Sci Data 6:41.

Hirasawa T, Kuratani S. 2013. A new scenario of the evolutionary derivation of the mammalian diaphragm from shoulder muscles. J Anat 222:504–517.

His W. 1868. Untersuchungen über die erste Anlage des Wirbelthierleibes. Die erste Entwickelung des Hühnchens im Ei. FCW Vogel.

Hongo H. 2017. Introduction of domestic animals to the Japanese archipelago. In: The Oxford Handbook of Zooarchaeology (Albarella U, et al. eds).

Hongo H, et al. 2009. The process of ungulate domestication at Cayönü, Southeastern Turkey: a multidisciplinary approach focusing on *Bos* sp. and *Cervus elaphus*. Anthropozoologica 44:63–78.

Honka J, et al. 2018. Over a thousand years of evolutionary history of domestic geese from Russian archaeological sites, analysed using ancient DNA. Genes 9:367.

Horlacher WR. 1930. Studies on inheritance in pigeons. VII. Inheritance of red and black color patterns in pigeons. Genetics 15:312–346.

Horschler DJ, et al. 2019. Absolute brain size predicts dog breed differences in executive function. Anim Cogn 22:187–198.

Houle D. 1998. High enthusiasm and low R-squared. Evolution 52:1872–1876.

Hoyt AM. 1982. History of Texas Longhorns. Texas Longhorn J:1–48.

Hsiao EY, et al. 2013. Microbiota modulate behavioral and physiological abnormalities associated with neurodevelopmental disorders. Cell 155:1451–1463.

Hubrecht RC, Kirkwood J. 2010. The UFAW Handbook on the Care and Management of Laboratory and Other Research Animals. John Wiley & Sons.

Huidekoper RS. 1891. Age of the Domestic Animals: Being a Complete Treatise on the Dentition of the Horse, Ox, Sheep, Hog, and Dog, and on the Various Other Means of Determining the Age of These Animals. FA Davis.

Hunter P. 2007. The silence of genes. Is genomic imprinting the software of evolution or just a battleground for gender conflict? EMBO Rep 8:441–443.

Hüppi E, et al. 2019. Development of the chondrocranium in the domesticated fowl (*Gallus gallus* f. *domestica*), with a study on the variation of the hypoglossal foramina. Vertebr Zool 69:299–310.

Hüppi E, Werneburg I, Sánchez-Villagra MR. 2021. Evolution and development of the bird chondrocranium. Front Zool 18:21.

Huxley J. 1932. Problems of Relative Growth. Methuen & Co.

Ikebuchi M, et al. 2017. Chick development and asynchroneous hatching in the Zebra Finch (*Taeniopygia guttata castanotis*). Zool Sci 34:369–376.

Imsland F, et al. 2012. The Rose-comb mutation in chickens constitutes a structural rearrangement causing both altered comb morphology and defective sperm motility. PLoS Genet 8:e1002775.

Ingold T. 1980. Hunters, Pastoralists and Ranchers: Reindeer Economies and their Transformations. Cambridge University Press.

Ingold T. 2000. The Perception of the Environment: Essays in Livelihood, Dwelling and Skill. Routledge.

Irie N, Kuratani S. 2011. Comparative transcriptome analysis reveals vertebrate phylotypic period during organogenesis. Nat Commun 2:248.

Irving-Pease EK, et al. 2018. Rabbits and the specious origins of domestication. Trends Ecol Evol 33:149–152.

Ishii M, et al. 2003. Msx2 and Twist cooperatively control the development of the neural crest-derived skeletogenic mesenchyme of the murine skull vault. Development 130:6131–6142.

Ito F, Matsumoto T, Hirata T. 2019. Frequent nonrandom shifts in the temporal sequence of developmental landmark events during teleost evolutionary diversification. Evol Dev 21:120–134.

Iwaniuk AN, Hurd PL. 2005. The evolution of cerebrotypes in birds. Brain Behav Evol 65:215–230.

Jablonski D. 2017. Approaches to macroevolution: 1. General concepts and origin of variation. Evol Biol 44:427–450.

Jabot G, et al. 2009. Intracranial lipomas: clinical appearances on neuroimaging and clinical significance. J Neurol 256:851–855.

Jacob F. 1981. Le Jeu des possibles. Fayard.

Janis CM. 1990. Correlation of cranial and dental variables with body size in ungulates and macropodoids. In: Body Size in Mammalian Paleobiology: Estimation and Biological Implications (Damuth JD, MacFadden BJ eds). Cambridge University Press, 255–299.

Jansen T, et al. 2002. Mitochondrial DNA and the origins of the domestic horse. PNAS 99:10905–10910.

Janssens LAA, Peeters S. 1997. Comparisons between stress incontinence in women and sphincter mechanism incompetence in the female dog. Vet Rec 141:620–625.

Janssens L, et al. 2016. Can orbital angle morphology distinguish dogs from wolves? Zoomorphology 135:149–158.

Janssens L, et al. 2019a. An evaluation of classical morphologic and morphometric parameters reported to distinguish wolves and dogs. J Archaeol Sci Rep 23:501–533.

Janssens LA, et al. 2019b. Bony labyrinth shape differs distinctively between modern wolves and dogs. Zoomorphology 138:409–417

Jeannin S. 2018. Cognition and emotions in dog domestication. In: Hybrid Communities: Biosocial Approaches to Domestication and Other Trans-species Relationships (Stépanoff C, Vigne J-D eds). Routledge, 324.

Jeffery JE, et al. 2002. Analyzing evolutionary patterns in amniote embryonic development. Evol Dev 4:292–302.

Jensen P. 2014. Behavior genetics and the domestication of animals. Annu Rev Anim Biosci 2:85–104.

Jensen P ed. 2017. The Ethology of Domestic Animals. An Introductory Text. CABI Publishing.

Jerison HJ. 1973. Evolution of the Brain and Intelligence. Academic Press.

Jetz W, et al. 2012. The global diversity of birds in space and time. Nature 491:444–448.

Jezyk PF. 1985. Constitutional disorders of the skeleton in dogs and cats. In: Textbook of Small Animal Orthopaedics (Newton CD, Nunamaker DM eds). International Veterinary Information Service.

Johnsson M, Henriksen R, Wright D. 2021. The neural crest cell hypothesis: no unified explanation for domestication. Genetics iyab097, https://doi.org/10.1093/genetics/iyab097.

Johnston RF, Johnson SG. 1989. Nonrandom mating in feral pigeons. Condor 91:23–29.

Joller S, et al. 2018. Crossed beaks in a local Swiss chicken breed. BMC Vet Res 14:68.

Jones MEH, et al. 2019. Digital dissection of the head of the rock dove (*Columba livia*) using contrast-enhanced computed tomography. Zool Lett 5:17.

Jones O'Day S, Van Neer W, Ervynck A eds. 2004. Behaviour behind Bones: The Zooarchaeology of Ritual, Religion, Status and Identity. Oxbow Books.

Joshi MB, et al. 2004. Phylogeography and origin of Indian domestic goats. Mol Biol Evol 21:454–462.

Julian LM, et al. 1957. Premature closure of the spheno-occipital synchondrosis in the horned Hereford dwarf of the "short-headed" variety. Am J Anat 100:269–287.

Kadletz M. 1932. Anatomischer Atlas der Extremitätengelenke von Pferd und Hund. Urban & Schwarzenberg.

Kadwell M, et al. 2001. Genetic analysis reveals the wild ancestors of the llama and the alpaca. Proc R Soc Lond B 268:2575–2584.

Kaiser G. 1971. Die Reproduktionsleistung der Haushunde in ihrer Beziehung zur Körpergröße und zum Gewicht der Rassen. Z Tierz Züchtungsbio 88:118–168.

Kamalakkannan R, et al. 2020. The complete mitochondrial genome of Indian gaur, *Bos gaurus* and its phylogenetic implications. Sci Rep 10:11936.

Kaplan BS, Webby RJ. 2013. The avian and mammalian host range of highly pathogenic avian H5N1 influenza. Virus Res 178:3–11.

Katajamaa R, et al. 2018. Activity, social and sexual behaviour in red junglefowl selected for divergent levels of fear of humans. PLoS ONE 13:e0204303.

Katajamaa R, et al. 2021. Cerebellum size is related to fear memory and domestication of chickens. Biol Lett 17:20200790.

Kause A, et al. 2003. Big and beautiful? Quantitative genetic parameters for appearance of large rainbow trout. J Fish Biol 62:610–622.

Kavanagh K. 2020. Developmental plasticity associated with early structural integration and evolutionary patterns: examples of developmental bias and developmental facilitation in the skeletal system. Evol Dev 22:196–204.

Kavanagh KD, Bailey CS, Sears KE. 2020. Evidence of five digits in embryonic horses and developmental stabilization of tetrapod digit number. Proc R Soc Lond B 287:20192756.

Kawabe S, et al. 2013. Variation in avian brain shape: relationship with size and orbital shape. J Anat 223:495–508.

Kawabe S, et al. 2015. Ontogenetic shape change in the chicken brain: implications for paleontology. PLoS ONE 10:e0129939.

Kawabe S, et al. 2017. Morphological variation in brain through domestication of fowl. J Anat 231:287–297.

Kawahara AY. 2007. Thirty-foot telescopic nets, bug-collecting video games, and beetle pets: entomology in modern Japan. Am Entomol 53:160–172.

Kays R, et al. 2020. The small home ranges and large local ecological impacts of pet cats. Anim Conserv.

Kealy JK, McAllister H, Graham JP. 2011. Diagnostic Radiology and Ultrasonography of the Dog and Cat. Saunders Elsevier.

Keibel F. 1897a. Normentafeln zur Entwicklungsgeschichte der Wirbelthiere (Vol. 1). G. Fischer.

Keibel F. 1897b. Normentafel zur Entwicklungsgeschichte des Schweines, *Sus scrofa domesticus*, vol. 1. G. Fischer.

Keibel F. 1905. Normentafeln zur Entwicklungsgeschichte der Wirbelthiere (Vol. 5). G. Fischer.

Keibel F. 1906. Die Entwickelung der äußeren Körperform der Wirbeltierembryonen, insbesondere der menschlichen Embryonen aus den ersten 2 Monaten. In: Handbuch der vergleichenden und experimentellen Entwickelungsgeschichte der Wirbeltiere Vol 1/2, (Hertwig O ed). G. Fischer, 1–176.

Keibel F, Abraham K. 1900. Normentafel zur Entwicklungsgeschichte des Huhnes. Heft 2 der Normentafeln zur Entwicklungsgeschichte der Wirbeltiere. G. Fischer.

Kelekna P. 2009. The Horse in Human History. Cambridge University Press.

Kelly RL. 2013. The Lifeways of Hunter-Gatherers: The Foraging Spectrum. Cambridge University Press.

Kelm H. 1938. Die postembryonale Schädelentwicklung des Wild-und Berkshire-Schweins. Z Anat Entwicklungsgesch 108:499–559.

Kempf A, et al. 2019. A potassium channel β-subunit couples mitochondrial electron transport to sleep. Nature 568:230–234.

Kerje S, et al. 2004. The *Dominant white*, *Dun* and *Smoky* color variants in chicken are associated with insertion/deletion polymorphisms in the PMEL17 gene. Genetics 168:1507–1518.

Keyte A, Hutson MR. 2012. The neural crest in cardiac congenital anomalies. Differentiation 84:25–40.

Khanal P, Nielsen MO. 2017. Impacts of prenatal nutrition on animal production and performance: a focus on growth and metabolic and endocrine function in sheep. J Anim Sci Biotechnol 8:75.

Kihlström JE. 1972. Period of gestation and body weight in some placental mammals. Comp Biochem Phys A 43A:673–680.

Kihslinger R, Lema SC, Nevitt G. 2006. Environmental rearing conditions produce forebrain differences in wild Chinook salmon *Oncorhynchus tshawytscha*. Comp Biochem Physiol Part A Mol Integr Physiol 145:145–151.

Kihslinger RL, Nevitt GA. 2006. Early rearing environment impacts cerebellar growth in juvenile salmon. J Exp Biol 209:504–509.

Kijas JW, et al. 2012. Genome-wide analysis of the world's sheep breeds reveals high levels of historic mixture and strong recent selection. PLoS Biol 10:e1001258.

Kijas JW, et al. 2013. Genetic diversity and investigation of polledness in divergent goat populations using 52 088 SNPs. Anim Genet 44:325–335.

Kilbourne BM, Hoffman LC. 2013. Scale effects between body size and limb design in quadrupedal mammals. PLoS ONE 8:e78392.

Kim B. 1933. Rassenunterschiede am embryonalen Schweineschädel und ihre Entstehung. Z Morph Anthrop 32:486–523.

Kimmel CB, et al. 1995. Stages of embryonic development of the zebrafish. Dev Dyn 203:253–310.

Kimura B, et al. 2013. Donkey domestication. Afr Archaeol Rev 30:83–95.

Kindler C, et al. 2017. Hybridization patterns in two contact zones of grass snakes reveal a new Central European snake species. Sci Rep 7:7378.

King HD. 1939. Life processes in gray Norway rats during fourteen years in captivity. Am Anat Mem 17.

King HD, Donaldson HH. 1929. Life processes and size of the body and organs of the gray Norway rat during ten generations in captivity. Am Anat Mem 14.

King JWB, Roberts RC. 1960. Carcass length in the bacon pig; its association with vertebrae numbers and prediction from radiographs of the young pig. Anim Sci J 2:59–65.

Kinne J, et al. 2010. Is there a two-humped stage in the embryonic development of the dromedary? Anat Histol Embryol 39:479–480.

Kirkwood TBL. 1977. Evolution of ageing. Nature 270:301–304.

Kirkwood JK. 1985. The influence of size on the biology of the dog. J Small Anim Pract 26:97–110.

Kissel P, Andre JM, Jacquier A. 1981. The Neurocristopathies. Masson Publishing.

Kistner TM, et al. 2021. Geometric morphometric investigation of craniofacial morphological change in domesticated silver foxes. Sci Rep 11:2582.

Klatt B. 1910. Zur Anatomie der Haubenhühner. Zool Anz 36:282–288.

Klatt B. 1911. Zur Frage der Hydrocephalie bei den Haubenhühnern. Sber Ges naturf Freunde Berl 2:75–84.

Klatt B. 1913. Über den Einfluss der Gesamtgrösse auf das Schädelbild nebst Bemerkungen über die Vorgeschichte der Haustiere. Arch Entwicklungsmechanik Organismen 36:387–471.

Klatt B. 1955. Noch einmal: Hirngröße und Körpergröße. Zool Anz 155:215–232.

Klein A-M, et al. 2007. Importance of pollinators in changing landscapes for world crops. Proc R Soc Lond B 274:303–313.

Klevezal GA. 1996. Recording Structures of Mammals. Determination of Age and Reconstruction of Life History. A.A. Balkema.

Klingenberg CP. 1998. Heterochrony and allometry: the analysis of evolutionary change in ontogeny. Biol Rev 73:79–123.

Klingenberg CP. 2005. Developmental constraints, modules, and evolvability. In: Variation (Hallgrímsson B, Hall BK eds). Academic Press, 219–247.

Klingenberg CP, Marugán-Lobón J. 2013. Evolutionary covariation in geometric morphometric data: analyzing integration, modularity, and allometry in a phylogenetic context. Syst Biol 62:591–610.

Knospe C. 2002. Periods and stages of the prenatal development of the domestic cat. Anat Histol Embryol 31:37–51.

Koch H, Schmid-Hempel P. 2011. Socially transmitted gut microbiota protect bumble bees against an intestinal parasite. PNAS 108:19288–19292.

Koecke HU. 1958. Normalstadien der Embryonalenentwicklung bei der Hausente (*Anas boschas domestica*). Embryologia 4:55–78.

Kohane MJ, Parsons PA. 1988. Domestication. Evol Biol 23:31–48.

Köhler M, Moyà-Solà S. 2009. Physiological and life history strategies of a fossil large mammal in a resource-limited environment. PNAS 106:20354–20358.

Kolb C, et al. 2015. Mammalian bone palaeohistology: a survey and new data with emphasis on island forms. PeerJ 3:e1358.

Kon T, et al. 2020. The genetic basis of morphological diversity in domesticated goldfish. Curr Biol 30:1–15.

König B, Lindholm AK. 2012. The complex social environment of female house mice (*Mus domesticus*). In: Evolution of the House Mouse (Macholán M, et al. eds). Cambridge University Press, 114–134.

König B, et al. 2015. A system for automatic recording of social behavior in a free-living wild house mouse population. Anim Biotelemetry 3:39.

Korn AK, Brandt HR, Erhardt G. 2016. Genetic and environmental factors influencing tooth and jaw malformations in rabbits. Vet Rec 178:341–341.

Kotrschal K, Van Staaden MJ, Huber R. 1998. Fish brains: evolution and environmental relationships. Rev Fish Biol Fish 8:373–408.

Koungoulos L. 2020. Old dogs, new tricks: 3D geometric analysis of cranial morphology supports ancient population substructure in the Australian dingo. Zoomorphology 139:263–275.

Koutsogiannouli EA, et al. 2010. Detection of hybrids between wild boars (*Sus scrofa scrofa*) and domestic pigs (*Sus scrofa* f. *domestica*) in Greece, using the PCR-RFLP method on melanocortin-1 receptor (MC1R) mutations. Mamm Biol 75:69–73.

Koyabu D, Maier W, Sánchez-Villagra MR. 2012. Paleontological and developmental evidence resolve the homology and dual embryonic origin of a mammalian skull bone, the interparietal. PNAS 109:14075–14080.

Koyabu D, et al. 2014. Mammalian skull heterochrony reveals modular evolution and a link between cranial development and brain size. Nat Commun 5:3625.

Kozák J. 2019. Variations of geese under domestication. Worlds Poult Sci J 75:247–260.

Kraatz BP, et al. 2015. Ecological correlates to cranial morphology in Leporids (Mammalia, Lagomorpha). PeerJ 3:e844.

Krämer E-M. 2009. Der grosse Kosmos Hundeführer. Franckh-Kosmos Verlag.

Kratochvíl Z. 1973. Schädelkriterien der Wild-und Hauskatze (*Felis silvestris silvestris* Schreb. 1777 und *Felis* s. f. *catus* L. 1758). Acta Sc Nat Brno 7:1–50.

Kratochwil CF, et al. 2018. Agouti-related peptide 2 facilitates convergent evolution of stripe patterns across cichlid fish radiations. Science 362:457–460.

Kraus C, Pavard S, Promislow DE. 2013. The size–life span trade-off decomposed: why large dogs die young. Am Nat 181:492–505.

Krautwald F. 1910. Die Haube der Hühner und Enten [Doctoral dissertation]. Universität Bern, Schweiz.

Krölling O. 1924. Die Form-und Organentwicklung des Hausrindes (*Bos taurus* L.) im ersten Embryonalmonat. Z Anat Entwicklungs 72:1–54.

Krölling O. 1925. Die Entwicklung des äußeren Ohres beim Hausrind (*Bos taurus* L.). Z Anat Entwicklungs 76:548–560.

Krölling O. 1942. Zur Frühentwicklung der Extremitäten beim Pferd. Z Anat Entwicklungs 111:490–507.

Kruska DCT. 1973. Cerebralisation, Hirnevolution und domestikationsbedingte Hirngrößenänderungen innerhalb der Ordnung Perissodactyla Owen, 1848 und ein Vergleich mit der Ordnung Artiodactyla Owen, 1848. J Zool Syst Evol Res 11:81–103.

Kruska DCT. 2005. On the evolutionary significance of encephalization in some eutherian mammals: effects of adaptive radiation, domestication, and feralization. Brain Behav Evol 65:73–108.

Kruska DCT. 2007. The effects of domestication on brain size. In: Evolution of Nervous Systems. Vol 3. The Evolution of Nervous Systems in Mammals (Krubitzer L, Kaas J eds). Elsevier, 143–153.

Kruska DCT, Röhrs M. 1974. Comparative–quantitative investigations on brains of feral pigs from the Galapagos Islands and of European domestic pigs. Z Anat Entwicklungsgesch 144:61–73.

Kruska DCT, Sidorovich VE. 2003. Comparative allometric skull morphometries in mink (*Mustela vison* Schreber, 1777) of Canadian and Belarus origin; taxonomic status. Mamm Biol 68:257–276.

Kruska DC, Steffen K. 2013. Comparative allometric investigations on the skulls of wild cavies (*Cavia aperea*) versus domesticated guinea pigs (*C. aperea f. porcellus*) with comments on the domestication of this species. Mamm Biol 78:178–186.

Kruska DCT, Stephan H. 1973. Volumenvergleich allokortikaler Hirnzentren bei Wild- und Hausschweinen. Acta Anat 84:387–415.

Kuhn M. 2001. Zur Morphogenese der Ohrregion spätfetaler Stadien von *Bos taurus* (Artiodactyla, Mammalia) [Diplomarbeit]. Universität Tübingen.

Kukekova AV, et al. 2011. Mapping loci for fox domestication: deconstruction/reconstruction of a behavioral phenotype. Behav Genet 41:593–606.

Kumar V, et al. 2015. Genetic diversity and phylogenetic relationship analysis between red jungle fowl and domestic chicken using AFLP markers. J Poult Sci 52:94–100.

Künzel E, et al. 1962. Die Entwicklung des Hühnchens im Ei. ZBL Veterinärmedizin 9:371–396.

Künzel W, Breit S, Oppel M. 2003. Morphometric investigations of breed-specific features in feline skulls and considerations on their functional implications. Anat Histol Embryol 32:218–223.

Künzl C, Sachser N. 1999. The behavioral endocrinology of domestication: a comparison between the domestic guinea pig (*Cavia aperea* f. *porcellus*) and its wild ancestor, the cavy (*Cavia aperea*). Horm Behav 35:28–37.

Kuo ZY. 1932. Ontogeny of embryonic behavior in Aves. I. The chronology and general nature of the behavior of the chick embryo. J Exp Zool 61:395–430.

Kupczik K, et al. 2017. The dental phenotype of hairless dogs with FOXI3 haploinsufficiency. Sci Rep 7:5459.

Kupczyńska M, et al. 2009. Dentition in brachycephalic dogs. Med Weter 65:334–339.

Kurushima J, et al. 2013. Variation of cats under domestication: genetic assignment of domestic cats to breeds and worldwide random-bred populations. Anim Genet 44:311–324.

Kwong WK, Moran NA. 2016. Gut microbial communities of social bees. Nat Rev Microbiol 14:374–384.

Laidlaw Jr HH. 1944. Artificial insemination of the queen bee (*Apis mellifera* L.): morphological basis and results. J Morphol 74:429–465.

Laland KN, Brown GR. 2011. Sense & Nonsense: Evolutionary Perspectives on Human Behaviour. 2nd ed. Oxford University Press.

Lamberson WR, et al. 1991. Direct responses to selection for increased litter size, decreased age at puberty, or random selection following selection for ovulation rate in swine. J Anim Sci 69:3129–3143.

Landau SY, Provenza FD. 2020. Of browse, goats, and men: contribution to the debate on animal traditions and cultures. Appl Anim Behav Sci 232:105127.

Landauer W, Chang TK. 1949. The Ancon or Otter sheep: history and genetics. J Hered 40:105–112.

Lande R. 1978. Evolutionary mechanisms of limb loss in tetrapods. Evolution 32:73–92.

Lande R, Arnold SJ. 1983. The measurement of selection on correlated characters. Evolution 37:1210–1226.

Lange A, Müller G. 2017. Polydactyly in development, inheritance, and evolution. Q Rev Biol 92:1–38.

Lange A, Nemeschkal HL, Müller GB. 2014. Biased polyphenism in polydactylous cats carrying a single point mutation: the Hemingway model for digit novelty. Evol Biol 41:262–275.

Lansing JS, Kremer JN. 2011. Rice, fish, and the planet. PNAS 108:19841–19842.

Larsen CS. 2006. The agricultural revolution as environmental catastrophe: implications for health and lifestyle in the Holocene. Quat Int 150:12–20.

Larson G, et al. 2005. Worldwide phylogeography of wild boar reveals multiple centers of pig domestication. Science 307.

Larson G, Burger J. 2013. A population genetics view of animal domestication. Trends Genet 29:197–205.

Larson G, Fuller DQ. 2014. The evolution of animal domestication. Annu Rev Ecol Evol Syst 45:115–136.

Larson G, et al. 2005. Worldwide phylogeography of wild boar reveals multiple centers of pig domestication. Science 307:1618–1621.

Larson G, et al. 2012. Rethinking dog domestication by integrating genetics, archeology, and biogeography. PNAS 109:8878–8883.

Law R. 1979. Optimal life histories under age-specific predation. Am Nat 114:399–417.

Lawal RA, et al. 2020. The wild species genome ancestry of domestic chickens. BMC Biol 18:1–18.

Lawling AM, Polly PD. 2010. Geometric morphometrics: recent applications to the study of evolution and development. J Zool 280:1–7.

Lawrence TLJ, Fowler VR, Novakofski JE. 2012. Growth of Farm Animals. 3rd ed. CABI.

Lazard J. 2009. La pisciculture des tilapias. Cah Agric 18:174–182.

Lázaro J, et al. 2018. Profound seasonal changes in brain size and architecture in the common shrew. Brain Struct Funct 223:2823–2840.

Le Douarin N. 2005. The Nogent Institute—50 years of embryology. Int J Dev Biol 49:85–103.

Le Douarin N, McLaren A eds. 1984. Chimeras in Developmental Biology. Academic Press.

Leach HM. 2003. Human domestication reconsidered. Curr Anthropol 44:349–368.

Leary RF, Allendorf FW, Knudsen KL. 1985. Developmental instability as an indicator of reduced genetic variation in hatchery trout. Trans Am Fish Soc 114:230–235.

Leathlobhair MN, et al. 2018. The evolutionary history of dogs in the Americas. Science 361:81–85.

Lebas F, et al. 1997. The Rabbit: husbandry, health, and production. Food and Agriculture Organization of the United Nations, Rome.

LeBlanc M, Festa-Bianchet M, Jorgenson JT. 2001. Sexual size dimorphism in bighorn sheep (*Ovis canadensis*): effects of population density. Can J Zool 79:1661–1670.

Lecocq T. 2018. Insects: the disregarded domestication histories. In: Animal Domestication (Teletchea F ed). Intech Open.

Lecointre G, Schnell NK, Teletchea F. 2020. Hierarchical analysis of ontogenetic time to describe heterochrony and taxonomy of developmental stages. Sci Rep 10:19732.

Lee JC. 1996. The Amphibians and Reptiles of the Yucatan Peninsula. Cornell University Press.

LeFebvre MJ, DeFrance SD. 2014. Guinea pigs in the pre-Columbian West Indies. J Island Coast Archaeol 9:16–44.

LeGrand EK, Brown CC. 2002. Darwinian medicine: applications of evolutionary biology for veterinarians. Can Vet J 43:556–559.

Lehman N, et al. 1991. Introgression of coyote mitochondrial DNA into sympatric North American gray wolf populations. Evolution 45:104–119.

Leight H. 1960. Pre-Inca Art and Culture. Orion Press.

Lemoine X, et al. 2014. A new system for computing dentition-based age profiles in *Sus scrofa*. J Archaeol Sci 47:179–193.

Lemus D. 1995. Contributions of heterospecific tissue recombinations to odontogenesis. Int J Dev Biol 39:291–297.

Lemus D, et al. 1986. Odontogenesis and amelogenesis in interacting lizard—quail tissue combinations. J Morphol 189:121–129.

Leonard JA, et al. 2002. Ancient DNA evidence for Old World origin of New World dogs. Science 298:1613–1616.

Leonard JA, et al. 2007. Megafaunal extinctions and the disappearance of a specialized wolf ecomorph. Curr Biol 17:1146–1150.

Leonard SP, et al. 2020. Engineered symbionts activate honey bee immunity and limit pathogens. Science 367:573–576.

Lerner IM, Dempster ER. 1951. Attenuation of genetic progress under continued selection in poultry. Heredity 5:75–94.

Leroi AM. 2003. Mutants: On Genetic Variety and the Human Body. Penguin Books.

Leroi AM. 2014. The Lagoon: How Aristotle Invented Science. Bloomsbury Publishing.

Lesbouyries G. 1949. Reproduction des mammifères domestiques: Sexualité. Vigot.

Levins R, Lewontin RC. 1985. The Dialectical Biologist. Harvard University Press.

Lewis JE, et al. 2011. The mismeasure of science: Stephen Jay Gould versus Samuel George Morton on skulls and bias. PLoS Biol 9:e1001071.

Levy MZ, et al. 2015. Bottlenecks in domestic animal populations can facilitate the emergence of *Trypanosoma cruzi*, the aetiological agent of Chagas disease. Proc R Soc Lond B 282:20142807.

Li Y, et al. 2006. cholesterol modification restricts the spread of Shh gradient in the limb bud. PNAS 103:6548–6553.

Li Q, et al. 2011. From organic waste to biodiesel: black soldier fly, *Hermetia illucens*, makes it feasible. Fuel 90:1545–1548.

Li IJ, et al. 2019a. Embryonic and postembryonic development of the ornamental twin-tail goldfish. Dev Dyn 248:251–283.

Li X, et al. 2019b. A breakthrough in the artificial cultivation of Chinese cordyceps on a large-scale and its impact on science, the economy, and industry. Crit Rev Biotechnol 39:181–191.

Li J, et al. 2020. Mutations upstream of the TBX5 and PITX1 transcription factor genes are associated with feathered legs in the domestic chicken. Mol Biol Evol 37:2477–2486.

Librado P, et al. 2015. Tracking the origins of Yakutian horses and the genetic basis for their fast adaptation to subarctic environments. PNAS 112:E6889-E6897.

Librado P, et al. 2017. Ancient genomic changes associated with domestication of the horse. Science 356:442–445.

Lieberman DE. 1993. Life history variables preserved in dental cementum microstructure. Science 261:1162–1164.

Lien ME. 2015. Becoming Salmon: Aquaculture and the Domestication of a Fish. University of California Press.

Lill A, Wood-Gush DGM. 1965. Potential ethological isolating mechanisms and assortative mating in the domestic fowl. Behaviour 25:16–44.

Lillie FR. 1952. Lillie's Development of the Chick: An Intoduction to Embryology (3rd Rev. ed.). Holt, Rinehart and Winston.

Lin J, et al. 2002. Myostatin knockout in mice increases myogenesis and decreases adipogenesis. Biochem Biophys Res Commun 291:701–706.

Lind L ed. 1963. Aldrovandi on Chickens: The Ornithology of Ulisse Aldrovandi (1600). University of Oklahoma Press.

Lindblad-Toh K, et al. 2005. Genome sequence, comparative analysis and haplotype structure of the domestic dog. Nature 438:803–819.

Linde-Medina M. 2016. Testing the cranial evolutionary allometric 'rule' in Galliformes. J Evol Biol 29:1873–1878.

Linderholm A, Larson G. 2013. The role of humans in facilitating and sustaining coat colour variation in domestic animals. Semin Cell Dev Biol. 24(6–7):587–593.

Linderholm A, et al. 2016. A novel MC1R allele for black coat colour reveals the Polynesian ancestry and hybridization patterns of Hawaiian feral pigs. R Soc Open Sci 3:160304.

Linnaeus C. 1758. Systema Naturae. 10th ed. Holmiae Impensis Direct. Laurentii Salvii.

Lipinski MJ, et al. 2008. The ascent of cat breeds: genetic evaluations of breeds and worldwide random-bred populations. Genomics 91:12–21.

Lloyd GT. 2016. Estimating morphological diversity and tempo with discrete character-taxon matrices: implementation, challenges, progress, and future directions. Biol J Linn Soc 118: 131–151.

Lobprise HB, Dodd JRB. 2019. Wiggs's Veterinary Dentistry: Principles and Practice. John Wiley & Sons.

Loog L, et al. 2017. Inferring allele frequency trajectories from ancient DNA indicates that selection on a chicken gene coincided with changes in medieval husbandry practices. Mol Biol Evol 34:1981–1990.

Lopes RJ, et al. 2016. Genetic basis for red coloration in birds. Curr Biol 26:1427–1434.

López-Aguirre C, et al. 2019. Prenatal allometric trajectories and the developmental basis of postcranial phenotypic diversity in bats (Chiroptera). J Exp Zool Part B 332:36–49.

Lord K, et al. 2013. Variation in reproductive traits of members of the genus *Canis* with special attention to the domestic dog (*Canis familiaris*). Behav Processes 92:131–142.

Lord K, Schneider RA, Coppinger R. 2016. Evolution of working dogs. In: The Domestic Dog: Its Evolution, Behavior and Interactions with People (Serpell J ed). Cambridge University Press, 42–66.

Lord E, et al. 2020a. Ancient DNA of guinea pigs (*Cavia* spp.) indicates a probable new center of domestication and pathways of global distribution. Sci Rep 10:8901.

Lord KA, et al. 2020b. The history of farm foxes undermines the animal domestication syndrome. Trends Ecol Evol 35:125–136.

Lord KA, Larson G, Karlsson EK. 2020c. Brain size does not rescue domestication syndrome. Trends Ecol Evol 35:1061–1062.

Lorenz K. 1935. Der Kumpan in der Umwelt des Vogels. J Ornithol 83:289–413.

Lorenz K. 1940. Durch Domestikation verursachte Störungen arteigenen Verhaltens. Z angew Psychol Charakterkd 59:2–81.

Lorenzen K, Beveridge MC, Mangel M. 2012. Cultured fish: integrative biology and management of domestication and interactions with wild fish. Biol Rev 87:639–660.

Losey RJ, et al. 2020. Domestication as enskilment: harnessing reindeer in Arctic Siberia. J Archaeol Method Theory.

Loss SR, Will T, Marra PP. 2013. The impact of free-ranging domestic cats on wildlife of the United States. Nat Commun 4:1396.

Ludwig A, et al. 2009. Coat color variation at the beginning of horse domestication. Science 324:485.

Luff R. 2000. Ducks. In: The Cambridge World History of Food (Kiple KF, Ornelas KC eds). Cambridge University Press, 517–524.

Luikart G, et al. 2001. Multiple maternal origins and weak phylogeographic structure in domestic goats. PNAS 98:5927–5932.

Lumer H. 1940. Evolutionary allometry in the skeleton of the domesticated dog. Am Nat 74:439–467.

Lüps P. 1974. Biometrische Untersuchungen an der Schädelbasis des Haushundes. Zoologischer Anzeiger Jena 5/6:383–413.

Lyne AG, Heideman MJ. 1959. The pre-natal development of skin and hair in cattle (*Bos taurus* L.). Aust J Biol Sci 12:72–95.

Maas SA, Suzuki T, Fallon JF. 2011. Identification of spontaneous mutations within the long-range limb-specific *Sonic Hedgehog* enhancer (ZRS) that alter *Sonic Hedgehog* expression in the chicken limb mutants *oligozeugodactyly* and Silkie breed. Dev Dyn 240:1212–1222.

Macdonald DW, Carr GM. 1995. Variation in dog society: between resource dispersion and social flux. In: The Domestic Dog: Its Evolution, Behaviour, and Interactions with People (Serpell J ed). Cambridge University Press, 199–216.

MacFadden BJ, et al. 2012. Fossil horses, orthogenesis, and communicating evolution in museums. Evol Educ Outreach 5:29–37.

Macho T. 2015. Schweine. Naturkunden. Matthes & Seitz.

Macintosh AA, Pinhasi R, Stock JT. 2016. Early life conditions and physiological stress following the transition to farming in central/southeast Europe: skeletal growth impairment and 6000 years of gradual recovery. PLoS ONE 11:e0148468

MacLachlan SS, et al. 2020. Influence of later exposure to perches and nests on flock level distribution of hens in an aviary system during lay. Poult Sci 99:30–38.

MacLean EL. 2016. Unraveling the evolution of uniquely human cognition. PNAS 113:6348–6354.

MacLean EL, et al. 2012. How does cognition evolve? Phylogenetic comparative psychology. Anim Cogn 15:223–238.

MacNeish RS, Vierra RK. 1983. The preceramic way of life in the thorn forest riverine ecozone. Prehistory of the Ayacucho Basin, Peru 4:48–129.

MacPhee RDE. 2018. End of the Megafauna: The Fate of the World's Hugest, Fiercest, and Strangest Animals. WW Norton & Co.

Madden RH. 2014. Hypsodonty in Mammals. Cambridge University Press.

Maes LD, et al. 2008. Steady locomotion in dogs: temporal and associated spatial coordination patterns and the effect of speed. J Exp Biol 211:138–149.

Maier W. 1993. Cranial morphology of the therian common ancestor, as suggested by the adaptations of neonate marsupials. In: Mammal Phylogeny: Mesozoic Differentiation, Multituberculates, Monotremes, Early Therians and Marsupials (Szalay FS, Novacek MJ, McKenna MC eds). Springer, 165–81.

Mainland I, Schutkowski H, Thomson AF. 2007. Macro- and micromorphological features of lifestyle differences in pigs and wild boar. Anthropozoologica 42:89–106.

Makino T, et al. 2018. Elevated proportions of deleterious genetic variation in domestic animals and plants. Genome Biol Evol 10:276–290.

Mallarino R, et al. 2011. Two developmental modules establish 3D beak-shape variation in Darwin's finches. PNAS 108:4057–4062.

Mallarino R, Abzhanov A. 2012. Paths less traveled: evo-devo approaches to investigating animal morphological evolution. Annu Rev Cell Dev Biol 28:743–763.

Mallet C, Cornette R, Guadelli JL. 2019. Morphometrical distinction between sheep (*Ovis aries*) and goat (*Capra hircus*) using the petrosal bone: application on French protohistoric sites. Int J Osteoarchaeol 29:525–537.

Malmkvist J, Hansen SW. 2002. Generalization of fear in farm mink, *Mustela vison*, genetically selected for behaviour towards humans. Anim Behav 64:487–501.

Malomane DK, et al. 2019. The SYNBREED chicken diversity panel: a global resource to assess chicken diversity at high genomic resolution. BMC Genomics 20:345.

Mann CC. 2005. 1491: New Revelations of the Americas before Columbus. Knopf.

Mann CC. 2011. 1493: Uncovering the New World Columbus Created. Alfred A. Knopf.

Mannen H, Nagata Y, Tsuji S. 2001. Mitochondrial DNA reveal that domestic goat (*Capra hircus*) are genetically affected by two subspecies of bezoar (*Capra aegagurus*). Biochem Genet 39:145–154.

Mannermaa K. 2014. Goose: domestication. In: Encyclopedia of Global Archaeology (Smith C ed). Springer, 3096–3098.

Manser A, et al. 2011. Polyandry and the decrease of a selfish genetic element in a wild house mouse population. Evolution 65:2435–2447.

Manzano MG, Návar J. 2000. Processes of desertification by goats overgrazing in the Tamaulipan thornscrub (matorral) in north-eastern Mexico. J Arid Environ 44:1–17.

Marchant TW, et al. 2017. Canine brachycephaly is associated with a retrotransposon-mediated misssplicing of SMOC2. Curr Biol 27:1573–1584. e1576.

Marchetti MP, Nevitt GA. 2003. Effects of hatchery rearing on brain structures of rainbow trout, *Oncorhynchus mykiss*. Environ Biol Fishes 66:9–14.

Marchini M, Rolian C. 2018. Artificial selection sheds light on developmental mechanisms of limb elongation. Evolution 72:825–837.

Marcucio RS, et al. 2011. Mechanisms that underlie co-variation of the brain and face. Genesis 49:177–189.

Marcy AE, et al. 2018. Low resolution scans can provide a sufficiently accurate, cost- and time-effective alternative to high resolution scans for 3D shape analyses. PeerJ 6:e5032.

Marín JC, et al. 2017. Y-chromosome and mtDNA variation confirms independent domestications and directional hybridization in South American camelids. Anim Genet 48:591–595.

Marín-Moratalla N, Jordana X, Köhler M. 2013. Bone histology as an approach to providing data on certain key life history traits in mammals: implications for conservation biology. Mamm Biol 78:422–429.

Marinov M, et al. 2018. Mitochondrial diversity of Bulgarian native dogs suggests dual phylogenetic origin. PeerJ 6:e5060.

Marra PP, Santella C. 2016. Cat Wars: The Devastating Consequences of a Cuddly Killer. Princeton University Press.

Marroig G, et al. 2009. The evolution of modularity in the mammalian skull II: evolutionary consequences. Evol Biol 36:136–148.

Marroig G, Cheverud J. 2010. Size as a line of least resistance II: direct selection on size or correlated response due to constraints? Evolution 64:1470–1488.

Marsden CD, et al. 2016. Bottlenecks and selective sweeps during domestication have increased deleterious genetic variation in dogs. PNAS 113:152–157.

Marshall F, Weissbrod L. 2011. Domestication processes and morphological change: through the lens of the donkey and African pastoralism. Curr Anthropol 52:S397-S413.

Marshall FB, et al. 2014. Evaluating the roles of directed breeding and gene flow in animal domestication. PNAS 111:6153–6158.

Martik ML, Bronner ME. 2017. Regulatory logic underlying diversification of the neural crest. Trends Genet 33:715–727.

Martin LF, et al. 2019. The way wear goes: phytolith-based wear on the dentine–enamel system in guinea pigs (Cavia porcellus). Proc R Soc Lond B 286:20191921.

Martini P, Schmid P, Costeur L. 2018. Comparative morphometry of Bactrian camel and dromedary. J Mamm Evol 25:407–425.

Martini A, et al. 2020. Plasticity of the skeleton and skeletal deformities in zebrafish (Danio rerio) linked to rearing density. J Fish Biol 98:971–986.

Martins AF, et al. 2015. R^{2}OBBIE-3D, a fast robotic high-resolution system for quantitative phenotyping of surface geometry and colour-texture. PLoS ONE 10:e0126740.

Masai H, Takatsuji K, Sato Y. 1982. Morphological variability of the brains under domestication from the crucian carp to the goldfish. J Zool Syst Evol Res 20:112–118.

Maselli V, et al. 2014. A dysfunctional sense of smell: the irreversibility of olfactory evolution in free-living pigs. Evol Biol 41:229–239.

Mason IL ed. 1984. Evolution of Domesticated Animals. Longman Group.

Masseti M. 2012. Atlas of Terrestrial Mammals of the Ionian and Aegean Islands. Walter de Gruyter.

Maturana-Romesín H, Mpodozis J. 2000. The origin of species by means of natural drift. Rev Chil Hist Nat 73:261–310.

Maxwell EE. 2008. Comparative embryonic development of the skeleton of the domestic turkey (Meleagris gallopavo) and other galliform birds. Zoology 111:242–257.

Maxwell EE, Harrison LB. 2009. Methods for the analysis of developmental sequence data. Evol Dev 11:109–119.

Mayyas AS, et al. 2012. Beeswax preserved in archaeological ceramics: function and use. Annals of Faculty of Arts Ain Shams University 40:343–371.

Mazengenya P, et al. 2017. Putative adult neurogenesis in two domestic pigeon breeds (Columba livia domestica): Racing Homer versus utility Carneau pigeons. Neural Regen Res 12:1086–1096.

Mbayahaga J, et al. 1998. Body weight, oestrous and ovarian activity in local Burundian ewes and goats after parturition in the dry season. Animal Reprod Sci 51:289–300.

McAndrew B, et al. 1988. The genetics and histology of red, blond and associated colour variants in Oreochromis niloticus. Genetica 76:127–137.

McCarthy EM. 2006. Handbook of Avian Hybrids of the World. Oxford University Press.

McGinnity P, et al. 1997. Genetic impact of escaped farmed Atlantic salmon (Salmo salar L.) on native populations: use of DNA profiling to assess freshwater performance of wild, farmed, and hybrid progeny in a natural river environment. ICES J Mar Sci 54:998–1008.

McGreevy PD, et al. 2018. Labrador retrievers under primary veterinary care in the UK: demography, mortality and disorders. Canine Genet Epidemiol 5:8.

McGrosky A, et al. 2016. Gross intestinal morphometry and allometry in Carnivora. Eur J Wildl Res 62:395–405.

McKellar AE, Hendry AP. 2009. How humans differ from other animals in their levels of morphological variation. PLoS ONE 4:e6876.

McKeown M. 1975. Craniofacial variability and its relationship to disharmony of the jaws and teeth. J Anat 119:579.

McLain N. 1885. Artificial fertilization. Report on experiments in apiculture. In: Report of the Entomologist Annual Report of the (U S) Commissioner of Agriculture, 339–342.

McLaren A. 1976. Growth from fertilization to birth in the mouse. In: Proceedings of the Embryogenesis in Mammals Ciba Foundation Symposium, 47–51.

McPhee ME. 2004. Generations in captivity increases behavioral variance: considerations for captive breeding and reintroduction programs. Biol Conserv 115:71–77.

McPhee ME, Carlstead K. 2012. The importance of maintaining natural behaviors in captive mammals. In: Wild Mammals in Captivity: Principles and Techniques for Zoo Management, 2nd ed. (Kleiman DG, Thompson KV, Baer CK, eds). University of Chicago Press, 303–313.

McPherron AC, Lawler AM, Lee S-J. 1997. Regulation of skeletal muscle mass in mice by a new TGF-p superfamily member. Nature 387:83–90.

Mead CS. 1909. The chondrocranium of an embryo pig, *Sus scrofa*. Am J Anat 9:167–215.

Meadow RH. 1989. Osteological evidence for the process of animal domestication. In: The Walking Larder: Patterns of Domestication, Pastoralism, and Predation (Clutton-Brock J ed). Unwin Hyman, 80–90.

Meadow R, Patel AK. 2003. Prehistoric pastoralism in northwestern South Asia from the Neolithic through the Harappan period. In: Indus Ethnobiology New Perspectives from the Field (Weber S, Belcher W eds). Lexington Books, 65–94.

Meadows J, Hiendleder S, Kijas J. 2011. Haplogroup relationships between domestic and wild sheep resolved using a mitogenome panel. Heredity 106:700–706.

Meagher R, et al. 2012. Decreased litter size in inactive female mink (*Neovison vison*): mediating variables and implications for overall productivity. Can J Anim Sci 92:92131–92141.

Medawar PB. 1952. An Unsolved Problem of Biology. H. K. Lewis.

Medina FM, et al. 2011. A global review of the impacts of invasive cats on island endangered vertebrates. Glob Change Biol 17:3503–3510.

Mekonnen M, Moges N. 2016. A review on dystocia in cows. Eur J Biol Sci 8:91–100.

Menegaz RA, et al. 2009. Phenotypic plasticity and function of the hard palate in growing rabbits. Anat Rec 292:277–284.

Mengoni Goñalons GL, Yacobaccio HD. 2006. The domestication of South American camelids: a view from the South-Central Andes. In: Documenting Domestication: New Genetic and Archaeological Paradigms (Zeder MA, et al. eds), 228–244. University of California Press.

Menotti-Raymond M, et al. 2008. Patterns of molecular genetic variation among cat breeds. Genomics 91:1–11.

Meredith AL, Prebble JL, Shaw DJ. 2015. Impact of diet on incisor growth and attrition and the development of dental disease in pet rabbits. J Small Anim Pract 56:377–382.

Merilä J, Hendry AP. 2014. Climate change, adaptation, and phenotypic plasticity: the problem and the evidence. Evol Appl 7:1–14.

Merrill AE, et al. 2008. Mesenchyme-dependent BMP signaling directs the timing of mandibular osteogenesis. Development 135:1223–1234.

Mess A. 1999. The evolutionary differentiation of the rostral nasal skeleton within Glires. A review with new data on lagomorph ontogeny. Zoosyst Evol 75:217–228.

Metscher BD. 2009. MicroCT for comparative morphology: simple staining methods allow high-contrast 3D imaging of diverse non-mineralized animal tissues. BMC Physiol 9:11.

Metzger J, Pfahler S, Distl O. 2016. Variant detection and runs of homozygosity in next generation sequencing data elucidate the genetic background of Lundehund syndrome. BMC Genomics 17:535.

Michl E. 1920. Beitrag zur Entwicklungsgeschichte von *Bos taurus* L. Anat Anz 53:193–215.

Mielke B, Lam R, Ter Haar G. 2017. Computed tomographic morphometry of tympanic bulla shape and position in brachycephalic and mesaticephalic dog breeds. Vet Radiol Ultrasound 58:552–558.

Miglino MA, et al. 2006. The carnivore pregnancy: the development of the embryo and fetal membranes. Theriogenology 66:1699–1702.

Mignon-Grasteau S, et al. 2005. Genetics of adaptation and domestication in livestock. Livest Prod Sci 93:3–14.

Miles AEW, Grigson C eds. 1990. Colyer's Variations and Diseases of the Teeth of Animals. Cambridge University Press.

Milinkovitch MC, Tzika A. 2007. Escaping the mouse trap: the selection of new evo-devo model species. J Exp Zool Part B 308:337–346.

Milocco L, Salazar-Ciudad I. 2020. Is evolution predictable? Quantitative genetics under complex genotype-phenotype maps. Evolution 74:230–244.

Minelli A. 2003. The Development of Animal Form: Ontogeny, Morphology, and Evolution. Cambridge University Press.

Minelli A. 2017. Evolvability and its evolvability. In: Challenging the Modern Synthesis: Adaptation, Development, and Inheritance (Huneman P, Walsh D eds). Oxford University Press.

Mishra G, Omkar. 2017. Insects as food. In: Industrial Entomology (Omkar ed). Springer, 413–434.

Mitchell P. 2018. The donkey in human history: an archaeological perspective. Oxford University Press.

Mitgutsch C, et al. 2008. Heterochronic shifts during early cranial neural crest cell migration in two ranid frogs. Acta Zool 89:69–78.

Mitgutsch C, et al. 2011. Timing of ossification in duck, quail, and zebra finch: intraspecific variation, heterochronies, and life history evolution. Zool Sci 28:491–500.

Mitsiadis TA, et al. 2003. Development of teeth in chick embryos after mouse neural crest transplantations. PNAS 100:6541–6545.

Mitteroecker P, et al. 2004. Comparison of cranial ontogenetic trajectories among great apes and humans. J Hum Evol 46:679–698.

Mitteroecker P, Huttegger SM. 2009. The concept of morphospaces in evolutionary and developmental biology: mathematics and metaphors. Biol Theory 4:54–67.

Mivart SG. 1871. On the Genesis of Species. D. Appleton and Company.

Mockler BK, et al. 2018. Microbiome structure influences infection by the parasite *Crithidia bombi* in bumble bees. Appl Environ Microbiol 84.

Mohamad K, et al. 2009. On the origin of Indonesian cattle. PLoS ONE 4:e5490.

Moiseyeva IG, et al. 2003. Evolutionary relationships of red jungle fowl and chicken breeds. Genet Sel Evol 35:403–423.

Møller AP, Sanotra GS, Vestergaard KS. 1995. Developmental stability in relation to population density and breed of chickens *Gallus gallus*. Poult Sci 74:1761–1771.

Mongin P, Plouzeau M. 1984. Guinea fowl. In: Evolution of Domesticated Animals (Mason IL ed). Longman, 322–325.

Monson TA, Hlusko LJ. 2018. The evolution of dental eruption sequence in artiodactyls. J Mamm Evol 25:15–26.

Montague MJ, et al. 2014. Comparative analysis of the domestic cat genome reveals genetic signatures underlying feline biology and domestication. PNAS 111:17230–17235.

Moore CE. 2019. Changes in antibiotic resistance in animals. Science 365:1251–1252.

Moore WJ. 1981. The Mammalian Skull. Cambridge University Press.

Moore WJ, Spence TF. 1969. Age changes in the cranial base of the rabbit (*Oryctolagus cuniculus*). Anat Rec 165:355–361.

Morand S, McIntyre KM, Baylis M. 2014. Domesticated animals and human infectious diseases of zoonotic origins: domestication time matters. Infect Genet Evol 24:76–81.

Moreira A, Marques Moreira H, Silva Hilsdorf A. 2005. Comparative growth performance of two Nile tilapia (Chitralada and Red-Stirling), their crosses and the Israeli tetra hybrid ND-56. Aquac Res 36:1049–1055.

Moreira JR, et al. 2013. Capybara: Biology, Use and Conservation of an Exceptional Neotropical Species. Springer Science & Business Media.

Morey DF. 1994. The early evolution of the domestic dog. Am Sci 82:336–347.

Morey DF. 2014. In search of Paleolithic dogs: a quest with mixed results. J Archaeol Sci 52:300–307.

Morey DF, Wiant MD. 1992. Early Holocene domestic dog burials from the North American Midwest. Curr Anthropol 33:224–229.

Morrison D. 2012. The first phylogenetic network. http://phylonetworks.blogspot.com/2012/02 /first-phylogenetic-network-1755.html.

Morton J. 1984. The domestication of the savage pig: the role of peccaries in tropical South and Central America and their relevance for the understanding of pig domestication in Melanesia. Canberra Anthropol 7:20–70.

Mosher DS, et al. 2007. A mutation in the myostatin gene increases muscle mass and enhances racing performance in heterozygote dogs. PLoS Genet 3:e79.

Mueller UG, et al. 2005. The evolution of agriculture in insects. Annu Rev Ecol Evol Syst 36:563–595.

Müller F. 1972. Evolutionary changes in the ontogenesis of Eutheria. Comparative morphological study of Marsupialia and Eutheria. Rev Suisse Zool 79:1599–1685.

Müller GB. 2010. Epigenetic innovation. In: Evolution: The Extended Synthesis. MIT Press, 307–332.

Müller J, et al. 2014. Growth and wear of incisor and cheek teeth in domestic rabbits (*Oryctolagus cuniculus*) fed diets of different abrasiveness. J Exp Zool A 321:283–298.

Müller J, et al. 2015. Tooth length and incisal wear and growth in guinea pigs (*Cavia porcellus*) fed diets of different abrasiveness. J Anim Physiol Anim Nutr 99:591–604.

Mun AM, Kosin IL. 1960. Developmental stages of the broad breasted bronze turkey embryo. Biol Bull 119:90–97.

Mundinger PC. 1995. Behaviour-genetic analysis of canary song: inter-strain differences in sensory learning, and epigenetic rules. Anim Behav 50:1491–1511.

Mundinger PC. 2010. Behaviour genetic analysis of selective song learning in three inbred canary strains. Behaviour 147:705–723.

Murawska D. 2012. The effect of age on the growth rate of tissues and organs and the percentage content of edible and nonedible carcass components in Pekin ducks. Poult Sci 91:2030–2038.

Murawska D, et al. 2005. Age-related changes in the carcass tissue composition and distribution of meat and fat with skin in carcasses of laying-type cockerels. Arch für Geflügelkunde 69:135–139.

Murawska D, et al. 2011. Age-related changes in the percentage content of edible and non-edible components in broiler chickens. Asian-Australas J Anim Sci 24:532–539.

Murawska D, et al. 2015. Age-related changes in the tissue composition of carcass parts and in the distribution of lean meat, fat with skin and bones in turkey carcasses. Eur Poult Sci 79:1–14.

Murawska D, et al. 2016. Selected growth parameters of farm-raised mallard (*Anas platyrhynchos* L.) ducklings. Can J Anim Sci 96:504–511.

Murray P, Drachman DB. 1969. The role of movement in the development of joints and related structures: the head and neck in the chick embryo. Development 22:349–371.

Murray JR, et al. 2013. Embryological staging of the Zebra Finch, *Taeniopygia guttata*. J Morphol 274:1090–1110.

Naaktgeboren C. 1960. Das embryonale Wachstum des Rindes mit besonderer Berücksichtigung der für die Geburt wichtigen Körperteile. Zeitschrift für Morphologie und Ökologie der Tiere 48:447–460.

Naaktgeboren C, et al. 1971. Die Geburt beim Texelschaf und beim Heideschaf, ein Beitrag zur Kenntnis der Domestikationseinflüsse auf den Geburtsverlauf. Z Tierzuecht Zuechtungsbiol 88:169–182.

Nacarino-Meneses C, Jordana X, Köhler M. 2016. Histological variability in the limb bones of the Asiatic wild ass and its significance for life history inferences. PeerJ 4:e2580.

Naess A. 1989. Ecology, Community and Lifestyle. Translated and edited by David Rothenberg. Cambridge University Press.

Nahashon S, et al. 2006. Modeling growth characteristics of meat-type guinea fowl. Poult Sci 85:943–946.

Nakajima T, et al. 2019. Common carp aquaculture in Neolithic China dates back 8,000 years. Nat Ecol Evol 3:1415–1418.

Nakane Y, Tsudzuki M. 1999. Development of the skeleton in Japanese quail embryos. Dev Growth Differ 41:523–534.

Narinc D, et al. 2010. Analysis of fitting growth models in medium growing chicken raised indoor system. Trends Anim Vet Sci 1:12–18.

Nätt D, et al. 2012. Heritable genome-wide variation of gene expression and promoter methylation between wild and domesticated chickens. BMC Genomics 13:59.

Natterson-Horowitz B, Bowers K. 2013. Zoobiquity: the astonishing connection between human and animal health. Vintage.

Navalón G, et al. 2020. The consequences of craniofacial integration for the adaptive radiations of Darwin's finches and Hawaiian honeycreepers. Nat Ecol Evol 4:270–278.

Neaux D, et al. 2020. Examining the effect of feralization on craniomandibular morphology in pigs, *Sus scrofa* (Artiodactyla: Suidae). Biol J Linn Soc 131:870–879.

Ness AR, Brown GL. 1956. The response of the rabbit mandibular incisor to experimental shortening and to the prevention of its eruption. Proc R Soc Lond B 146:129–154.

Newbern J, et al. 2008. Mouse and human phenotypes indicate a critical conserved role for ERK2 signaling in neural crest development. PNAS 105:17115–17120.

Nguyen TT, Tomberlin JK, Vanlaerhoven S. 2015. Ability of black soldier fly (Diptera: Stratiomyidae) larvae to recycle food waste. Environ Entomol 44:406–410.

Nice MM. 1962. Development of behavior in precocial birds. Trans Linn Soc N Y 8:1–211.

Nickel R, Schummer A, Seiferle E. 1986. The Anatomy of the Domestic Animals. Vol 1. The Locomotor System of the Domestic Mammals. Verlag Paul Parey.

Nieberle K, Cohrs P. 1967. Textbook of the special pathological anatomy of domestic animals. Pergamon Press.

Nishibori M, et al. 2005. Molecular evidence for hybridization of species in the genus *Gallus* except for *Gallus varius*. Anim Genet 36:367–375.

Niu D, et al. 2002. The origin and genetic diversity of Chinese native chicken breeds. Biochem Genet 40:163–174.

Noddle B. 1974. Ages of epiphyseal closure in feral and domestic goats and ages of dental eruption. J Archaeol Sci 1:195–204.

Noden DM, La Hunta A. 1985. The Embryology of Domestic Animals. Developmental Mechanisms and Malformations. Williams & Wilkins.

Nogueira-Filho SLG, Nogueira SSC, Fragoso JMV. 2009. Ecological impacts of feral pigs in the Hawaiian Islands. Biodivers Conserv 18:3677–3683.

Nolan MB, et al. 2017. Artificially extended photoperiod administered to pre-partum mares via blue light to a single eye: observations on gestation length, foal birth weight and foal hair coat at birth. Theriogenology 100:126–133.

Norton HL, et al. 2019. Human races are not like dog breeds: refuting a racist analogy. Evolution: Education and Outreach 12:17.

Nourinezhad J, Gilanpour H, Radmehr B. 2013. Prenatal development of the fetal thoracic sympathetic trunk in sheep (*Ovis aries*). Auton Neurosci 177:154–162.

Novoa-Bravo M, Bernal-Pinilla E, García LF. 2021. Microevolution operating in domestic animals: evidence from the Colombian Paso horses. Mamm Biol 101:181–192.

Nowlan NC, et al. 2010. Mechanobiology of embryonic skeletal development: insights from animal models. Birth Defects Res C Embryo Today Rev 90:203–213.

Núñez-León D, et al. 2019. Morphological diversity of integumentary traits in fowl domestication: insights from disparity analysis and embryonic development. Dev Dyn 248:1044–1058.

Núñez-León D. 2021. Shifts in growth, but not differentiation, foreshadow the formation of exaggerated forms under chicken domestication. Proc R Soc Lond B: 288:20210392.

Nunn CL, Smith KK. 1998. Statistical analyses of developmental sequences: the craniofacial region in marsupial and placental mammals. Am Nat 152:82–101.

Nussbaumer M. 1982. Über die Variabilitat der dorso-basalen Schädelknickungen bei Haushunden. Zool Anz 209:1–32.

Nussbaumer MB. 2000. Barry vom Grossen St. Bernard. Naturhistorisches Museum der Burgergemeinde Bern.

Nwibo DD, et al. 2015. Current use of silkworm larvae (*Bombyx mori*) as an animal model in pharmaco-medical research. Drug Discov Ther 9:133–135.

Odling-Smee FJ, Laland KN, Feldman MW. 2003. Niche Construction: The Neglected Process in Evolution. Princeton University Press.

Okanoya K. 2015. Evolution of song complexity in Bengalese finches could mirror the emergence of human language. J Ornithol 156:65–72.

Olea GB, Sandoval MT. 2012. Embryonic development of *Columba livia* (Aves: Columbiformes) from an altricial-precocial perspective. Rev Colomb Cienc Pec 25:3–13.

Oliver J. 2005. Jamie's Italy. Michael Joseph.

Ollerton J, Winfree R, Tarrant S. 2011. How many flowering plants are pollinated by animals? Oikos 120:321–326.

Olmstead MP. 1911. Das Primordialcranium eines Hundeembryo. Ein Beitrag zur Morphologie des Säugetierschädels. Anat Hefte 43:335–375.

Olsen SJ. 1985. Origins of the Domestic Dog: The Fossil Record. University of Arizona Press.

Olsen SJ. 1990. Fossil ancestry of the yak, its cultural significance and domestication in Tibet. Proc Acad Nat Sci Phila 142:73–100.

Onstein RE, et al. 2018. To adapt or go extinct? The fate of megafaunal palm fruits under past global change. Proc R Soc Lond B 285:20180882.

Onzima RB, et al. 2018. Genome-wide characterization of selection signatures and runs of homozygosity in Ugandan goat breeds. Front Genet 9:318.

Oppenheimer EC, Oppenheimer JR. 1975. Certain behavioral features in the pariah dog (*Canis familiaris*) in West Bengal. Appl Anim Ethol 2:81–92.

Oralová V, et al. 2020. Multiple epithelia are required to develop teeth deep inside the pharynx. PNAS 117:11503–11512.

O'Regan HJ, Kitchener AC. 2005. The effects of captivity on the morphology of captive, domesticated and feral mammals. Mammal Rev 35:215–230.

Orlando L. 2016. Back to the roots and routes of dromedary domestication. PNAS 113:6588–6590.

Ostrander GK ed. 2000. The Laboratory Fish. Academic Press.

Ostrander EA, Wayne RK. 2005. The canine genome. Genome Res 15:1706–1716.

Ota KG, Abe G. 2016. Goldfish morphology as a model for evolutionary developmental biology. WIREs Dev Biol 5:272–295.

Ottoni C, et al. 2017. The palaeogenetics of cat dispersal in the ancient world. Nat Ecol Evol 1:0139.

Outram AK, et al. 2009. The earliest horse harnessing and milking. Science 323:1332–1335.

Ovodov ND, et al. 2011. A 33,000-year-old incipient dog from the Altai Mountains of Siberia: Evidence of the earliest domestication disrupted by the Last Glacial Maximum. PLoS ONE 6:e22821.

Owen R. 1853. Descriptive Catalogue of the Osteological Series Contained in the Museum. Taylor & Francis.

Owen J, et al. 2014. The zooarchaeological application of quantifying cranial shape differences in wild boar and domestic pigs (*Sus scrofa*) using 3D geometric morphometrics. J Archaeol Sci 43:159–167.

Oyama S, Griffiths PE, Gray RD. 2003. Cycles of Contingency: Developmental Systems and Evolution. MIT Press.

Packard JM, et al. 1985. Causes of reproductive failure in two family groups of wolves (*Canis lupus*). Ethology 68:24–40.

Padgett CS, Ivey WD. 1960. The normal embryology of the Coturnix quail. Anat Rec 137:1–11.

Padian K, Lamm E-T eds. 2013. Bone Histology of Fossil Tetrapods: Advancing Methods, Analysis, and Interpretation. University of California Press.

Pailhoux E, et al. 2001. A 11.7-kb deletion triggers intersexuality and polledness in goats. Nat Genet 29:453–458.

Pal SK. 2001. Population ecology of free-ranging urban dogs in West Bengal, India. Acta Theriol 46:69–78.

Palmer C. 2010. Animal Ethics in Context. Columbia University Press.

Pandey UB, Nichols CD. 2011. Human disease models in *Drosophila melanogaster* and the role of the fly in therapeutic drug discovery. Pharmacol Rev 63:411–436.

Pardue S, Ring N, Smyth Jr J. 1985. Pleiotropisms associated with alleles of the C locus in the domestic fowl. Poult Sci 64:1821–1828.

Parés-Casanova PM. 2018. Skull growth in equids beyond domestication. Anim Husb Dairy Vet Sci 2:1–3.

Parés-Casanova PM, Kucherova I. 2014. Comparación de modelos no lineales para describir curvas de crecimiento en la cabra catalana. Rev Investig Vet Perú 25:390–398.

Parés-Casanova PM, Medina A. 2019. Asimetría direccional en el cráneo del conejo de compañía. Rev Investig Vet Perú 30:1003–1008.

Parés-Casanova PM, et al. 2020. A comparison of traditional and geometric morphometric techniques for the study of basicranial morphology in horses: a case study of the Araucanian horse from Colombia. Animals 10:118.

Park K, et al. 2008. Canine polydactyl mutations with heterogeneous origin in the conserved intronic sequence of LMBR1. Genetics 179:2163–2172.

Park SD, et al. 2015. Genome sequencing of the extinct Eurasian wild aurochs, *Bos primigenius*, illuminates the phylogeography and evolution of cattle. Genome Biol 16:234.

Parker HG, et al. 2004. Genetic structure of the purebred domestic dog. Science 304:1160–1164.

Parker HG, et al. 2009. An expressed Fgf4 retrogene is associated with breed-defining chondrodysplasia in domestic dogs. Science 325:995–998.

Parker HG, et al. 2017. Genomic analyses reveal the influence of geographic origin, migration, and hybridization on modern dog breed development. Cell Rep 19:697–708.

Parsons KJ, et al. 2020. Skull morphology diverges between urban and rural populations of red foxes mirroring patterns of domestication and macroevolution. Proc R Soc Lond B 287:20200763.

Pasquet A. 2018. Effects of domestication on fish behaviour. In: Animal Domestication (Teletchea F ed). IntechOpen.

Patten BM. 1951. Early Embryology of the Chick, 4th Edition. McGraw-Hill.

Pavlicev M, et al. 2008. Genetic variation in pleiotropy: differential epistasis as a source of variation in the allometric relationship between long bone lengths and body weight. Evolution 62:199–213.

Pawlick TF. 2006. The End of Food: How the Food Industry Is Destroying Our Food Supply—And What You Can Do about It. Barricade Books.

Payne S. 1973. Kill-off patterns in sheep and goats: the mandibles from Aşvan Kale. Anatol Stud 23:281–303.

Payne CLR, Evans JD. 2017. Nested houses: domestication dynamics of human–wasp relations in contemporary rural Japan. J Ethnobiol Ethnomed 13:13.

Payne JL, Wagner A. 2019. The causes of evolvability and their evolution. Nat Rev Genet 20:24–38.

Pedrosa S, et al. 2005. Evidence of three maternal lineages in Near Eastern sheep supporting multiple domestication events. Proc R Soc Biol Sci Ser B 272:2211–2217.

Peiris JSM, et al. 2004. Re-emergence of fatal human influenza A subtype H5N1 disease. The Lancet 363:617–619.

Pendleton AL, et al. 2018. Comparison of village dog and wolf genomes highlights the role of the neural crest in dog domestication. BMC Biol 16:64.

Pereira Verdugo M, et al. 2019. Ancient cattle genomics, origins, and rapid turnover in the Fertile Crescent. Science 365:173–176.

Pérez B, et al. 2010. Computed tomographic anatomy of the larynx in mesaticephalic dogs. Arch Med Vet 42:91–99.

Perreault C. 2019. The Quality of the Archaeological Record. University of Chicago Press.

Perri A. 2016. A wolf in dog's clothing: initial dog domestication and Pleistocene wolf variation. J Archaeol Sci 68:1–4.

Perri A, et al. 2021. Dog domestication and the dual dispersal of people and dogs into the Americas. PNAS 118:e2010083118.

Perry JC, Rowe L. 2015. The evolution of sexually antagonistic phenotypes. Cold Spring Harb Perspect Biol 7:a017558.

Perry WB, et al. 2019. Evolutionary drivers of kype size in Atlantic salmon (*Salmo salar*): domestication, age and genetics. R Soc Open Sci 6:190021.

Peters J, von den Driesch A. 1997. The two-humped camel (*Camelus bactrianus*): new light on its distribution, management and medical treatment in the past. J Zool 242:651–679.

Peters J, et al. 2015. Questioning new answers regarding Holocene chicken domestication in China. PNAS 112:E2415-E2415.

Peters J, et al. 2016. Holocene cultural history of Red jungle fowl (*Gallus gallus*) and its domestic descendant in East Asia. Quat Sci Rev 142:102–119.

Petersen B. 2017. Basics of genome editing technology and its application in livestock species. Reprod Domest Anim 52:4–13.

Petersson E, et al. 1996. The effect of domestication on some life history traits of sea trout and Atlantic salmon. J Fish Biol 48:776–791.

Petitjean M. 1969. De quelques applications de l'insémination artificielle en aviculture. Essais d'hybridation interspécifique coq x pintade et pintade mäle x poule. Revue de l'Elevage 24:123–131.

Petkov CI, Jarvis ED. 2012. Birds, primates, and spoken language origins: behavioral phenotypes and neurobiological substrates. Front Evol Neurosci 4:12.

Petri C. 1935. Die Skelettentwicklung beim Meerschwein. Zugleich ein Beitrag zur vergleichenden Anatomie der Skelettentwicklung der Säuger. Vierteljahresschr Naturforsch Ges Zürich 80:157–240.

Peyer B. 1950. Goethes Wirbeltheorie des Schädels. Kommissionsverlag Gebr. Fretz AG.

Pielberg G, et al. 2002. Unexpectedly high allelic diversity at the KIT locus causing dominant white color in the domestic pig. Genetics 160:305–311.

Piperno DR, Pearsall DM. 1998. The Origins of Agriculture in the Lowland Neotropics. Academic Press.

Pitsillides AA. 2006. Early effects of embryonic movement: "a shot out of the dark." J Anat 208:417–431.

Pitt D, et al. 2019. Domestication of cattle: two or three events? Evol Appl 12:123–136.

Pittroff W, et al. 2008. Onset of puberty and the inflection point of the growth curve in sheep—Brody's Law revisited. J Agric Sci 146:239–250.

Plassará K. 2005. Greek Animals Disappearing. An Unknown Treasure on Its Way of Becoming Extinct. 2nd ed. ERGO Publisher.

Plate L. 1929. Über Nackthunde und Kreuzungen von Ceylon-Nackthund und Dackel. Jena Z Naturwissenschaft 64:227–282.

Platt DE, et al. 2017. Mapping post-glacial expansions: the peopling of Southwest Asia. Sci Rep 7:1–10.

Pocock MJ, Searle JB, White PC. 2004. Adaptations of animals to commensal habitats: population dynamics of house mice *Mus musculus domesticus* on farms. J Anim Ecol 73:878–888.

Podlesnykh AV, Brykov VA, Skurikhina LA. 2015. Polyphyletic origin of ornamental goldfish. Food Nutr Sci 6:1005–1013.

Poe S, Wake MH. 2004. Quantitative tests of general models for the evolution of development. Am Nat 164:415–422.

Pohle C. 1969. Ein Fall von Mopsköpfigkeit und einige Zahnanomalien beim Farmnerz, *Mustela vison* (Schreber, 1778). Saeugetierkd Mitt 17:129–131.

Pointer MA, et al. 2012. RUNX2 tandem repeats and the evolution of facial length in placental mammals. BMC Evol Biol 12:103.

Polák J, Frynta D. 2009. Sexual size dimorphism in domestic goats, sheep, and their wild relatives. Biol J Linn Soc 98:872–883.

Pollard RE, et al. 2015. Japanese Bobtail: vertebral morphology and genetic characterization of an established cat breed. J Feline Med Surg 17:719–726.

Polly PD. 2012. Measuring the evolution of body size in mammals. PNAS 109:4027–4028.

Pontzer H, et al. 2012. Hunter-gatherer energetics and human obesity. PLoS ONE 7:e40503.

Porter V. 1993. Pigs, a Handbook to the Breeds of the World. Helm Information.

Porter V. 1996. Goats of the World. Farming Press.

Porter V. 2002. Mason's World Dictionary of Liverstock Breeds, Types and Varieties. 5th ed. CABI Publishing.

Portmann A. 1951. Ontogenesetypus und Cerebralisation in der Evolution der Vögel und Säuger. Rev Suisse Zool 58:427–434.

Porto A, et al. 2009. The evolution of modularity in the mammalian skull I: morphological integration patterns and magnitudes. Evol Biol 36:118–135.

Porto A, et al. 2013. Size variation, growth strategies, and the evolution of modularity in the mammalian skull. Evolution 67:3305–3322.

Potts D. 2004. Camel hybridization and the role of *Camelus bactrianus* in the ancient Near East. J Econ Soc Hist Orient 47:143–165.

Prassack KA, et al. 2020. Dental microwear as a behavioral proxy for distinguishing between canids at the Upper Paleolithic (Gravettian) site of Předmostí, Czech Republic. J Archaeol Sci 115:105092.

Prentiss CW. 1906. Extra digits and digital reductions. Popular Science Monthly 68:335–348.

Price EO. 1984. Behavioral aspects of animal domestication. Q Rev Biol 59:1–32.

Price EO. 1999. Behavioral development in animals undergoing domestication. Appl Anim Behav Sci 65:245–271.

Price EO. 2002a. Animal Domestication and Behavior. CABI Publishing.

Price TD. 2002b. Domesticated birds as a model for the genetics of speciation by sexual selection. Genetica 116:311–327.

Proks P, et al. 2015. Congenital abnormalities of the vertebral column in ferrets. Vet Radiol Ultrasound 56:117–123.

Promislow DEL, Harvey PH. 1990. Living fast and dying young: a comparative analysis of life-history variation among mammals. J Zool 220:417–437.

Prondvai E, et al. 2020. Extensive chondroid bone in juvenile duck limbs hints at accelerated growth mechanism in avian skeletogenesis. J Anat 236:463–473.

Proops L, Burden F, Osthaus B. 2009. Mule cognition: a case of hybrid vigour? Anim Cogn 12:75–84.

Pruvost M, et al. 2011. Genotypes of predomestic horses match phenotypes painted in Paleolithic works of cave art. PNAS 108:18626–18630.

Pryor K. 2001. Cultural transmission of behavior in animals: How a modern training technology uses spontaneous social imitation in cetaceans and facilitates social imitation in horses and dogs. Behavioral and Brain Sciences 24:352–352.

Purohit G. 2010. Parturition in domestic animals: a review. Reproduction 1:WMC00748.

Purugganan MD, Fuller DQ. 2011. Archaeological data reveal slow rates of evolution during plant domestication. Evolution 65:171–183.

Qian Y, Chen W, Guo B. 2020. Zing-Yang Kuo and behavior epigenesis based on animal experiments. Protein & Cell 11:387–390.

Qin Q, et al. 2016. Autotriploid origin of *Carassius auratus* as revealed by chromosomal locus analysis. Sci China Life Sci 59:622–626.

Qiu Q, et al. 2015. Yak whole-genome resequencing reveals domestication signatures and prehistoric population expansions. Nat Commun 6:10283.

Qu S, et al. 1998. Mutations in mouse Aristaless-like4 cause Strong's luxoid polydactyly. Development 125:2711–2721.

Quammen D. 2012. Spillover: Animal Infections and the Next Human Pandemic. WW Norton & Co.

Quintana Cardona J, Ramis Bernat D, Bover Arbós P. 2016. Primera datació d'un mamífer no autòcton (*Oryctolagus cuniculus* [Linnaeus, 1758], Mammalia: Lagomorpha) del jaciment holocènic del Pas d'en Revull (barranc d'Algendar, Ferreries). Revista de Menorca 95:185–200.

Räber H. 1993. Enzyklopädie der Rassehunde. Franckh-Kosmos.

Rabosky DL, et al. 2013. Rates of speciation and morphological evolution are correlated across the largest vertebrate radiation. Nat Commun 4:1–8.

Rabosky DL, et al. 2018. An inverse latitudinal gradient in speciation rate for marine fishes. Nature 559:392–395.

Radinsky L. 1983. Allometry and reorganization in horse skull proportions. Science 221: 1189–1191.

Radinsky L. 1984. Ontogeny and phylogeny in horse skull evolution. Evolution 38:1–15.

Rager L, et al. 2014. Timing of cranial suture closure in placental mammals: phylogenetic patterns, intraspecific variation, and comparison with marsupials. J Morphol 275:125–140.

Randau M, Goswami A. 2017. Unravelling intravertebral integration, modularity and disparity in Felidae (Mammalia). Evol Dev 19:85–95.

Rasali D, Shrestha J, Crow G. 2006. Development of composite sheep breeds in the world: a review. Can J Anim Sci 86:1–24.

Ratcliffe NA, et al. 2011. Insect natural products and processes: new treatments for human disease. Insect Biochem Mol Biol 41:747–769.

Ravinet M, et al. 2018. Signatures of human-commensalism in the house sparrow genome. Proc R Soc Lond B 285:20181246.

Ravn-Mølby E-M, et al. 2019. Breeding French bulldogs so that they breathe well—A long way to go. PLoS ONE 14:e0226280.

Recht J, Schuenemann VJ, Sánchez-Villagra MR. 2020. Host diversity and origin of zoonoses: the ancient and the new. Animals 10:1672.

Reedy MV, Faraco CD, Erickson CA. 1998. Specification and migration of melanoblasts at the vagal level and in hyperpigmented Silkie chickens. Dev Dyn 213:476–485.

Regodón S, et al. 1993. Craniofacial angle in dolicho-, meso- and brachycephalic dogs: radiological determination and application. Ann Anat 175:361–363.

Rehkämper G, Haase E, Frahm HD. 1988. Allometric comparison of brain weight and brain structure volumes in different breeds of the domestic pigeon, Columba livia f. d. (fantails, homing pigeons, strassers). Brain Behav Evol 31:141–149.

Rehkämper G, et al. 2003. Discontinuous variability of brain composition among domestic chicken breeds. Brain Behav Evol 61:59–69.

Rehkämper G, Frahm HD, Cnotka J. 2008. Mosaic evolution and adaptive brain component alteration under domestication seen on the background of evolutionary theory. Brain Behav Evol 71:115–126.

Reitz EJ, Wing ES. 2008. Zooarchaeology. 2nd ed. Cambridge University Press.

Rempe U. 1970. Morphometrische Untersuchungen an Iltisschädeln zur Klärung der Verwandtschaft von Steppeniltis, Waldiltis und Frettchen. Analyse eines "Grenzfalles" zwischen Unterart und Art. Z Wiss Zool Abt A 180:185–367.

Rempel AG, Eastlick HL. 1957. Developmental stages of normal White Silkie fowl embryos. Northwest Sci 31:1–13.

Rensch B. 1950. Die Abhängigkeit der relativen Sexualdifferenz von der Körpergrösse. Bonn Zool Beitr 1:58–69.

Requate H. 1959. Federhauben bei Vögeln: Eine genetische und entwicklungsphysiologische Studie zum Problem der Parallelbildungen. Z wiss Zool 162:192–313.

Reznick DN. 2010. The Origin Then and Now: An Interpretive Guide to the Origin of Species. Princeton University Press.

Reznick DN, et al. 1997. Evaluation of the rate of evolution in natural populations of guppies (Poecilia reticulata). Science 275:1934–1937.

Rice WR. 1996. Sexually antagonistic male adaptation triggered by experimental arrest of female evolution. Nature 381:232–234.

Richards KW. 1987. Alfalfa leafcutter bee management in Canada. Bee World 68:168–178.

Richardson MK, et al. 2009. Heterochrony in limb evolution: developmental mechanisms and natural selection. J Exp Zool B 312B:639–664.

Richmond ML. 2001. Women in the early history of genetics: William Bateson and the Newnham College Mendelians, 1900–1910. Isis 92:55–90.

Richter CP. 1952. Domestication of the Norway rat and its implication for the study of genetics in man. Am J Hum Genet 4:273–285.

Richter J, Götze R. 1978. Tiergeburtshilfe. Paul Parey.

Richtsmeier JT, et al. 2006. Phenotypic integration of neurocranium and brain. J Exp Zool Part B 306:360–378.

Riddle RD, et al. 1993. Sonic hedgehog mediates the polarizing activity of the ZPa. Cell 75:1401–1416.

Riedel A. 1993. The Austrian 'Blondvieh' cattle horncores. An Archaeozoological view. In: Skeletons in Her Cupboard: Festschrift for Juliet Clutton-Brock (Clason A, Payne S, Uerpmann H-P eds). Oxbow Books, 183–188.

Riedl R. 1975. Die Ordnung des Lebendigen. Verlag Paul Parey.

Rieppel O. 2017. Turtles as Hopeful Monsters: Origins and Evolution. Indiana University Press.

Rimbault M, et al. 2013. Derived variants at six genes explain nearly half of size reduction in dog breeds. Genome Res 23:1985–1995.

Roberts T, McGreevy P, Valenzuela M. 2010. Human induced rotation and reorganization of the brain of domestic dogs. PLoS ONE 5:e11946.

Roberts P, et al. 2018. Calling all archaeologists: guidelines for terminology, methodology, data handling, and reporting when undertaking and reviewing stable isotope applications in archaeology. Rapid Commun Mass Spectrom 32:361–372.

Robinson A, Gibson A. 1917. VIII.—Description of a reconstruction model of a horse embryo twenty-one days old. Earth Env Sci Trans R Soc 51:331–347.

Rodríguez LC, Méndez MA, Niemeyer HM. 2001. Direction of dispersion of cochineal (*Dactylopius coccus* Costa) within the Americas. Antiquity 75:73–77.

Røed KH, Bjørklund I, Olsen BJ. 2018. From wild to domestic reindeer—genetic evidence of a non-native origin of reindeer pastoralism in northern Fennoscandia. J Archaeol Sci Rep 19:279–286.

Roegner AF, et al. 2019. *Giardia* infection and *Trypanosoma cruzi* exposure in dogs in the Bosawás Biosphere Reserve, Nicaragua. EcoHealth 16:512–522.

Roffet-Salque M, et al. 2015. Widespread exploitation of the honeybee by early Neolithic farmers. Nature 527:226–230.

Rogell B, Dowling DK, Husby A. 2020. Controlling for body size leads to inferential biases in the biological sciences. Evol Lett 4:73–82.

Rohner N, et al. 2009. Duplication of fgfr1 permits Fgf signaling to serve as a target for selection during domestication. Curr Biol 19:1642–1647.

Röhrs M, Ebinger P. 1999. Verwildert ist nicht gleich wild: Die Hirngewichte verwilderter Haussäugetiere. Berl Munch Tierarztl Wochenschr 112:234–238.

Rolian C. 2020. Endochondral ossification and the evolution of limb proportions. Wiley Interdisciplinary Reviews: Developmental Biology 9:e373.

Rollo CD. 2002. Growth negatively impacts the life span of mammals. Evol Dev 4:55–61.

Ruscillo D. 2006. Recent Advances in Ageing and Sexing Animal Bones. Oxbow Books.

Romanoff AL, Romanoff AJ. 1972. Pathogenesis of the avian embryo: an analysis of causes of malformations and prenatal death. John Wiley and Sons.

Rong R, et al. 1985. Fertile mule in China and her unusual foal. J R Soc Med 78:821–825.

Rose AB, Platt KH. 1992. Snow tussock (*Chionochloa*) population responses to removal of sheep and European hares, Canterbury, New Zealand. N Z J Bot 30:373–382.

Rosenberg KFA. 1966. Die postnatale Proportionsänderung der Schädel zweier extremer Wuchsformen des Haushundes. Vergleichend allometrische Untersuchungen an Whippets und Pekingesen. Z Tierzuecht Zuechtungsbiol 82:1–36.

Rossel S, et al. 2008. Domestication of the donkey: timing, processes, and indicators. PNAS 105:3715–3720.

Roux W. 1894. Einleitung zum Archiv für Entwicklungsmechanik der Organismen. Archiv fur Entwicklungsmechanik der Organismen 1:1–142.

Rubin C-J, et al. 2010. Whole-genome resequencing reveals loci under selection during chicken domestication. Nature 464:587–591.

Ruf I. 2020. Ontogenetic transformations of the ethmoidal region in Muroidea (Rodentia, Mammalia): new insights from perinatal stages. Vertebr Zool 70:383–415.

Ruff CB ed. 2017. Skeletal Variation and Adaptation in Europeans: Upper Paleolithic to the Twentieth Century. Wiley-Blackwell.

Ruse M. 1975. Charles Darwin and artificial selection. J Hist Ideas 36:339–350.

Ruse M. 2010. Charles Darwin and the Origin of Species. In: Reznick DN, The Origin Then and Now. An Interpretative Guide to the Origin of Species, introduction. Princeton University Press.

Ruskin F ed. 1983. Little-known Asian Animals with a Promising Economic Future. National Academy Press.

Rüsse I, Sinowatz F. 1991. Lehrbuch der Embryologie der Haustiere. Parey.

Russell N. 2002. The wild side of animal domestication. Soc Anim 10:285–302.

Rütimeyer L. 1861. Die Fauna der Pfahlbauten der Schweiz. Bahnmaier (C. Detloff).

Rütimeyer L. 1866. Über Art und Race des zahmen europäischen Rindes. Archiv für Anthropologie 1:219–250.

Rütimeyer L. 1867. Versuch einer natürlichen Geschichte des Rindes in seinen Beziehungen zu den Wiederkäuern im Allgemeinen. Neue Denkschriften der Allg Schweizerischen Gesellschaft für die gesammten Naturwissenschaften 22:102.

Ryder ML. 1958. Follicle arrangement in skin from wild sheep, primitive domestic sheep and in parchment. Nature 182:781–783.

Sablin MV, Khlopachev GA. 2002. The earliest Ice Age dogs: evidence from Eliseevichi. Curr Anthropol 43:795–799.

Sætre G-P. 2013. Hybridization is important in evolution, but is speciation? J Evol Biol 26:256–258.

Sahoo T, et al. 2011. Copy number variants of schizophrenia susceptibility loci are associated with a spectrum of speech and developmental delays and behavior problems. Genet Med 13:868–880.

Saif R, et al. 2020. The LCORL locus is under selection in large-sized Pakistani goat breeds. Genes 11:168.

Salazar-Ciudad I. 2006. Developmental constraints vs. variational properties: how pattern formation can help to understand evolution and development. J Exp Zool Part B 306B:107–125.

Salazar-Ciudad I, Jernvall J. 2013. The causality horizon and the developmental bases of morphological evolution. Biol Theory 8:286–292.

Salmi A-K, Niinimäki S. 2016. Entheseal changes and pathological lesions in draught reindeer skeletons—four case studies from present-day Siberia. Int J Paleopathol 14:91–99.

Salque M, et al. 2013. Earliest evidence for cheese making in the sixth millennium BC in northern Europe. Nature 493:522–525.

Salzburger W. 2018. Understanding explosive diversification through cichlid fish genomics. Nat Rev Genet 19:705–717.

Sambraus H, Sander H. 1980. Imprinting of pigeons on to birds of their own colouring within the species. Arch für Geflügelkunde 44:200–207.

Sánchez-Villagra MR, Sultan F. 2002. The cerebellum at birth in therian mammals, with special reference to rodents. Brain Behav Evol 59:101–113.

Sánchez-Villagra MR, Geiger M, Schneider RA. 2016. The taming of the neural crest: a developmental perspective on the origins of morphological covariation in domesticated mammals. R Soc Open Sci 3:160107.

Sánchez-Villagra MR, Forasiepi AM. 2017. On the development of the chondrocranium and the histological anatomy of the head in perinatal stages of marsupial mammals. Zool Lett 3:1.

Sánchez-Villagra MR, et al. 2017. On the lack of a universal pattern associated with mammalian domestication: differences in skull growth trajectories across phylogeny. R Soc Open Sci 4:170876.

Sánchez-Villagra MR, van Schaik CP. 2019. Evaluating the self-domestication hypothesis of human evolution. Evol Anthropol 28:133–143.

Sander K. 1983. The evolution of patterning mechanisms: gleanings from insect embryogenesis and spermatogenesis. In: Development and Evolution, (Goodwin BC, Holder N, Wylie CC eds). Cambridge University Press, 123–159.

Santini F, et al. 2009. Did genome duplication drive the origin of teleosts? A comparative study of diversification in ray-finned fishes. BMC Evol Biol 9:194.

Santos M, et al. 2010. Playing Darwin. Part B. 20 years of domestication in *Drosophila subobscura*. Theory Biosci 129:97–102.

Santos ME, et al. 2014. The evolution of cichlid fish egg-spots is linked with a cis-regulatory change. Nat Commun 5:5149.

Sarkozy A, et al. 2009. Germline BRAF mutations in Noonan, LEOPARD, and cardiofaciocutaneous syndromes: molecular diversity and associated phenotypic spectrum. Hum Mutat 30:695–702.

Sarma K. 2006. Morphological and craniometrical studies on the skull of Kagani goat (*Capra hircus*) of Jammu region. Int J Morphol 24:449–455.

Sato JJ, et al. 2003. Phylogenetic relationships and divergence times among mustelids (Mammalia: Carnivora) based on nucleotide sequences of the nuclear interphotoreceptor retinoid binding protein and mitochondrial cytochrome b genes. Zool Sci 20:243–264.

Sauer CO. 1952. Agricultural Origins and Dispersals. American Geographical Society.

Saunders W, Gasseling MT. 1968. Ectodermal-mesenchymal interactions in the origins of limb symmetry. In: Epithelial-Mesenchymal Interactions (Fleischmeyer R, Billingham RE eds). Williams and Wilkins, 78–97.

Saunders FC, et al. 2013. Computed tomographic method for measurement of inclination angles and motion of the sacroiliac joints in German Shepherd dogs and Greyhounds. Am J Vet Res 74:1172–1182.

Sawin PB. 1937. Preliminary studies of hereditary variation in the axial skeleton of the rabbit. Anat Rec 69:407–428.

Sayol F, et al. 2018. Predictable evolution towards larger brains in birds colonizing oceanic islands. Nat Commun 9:2820.

Scally A. 2016. The mutation rate in human evolution and demographic inference. Curr Opin Genet Dev 41:36–43.

Schaller GB. 1977. Mountain Monarchs. Wild Sheep and Goats of the Himalaya. University of Chicago Press.

Schleifenbaum C. 1973. Untersuchungen zur Ontogenese des Gehirns von Großpudeln und Wölfen. Z Anat Entwicklungs 141:179–205.

Schliemann H. 1966. Zur Morphologie und Entwicklung des Craniums von *Canis lupus* f. *familiaris* L. Gegenbaurs Morphol Jahrb 109:501–603.

Schlueter C, et al. 2009. Brachycephalic feline noses: CT and anatomical study of the relationship between head conformation and the nasolacrimal drainage system. J Feline Med Surg 11:891–900.

Schmidt CJ, et al. 2009. Comparison of a modern broiler line and a heritage line unselected since the 1950s. Poult Sci 88:2610–2619.

Schmidt-Küntzel A, et al. 2005. Tyrosinase and Tyrosinase related protein 1 alleles specify domestic cat coat color phenotypes of the albino and brown loci. J Hered 96:289–301.

Schmidt-Nielsen K. 1959. The physiology of the camel. Sci Am 201:140–151.

Schmitt E, Wallace S. 2014. Shape change and variation in the cranial morphology of wild canids (*Canis lupus, Canis latrans, Canis rufus*) compared to domestic dogs (*Canis familiaris*) using geometric morphometrics. Int J Osteoarchaeol 24:42–50.

Schneider KM. 1950. Zur gewichtsmässigen Jugendentwicklung einiger gefangengehaltener Wildcaniden nebst einigen zeitlichen Bestimmungen über ihre Fortpflanzung. I. Der Wolf. Zool Anz Suppl. 145:867–910.

Schneider RA. 2005. Developmental mechanisms facilitating the evolution of bills and quills. J Anat 207:563–573.

Schneider RA. 2018a. Neural crest and the origin of species-specific pattern. Genesis 56:e23219.

Schneider RA. 2018b. Cellular control of time, size, and shape in development and evolution. In: Cells in Evolutionary Biology: Translating Genotypes into Phenotypes—Past, Present, Future (Hall BK, Moody SA eds). CRC Press.

Schnorr B, Kressin M. 2006. Embryologie der Haustiere. Enke Verlag.

Schowing J. 1968. Influence inductrice de l'encéphale embryonnaire sur le développement du crâne chez le Poulet: I. Influence de l'excision des territoires nerveux antérieurs sur le développement cranien. Development 19:9–22.

Schubert M, et al. 2014. Prehistoric genomes reveal the genetic foundation and cost of horse domestication. PNAS 111:E5661-E5669.

Schultz AH. 1956. Postembryonic age changes. In: Primatologia (Hofer H, Schultz A, Starck D eds). Karger, 887–964.

Schultz AH. 1960. Age changes in primates and their modification in man. In: Human Growth (Tanner JM ed). Pergamon Press, 1–20.

Schultz W. 1969. Zur Kenntis des Hallstromhunds (*Canis hallstromi*, Troughton 1957). Zool Anz 183:47–72.

Schulz E, et al. 2013. Dietary abrasiveness is associated with variability of microwear and dental surface texture in rabbits. PLoS ONE 8:e56167.

Schultze H-P, Arratia G. 1989. The composition of the caudal skeleton of teleosts (Actinopterygil: Osteichthyes). Zool J Linn Soc 97:189–231.

Schuppli C, van Schaik CP. 2019. Animal cultures: how we've only seen the tip of the iceberg. Evol Hum Sci 1:e2.

Schütz KE, et al. 2004. Major growth QTLs in fowl are related to fearful behavior: possible genetic links between fear responses and production traits in a red junglefowl × White Leghorn intercross. Behav Genet 34:121–130.

Schweizer AV, et al. 2017. Size variation under domestication: conservatism in the inner ear shape of wolves, dogs and dingoes. Sci Rep 7:13330.

Scipioni Ball R. 2006. Issues to consider for preparing ferrets as research subjects in the laboratory. ILAR J 47:348–357.

Scott JP. 1968. Evolution and domestication of the dog. In: Evol Biol (Dobzhansky T, Hecht MK, Steere WC eds). Appleton-Century-Crofts, 243–275.

Scott JC. 2017. Against the Grain. A Deep History of the Earliest States. Yale University Press.

Scott JP, Fuller JL. 1965. Genetics and the Social Behaviour of the Dog. University of Chicago Press.

Seal U, et al. 1979. Endocrine correlates of reproduction in the wolf. I. Serum progesterone, estradiol and LH during the estrous cycle. Biol Reprod 21:1057–1066.

Sealey WM, et al. 2011. Sensory analysis of rainbow trout, *Oncorhynchus mykiss,* fed enriched black soldier fly prepupae, *Hermetia illucens.* J World Aquacult Soc 42:34–45.

Sears K, et al. 2007. The correlated evolution of Runx2 tandem repeats, transcriptional activity, and facial length in carnivora. Evol Dev 9:555–565.

Seetah K, et al. 2014. A geometric morphometric re-evaluation of the use of dental form to explore differences in horse (*Equus caballus*) populations and its potential zooarchaeological application. J Archaeol Sci 41:904–910.

Seetah K, Cardini A, Barker G. 2016. A 'long-fuse domestication' of the horse? Tooth shape suggests explosive change in modern breeds compared with extinct populations and living Przewalski's horses. The Holocene 26:1326–1333.

Segura V, et al. 2022. Biological and cultural history of domesticated dogs in the Americas. Anthropozoologica: in press.

Segura V, Sánchez-Villagra MR. 2021 Human-canid relationship in the Americas: an examination of canid biological attributes and domestication. Mamm Biol 101: in press.

Selba MC, et al. 2020. The impact of selection for facial reduction in dogs: geometric morphometric analysis of canine cranial shape. Anat Rec 303:330–346.

Sell A. 1994. Breeding and Inheritance in Pigeons. Schober Verlags-GmbH.

Serjeantson D. 2002. Goose husbandry in Medieval England, and the problem of ageing goose bones. Acta Zool Cracov 45:39–54.

Serjeantson D. 2009. Birds (Cambridge Manuals in Archaeology). Cambridge University Press.

Serpell J. 1995. The Domestic Dog: Its Evolution, Behavior and Interactions with People. Cambridge University Press.

Shannon LM, et al. 2015. Genetic structure in village dogs reveals a Central Asian domestication origin. PNAS 112:13639–13644.

Shapiro B, Hofreiter M. 2014. A paleogenomic perspective on evolution and gene function: new insights from ancient DNA. Science 343:1236573.

Shapiro MD, et al. 2013. Genomic diversity and evolution of the head crest in the rock pigeon. Science 339:1063–1067.

Sharma P, et al. 2020. Altered gut microbiota and immunity defines *Plasmodium vivax* survival in *Anopheles stephensi*. Front Immunol 11.

Sharp HS, Sharp K. 2015. Hunting Caribou: Subsistence Hunting along the Northern Edge of the Boreal Forest. University of Nebraska Press.

Sharpe AE, et al. 2018. Earliest isotopic evidence in the Maya region for animal management and long-distance trade at the site of Ceibal, Guatemala. PNAS 115:3605–3610.

Shelton WL, Rothbard S. 2006. Exotic species in global aquaculture—a review. Israeli J Aquac Bamidgeh 58:3–28.

Sheppard DC, et al. 1994. A value added manure management system using the black soldier fly. Bioresour Technol 50:275–279.

Sheppard DC, et al. 2002. Rearing methods for the black soldier fly (Diptera: Stratiomyidae). J Med Entomol 39:695–698.

Shiomi M. 2009. Rabbit as a model for the study of human diseases. In: Rabbit Biotechnology (Houdebine L-M, Fan J eds). Springer, 49–63.

Signor III PW, Lipps JH eds. 1982. Sampling bias, gradual extinction patterns, and catastrophes in the fossil record. Geological Society of America Special Publication.

Sikes RS, Ylönen H. 1998. Considerations of optimal litter size in mammals. Oikos 83:452–465.

Silliman BR, Newell SY. 2003. Fungal farming in a snail. PNAS 100:15643–15648.

Simoens P, De Vos NR, Lauwers H. 1978. Variations des vertèbres lombales et sacrales de la chèvre. Anat Histol Embryol 7:214–225.

Simões P, Santos J, Matos M. 2009. Experimental evolutionary domestication. In: Experimental Evolution: Concepts, Methods, and Applications of Selection Experiments (Garland T, Rose MR eds). University of California Press, 89–110.

Singh N, et al. 2017. Facial shape differences between rats selected for tame and aggressive behaviors. PLoS ONE 12.

Shipman P. 2011. The Animal Connection: A New Perspective on What Makes Us Human. WW Norton & Company.

Sierts-Roth U. 1953. Geburts-und Aufzuchtgewichte von Rassehunden. Zeitschrift für Hundeforschung.

Siripan S, et al. 2019. Origin of prehistoric cattle excavated from four archaeological sites in central and northeastern Thailand. Mitochondrial DNA Part A 30:609–617.

Skaala Ø, et al. 2012. Performance of farmed, hybrid, and wild Atlantic salmon (*Salmo salar*) families in a natural river environment. Can J Fish Aquat Sci 69:1994–2006.

Skaala Ø, et al. 2019. An extensive common-garden study with domesticated and wild Atlantic salmon in the wild reveals impact on smolt production and shifts in fitness traits. Evol Appl 12:1001–1016.

Slack JM, Holland PW, Graham CF. 1993. The zootype and the phylotypic stage. Nature 361:490–492.

Slijper EJ. 1942. Biologic anatomical investigations on the bipedal gait and upright posture in mammals, with special reference to a little goat born without forelegs. Proceedings of the Koninklijke Nederlandse Akademie Van Wetenschappen 45:407–415.

Slijper E. 1946. Comparative biologic anatomical investigations on the vertebral column and spinal musculature of mammals. Verhandelingen der Koninklijke Nederlandse Akademie van Wetenschappen, Afdeling Natuurkunde 42:1–128.

Smartt J. 2001. Goldfish Varieties and Genetics: Handbook for Breeders. Blackwell Science.

Smith BD. 1998. The Emergence of Agriculture. Scientific American Library.

Smith BD. 2011a. General patterns of niche construction and the management of 'wild' plant and animal resources by small-scale pre-industrial societies. Philos Trans R Soc B 366:836–848.

Smith BD. 2011b. A cultural niche construction theory of initial domestication. Biol Theory 6:260–271.

Smith BH. 2000. "Schultz's Rule" and the evolution of tooth emergence and replacement patterns in primates and ungulates. In: Development, Function and Evolution of Teeth (Teaford MF, Smith MM, Ferguson MWJ eds). Cambridge University Press, 212–227.

Smith KK. 2001a. Heterochrony revisited: the evolution of developmental sequences. Biol J Linn Soc 73:169–186.

Smith KK. 2001b. Early development of the neural plate, neural crest and facial region of marsupials. J Anat 199:121–131.

Smith NJH. 1999. The Amazon River Forest. A Natural History of Plants, Animals, and People. Oxford University Press.

Smith T. 2006. Real Rudolph: A Natural History of the Reindeer. The History Press.

Smith B, Savolainen P. 2015. The origin and ancestry of the dingo. In: The Dingo Debate: Origins, Behaviour and Conservation (Smith B ed). Csiro Publishing, 55–79.

Smith FJ, et al. 2015. Divergence of craniofacial developmental trajectories among avian embryos. Dev Dyn 244:1158–1167.

Smith B, et al. 2018. Brain size/body weight in the dingo (*Canis dingo*): comparisons with domestic and wild canids. Aust J Zool 65:292–301.

Smith-Paredes D, et al. 2018. Dinosaur ossification centres in embryonic birds uncover developmental evolution of the skull. Nat Ecol Evol 2:1966.

Sniegowski PD, Murphy HA. 2006. Evolvability. Curr Biol 16:R831-R834.

Snipes RL, Snipes H. 1997. Quantitative investigation of the intestines in eight species of domestic animals. Z Saugetierkd.

Soares LC, et al. 2015. Growth and carcass traits of three Portuguese autochthonous chicken breeds: Amarela, Preta Lusitânica and Pedrês Portuguesa. Italian J Anim Sci 14:3566.

Sol D. 2008. Artificial selection, naturalization, and fitness: Darwin's pigeons revisited. Biol J Linn Soc 93:657–665.

Solberg MF, et al. 2013. Does domestication cause changes in growth reaction norms? A study of farmed, wild and hybrid Atlantic salmon families exposed to environmental stress. PLoS ONE 8:e54469.

Solberg MF, et al. 2014. Hatching time and alevin growth prior to the onset of exogenous feeding in farmed, wild and hybrid Norwegian Atlantic salmon. PLoS ONE 9:e113697.

Sommer. 1931. Untersuchungen über die Wachstumvorgänge am Hundeskelett. Archiv für Tierernaehrung 6:439–469.

Sommer RJ. 2009. The future of evo–devo: model systems and evolutionary theory. Nat Rev Genet 10:416–422.

Sommer-Trembo C, et al. 2017. Predation risk and abiotic habitat parameters affect personality traits in extremophile populations of a neotropical fish (*Poecilia vivipara*). Ecol Evol 7:6570–6581.

Sorbe D, Kruska D. 1975. Vergleichende allometrische Untersuchungen an den Schädeln von Wander- und Laborratten. Zool Anz 195:124–144.

Sossinka R. 1982. Domestication in birds. In: Avian Biol (Farner D, King J, Parkes K eds). Academic Press, 373–403.

Soumya M, et al. 2017. Silkworm (*Bombyx mori*) and its constituents: a fascinating insect in science and research. J Ent Zool Stud 5:1701–1705.

Speakman JR, et al. 2002. Living fast, dying when? The link between aging and energetics. J Nutr 132:1583S–1597S.

Speakman JR, van Acker A, Harper EJ. 2003. Age-related changes in the metabolism and body composition of three dog breeds and their relationship to life expectancy. Aging Cell 2:265–275.

Speakman JR, et al. 2004. Uncoupled and surviving: individual mice with high metabolism have greater mitochondrial uncoupling and live longer. Aging Cell 3:87–95.

Spotorno AE, et al. 2006. Ancient and modern steps during the domestication of guinea pigs (*Cavia porcellus* L.). J Zool 270:57–62.

Springer MS, et al. 2003. Placental mammal diversification and the Cretaceous–Tertiary boundary. PNAS 100:1056–1061.

Stahl PW. 2012. Interactions between humans and endemic canids in Holocene South America. J Ethnobiol 32:108–127.

Stahl PW. 2013. Early dogs and endemic South American canids of the Spanish main. J Anthropol Res 69:515–533.

Stahl PW. 2014. Perspectival ontology and animal non-domestication in the Amazon Basin. In: Antes de Orellana Actas del 3er Encuentro Internacional de Arqueología Amazónica (Rostain S ed). Instituto Francés de Estudios Andinos, 221–231.

Stange M, et al. 2018. Morphological disparity under domestication: how variable are chicken? R Soc Open Sci 5:180993.

Stanley CE, Kulathinal RJ. 2016. Genomic signatures of domestication on neurogenetic genes in *Drosophila melanogaster*. BMC Evol Biol 16:6.

Starck D. 1962. Der heutige Stand des Fetalisationsproblems. Verlag Paul Parey.

Starck JM. 1993. Evolution of avian ontogenies. Curr Ornithol 10:275–366.

Starck JM. 2018. From fertilization to independence. In: Ornithology: Foundation, Analysis, and Application (Morrison ML, et al. eds). Johns Hopkins University Press.

Starck JM, Ricklefs RE. 1998. Patterns of development: the altricial-precocial spectrum. In: Avian Growth and Development Evolution within the altricial precocial spectrum (Starck JM, Ricklefs RE eds). Oxford University Press, 3–30.

Starck JM, Sutter E. 2000. Patterns of growth and heterochrony in moundbuilders (Megapodiidae) and fowl (Phasianidae). J Avian Biol 31:527–547.

Statham MJ, et al. 2011. On the origin of a domesticated species: identifying the parent population of Russian silver foxes (*Vulpes vulpes*). Biol J Linn Soc 103:168–175.

Stearns S. 1992. The Evolution of Life Histories. Oxford University Press.

Steckel RH, Rose JC. 2002. The Backbone of History: Health and Nutrition in the Western Hemisphere. Cambridge University Press.

Steele TE. 2015. The contributions of animal bones from archaeological sites: the past and future of zooarchaeology. J Archaeol Sci 56:168–176.

Stelkens R, Seehausen O. 2009. Genetic distance between species predicts novel trait expression in their hybrids. Evolution 63:884–897.

Štěrba O. 1995. Staging and ageing of mammalian embryos and fetuses. Acta Vet Brno 64:83–89.

Sterelny K. 2005. Another view of life. Stud Hist Philos Sci C: Stud Hist Philos Biol Biomed Sci 36:585–593.

Stern DL. 2011. Evolution, Development, & the Predictable Genome. Roberts and Company.

Stern C. 2018. The chick model system: a distinguished past and a great future. Int J Dev Biol 62:1–4.

Stock JT, Pfeiffer SK. 2004. Long bone robusticity and subsistence behaviour among Later Stone Age foragers of the forest and fynbos biomes of South Africa. J Archaeol Sci 31:999–1013.

Stockard CR. 1941. The genetic and endocrine basis for differences in form and behaviour as elucidated by studies of contrasted pure-line dog breeds and their hybrids. With special contributions on behaviour by O. D. Anderson and W. T. James. American Anatomical Memoirs 19.

Stokstad E. 2015. Bringing back the aurochs. Science 350:1144–1147.

Stone HR, et al. 2016. Associations between domestic-dog morphology and behaviour scores in the dog mentality assessment. PLoS ONE 11:e0149403.

Storey AA, et al. 2007. Radiocarbon and DNA evidence for a pre-Columbian introduction of Polynesian chickens to Chile. PNAS 104:10335–10339.

Storey AA, et al. 2012. Investigating the global dispersal of chickens in prehistory using ancient mitochondrial DNA signatures. PLoS ONE 7:e39171.

Stotts C, Darrow M. 1955. Application of artificial insemination in turkey breeder flocks. Poult Sci 34:508–518.

Stover SM, et al. 1992. Histological features of the dorsal cortex of the third metacarpal bone mid-diaphysis during postnatal growth in thoroughbred horses. J Anat 181:455.

Straehl FR, et al. 2013. Evolutionary patterns of bone histology and bone compactness in xenarthran mammal long bones. PLoS ONE 8:e69275.

Strain GM. 2015. The genetics of deafness in domestic animals. Front Vet Sci 2:29.

Stringham SA, et al. 2012. Divergence, convergence, and the ancestry of feral populations in the domestic rock pigeon. Curr Biol 22:302–308.

Studer T. 1901. Die praehistorischen Hunde in ihrer Beziehung zu den gegenwärtig lebenden Rassen. Verlag Zürcher and Furrer.

Stuermer IW, et al. 2003. Intraspecific allometric comparison of laboratory gerbils with Mongolian gerbils trapped in the wild indicates domestication in *Meriones unguiculatus* (Milne-Edwards, 1867) (Rodentia: Gerbillinae). Zool Anz 242:249–266.

Stuermer IW, Wetzel W. 2006. Early experience and domestication affect auditory discrimination learning, open field behaviour and brain size in wild Mongolian gerbils and domesticated laboratory gerbils (*Meriones unguiculatus* forma domestica). Behav Brain Res 173:11–21.

Sulik KK, Johnston MC. 1983. Sequence of developmental alterations following acute ethanol exposure in mice: craniofacial features of the fetal alcohol syndrome. Am J Anat 166:257–269.

Sultan F. 2005. Why some bird brains are larger than others. Curr Biol 15:R649-R650.

Sultana S, Mannen H, Tsuji S. 2003. Mitochondrial DNA diversity of Pakistani goats. Anim Genet 34:417–421.

Sutter NB, et al. 2008. Morphometrics within dog breeds are highly reproducible and dispute Rensch's rule. Mamm Genome 19:713–723.

Svobodová J, et al. 2020. Differences in the growth rate and immune strategies of farmed and wild mallard populations. PLoS ONE 15: e0236583.

Sykes N. 2014. Beastly Questions: Animal Answers to Archaeological Issues. Bloomsbury.

Sykes N, Spriggs M, Evin A. 2019. Beyond curse or blessing: the opportunities and challenges of aDNA analysis. Wld Archaeol 51:503–516.

Syropoulos S. 2018. A Bestiary of Monsters in Greek Mythology. Archaeopress Archaeology.

Szulkin M, Munshi-South J, Charmantier A eds. 2020. Urban Evolutionary Biology. Oxford University Press.

Tafra V, et al. 2014. Primera caracterización morfoestructural y faneróptica del perro ovejero Magallánico, Chile. Arch Zootec 63:371–380.

Takai H, et al. 2018. Morphological and electrophysiological differences in tarsal chemosensilla between the wild silkmoth *Bombyx mandarina* and the domesticated species *Bombyx mori*. Arthropod Struct Dev 47:238–247.

Tamlin AL, Bowman J, Hackett DF. 2009. Separating wild from domestic American mink *Neovison vison* based on skull morphometries. Wildl Biol 15:266–277.

Tanner A. 1979. Bringing Home Animals: Religious Ideology and Mode of Production of the Mistassini Cree Hunters. C. Hurst and Company.

Taverne M, Van Der Weijden G. 2008. Parturition in domestic animals: targets for future research. Reprod Domest Anim 43:36–42.

Tchernov E, Horwitz LK. 1991. Body size diminution under domestication: unconscious selection in primeval domesticates. J Anthropol Archaeol 10:54–75.

Teletchea F. 2016. De la pêche à l'aquaculture. Demain, quels poissons dans nos assiettes? Belin.

ten Broek CMA, et al. 2012. Evo-devo of the human vertebral column: on homeotic transformations, pathologies and prenatal selection. Evol Biol 39:456–471.

ten Cate C, Verzijden MN, Etman E. 2006. Sexual imprinting can induce sexual preferences for exaggerated parental traits. Curr Biol 16:1128–1132.

Teng CS, et al. 2019. Resolving homology in the face of shifting germ layer origins: lessons from a major skull vault boundary. eLife 8:e52814.

Teotónio H, Rose MR. 2000. Variation in the reversibility of evolution. Nature 408:463–466.

Terry RJ. 1917. The primordial cranium of the cat. J Morphol 29:281–433.

Thalmann O, et al. 2013. Complete mitochondrial genomes of ancient canids suggest a European origin of domestic dogs. Science 342:871–874.

Theiler K. 1972. The House Mouse. Development and Normal Stages from Fertilization to 4 Weeks Age. Springer.

Theiler K. 1989. The House Mouse: Atlas of Embryonic Development. Springer Verlag.

Theofanopoulou C, et al. 2017. Self-domestication in *Homo sapiens*: insights from comparative genomics. PLoS ONE 12:e0185306.

Thodesen J, et al. 2011. Genetic improvement of tilapias in China: genetic parameters and selection responses in growth of Nile tilapia (*Oreochromis niloticus*) after six generations of multi-trait selection for growth and fillet yield. Aquaculture 322–323:51–64.

Thomas J, Kirby S. 2018. Self domestication and the evolution of language. Biol Philos 33:9.

Thompson DAW. 1917. On Growth and Form. Cambridge University Press.

Thompson KG, et al. 2005. Inherited chondrodysplasia in Texel sheep. N Z Vet J 53:208–212.

Thomson APD. 1951. A history of the ferret. J Hist Med Allied Sci VI:471–480.

Thomson PC. 1992. The behavioural ecology of dingoes in north-western Australia. II. Activity patterns, breeding season and pup rearing. Wildl Res 19:519–529.

Thomson JA, et al. 1998. Embryonic stem cell lines derived from human blastocysts. Science 282:1145–1147.

Thomson VA, Lebrasseur O, Austin JJ, et al. 2014. Using ancient DNA to study the origins and dispersal of ancestral Polynesian chickens across the Pacific. PNAS 111:4826–4831.

Thornton EK. 2016. Introduction to the special issue—Turkey husbandry and domestication: recent scientific advances. J Archaeol Sci Rep 10:514–519.

Thornton EK, et al. 2012. Earliest Mexican turkeys (*Meleagris gallopavo*) in the Maya region: implications for pre-Hispanic animal trade and the timing of turkey domestication. PLoS ONE 7:e42630.

Tickle CA, Summerbell D, Wolpert L. 1975. Positional signalling and specification of digits in chick limb morphogenesis. Nature 254:199–202.

Tidière M, et al. 2016. Comparative analyses of longevity and senescence reveal variable survival benefits of living in zoos across mammals. Sci Rep 6:36361.

Tiemann I, Rehkämper G. 2012. Evolutionary pets: offspring numbers reveal speciation process in domesticated chickens. PLoS ONE 7:e41453.

Tixier-Boichard M, Bed'hom B, Rognon X. 2011. Chicken domestication: from archeology to genomics. C R Biol 334:197–204.

Toledo Fonseca E, et al. 2013. Embryonic development of chicken (*Gallus gallus domesticus*) from 1st to 19th day—ectodermal structures. Microsc Res Tech 76:1217–1225.

Toli E, et al. 2017. Environmental enrichment, sexual dimorphism, and brain size in sticklebacks. Ecol Evol 7:1691–1698.

Tompkins DM, Mitchell RA, Bryant DM. 2006. Hybridization increases measures of innate and cell-mediated immunity in an endangered bird species. J Anim Ecol 75:559–564.

Toscano MJ. 2018. Skeletal problems in contemporary commercial laying hens. In: Advances in Poultry Welfare (Mench J ed). Woodhead Publishing, 151–174.

Townsend SW, et al. 2018. Compositionality in animals and humans. PLoS Biol 16:e2006425.

Trainor PA. 2016. Developmental biology: we are all walking mutants. Curr Top Dev Biol 117: 523–538.

Troy CS, et al. 2001. Genetic evidence for Near-Eastern origins of European cattle. Nature 410:1088–1091.

True JR, Haag ES. 2001. Developmental system drift and flexibility in evolutionary trajectories. Evol Dev 3:109–119.

Trut L. 1999. Early canid domestication: the farm-fox experiment. Am Sci 87:160–169.

Trut L, Oskina I, Kharlamova A. 2009. Animal evolution during domestication: the domesticated fox as a model. Bioessays 31:349–360.

Tsukaguchi R. 1912. Zur Entwickelungsgeschichte der Ziege (*Capra hircus*). Anat Hefte 46:413–492.

Tucker AS, Lumsden A. 2004. Neural crest cells provide species-specific patterning information in the developing branchial skeleton. Evol Dev 6:32–40.

Turcotte MM, et al. 2017. The eco-evolutionary impacts of domestication and agricultural practices on wild species. Philos Trans R Soc B 372:20160033.

Ueck M. 1961. Abstammung und Rassebildung der vorkolumbianischen Haushunde in Südamerika. Z Säugetierkd 26:157–176.

Uerpmann H-P, Uerpmann M. 2002. The appearance of the domestic camel in south-east Arabia. J Oman Stud 12:235–260.

Uerpmann H-P, Uerpmann M. 2017. The "commodification" of animals. In: Ancient West Asian Civilization: Geoenvironment and Society in the Pre-Islamic Middle East (Tsuneki A, Yamada S, Hisada K-i eds). Springer Singapore, 99–113.

Ündar L, Akpinar MA, Yanikoğlu A. 1990. "Doctor fish" and psoriasis. Lancet 335:470–471.

Uno Y, et al. 2013. Homoeologous chromosomes of *Xenopus laevis* are highly conserved after whole-genome duplication. Heredity 111:430–436.

Upex B, Dobney K. 2012. More than just mad cows: exploring human-animal relationships through animal paleopathology. In: A Companion to Paleopathology (Grauer AL ed). Blackwell, 191–213.

Upham NS, Esselstyn JA, Jetz W. 2019. Inferring the mammal tree: species-level sets of phylogenies for questions in ecology, evolution, and conservation. PLoS Biol 17:e3000494.

Usherwood JR, Wilson AM. 2005. No force limit on greyhound sprint speed. Nature 438:753–754.

Usui K, Tokita M. 2018. Creating diversity in mammalian facial morphology: a review of potential developmental mechanisms. EvoDevo 9:15.

Vaglia JL, Smith KK. 2003. Early differentiation and migration of cranial neural crest in the opossum, *Monodelphis domestica*. Evol Dev 5:121–135.

Valadez Azúa R, et al. 2002. Híbridos de lobos y perros (tercer acto): hallazgos en la pirámide de Quetzalcóatl de la antigua ciudad de Teotihuacan (segunda y última de dos partes). AMMVEPE 13:219–231.

van Alphen JJM, Fernhout BJ. 2020. Natural selection, selective breeding, and the evolution of resistance of honeybees (*Apis mellifera*) against *Varroa*. Zool Lett 6:6.

Van Boeckel TP, et al. 2019. Global trends in antimicrobial resistance in animals in low-and middle-income countries. Science 365:eaaw1944.

Van Dam AR, et al. 2015. Range wide phylogeography of *Dactylopius coccus* (Hemiptera: Dactylopiidae). Ann Entomol Soc Am 108:299–310.

van der Geer AAE. 2019. Effect of isolation on coat colour polymorphism of Polynesian rats in Island Southeast Asia and the Pacific. PeerJ 7:e6894.

van der Geer A, et al. 2010. Evolution of Island Mammals: Adaptation and Extinction of Placental Mammals on Islands. Wiley-Blackwell.

van der Geer AAE, Galis F. 2017. High incidence of cervical ribs indicates vulnerable condition in Late Pleistocene woolly rhinoceroses. PeerJ 5:e3684.

Van Grouw K. 2018. Unnatural Selection. Princeton University Press.

van Huis A. 2013. Potential of insects as food and feed in assuring food security. Annu Rev Entomol 58:563–583.

van Huis A, et al. 2013. Edible Insects: Future Prospects for Food and Feed Security. Food and Agriculture Organization of the United Nations.

van Huis A. 2020. Insects as food and feed, a new emerging agricultural sector: a review. Journal of Insects as Food and Feed 6:27–44.

Van Valen L. 1962. A study of fluctuating asymmetry. Evolution 16:125–142.

Van Vuren D, Coblentz BE. 1988. Dental anomalies of feral goats (*Capra hircus*) on Aldabra Atoll. J Zool 216:503–506.

Van Vuure C. 2005. Retracing the Aurochs. History, Morphology and Ecology of an Extinct Wild Ox. Pensoft.

VandeHaar MJ, St-Pierre N. 2006. Major advances in nutrition: relevance to the sustainability of the dairy industry. J Dairy Sci 89:1280–1291.

Vandeputte M, et al. 2009. Amélioration génétique des poissons: quelles réalisations et quels défis pour demain? Cah Agric 18:262–269.

vanEngelsdorp D, Meixner MD. 2010. A historical review of managed honey bee populations in Europe and the United States and the factors that may affect them. J Invertebr Pathol 103:S80-S95.

Vargas-Ramírez M, et al. 2012. Extremely low genetic diversity and weak population differentiation in the endangered Colombian river turtle *Podocnemis lewyana* (Testudines: Podocnemididae). Conserv Genet 13:65–77.

Vázquez-Gómez M, et al. 2020. Differential effects of litter size and within-litter birthweight on postnatal traits of fatty pigs. Animals 10:870.

Veasey JS. 2017. In pursuit of peak animal welfare; the need to prioritize the meaningful over the measurable. Zoo Biol 36:413–425.

Veitschegger K, Sánchez-Villagra MR. 2016. Tooth eruption sequences in cervids and the effect of morphology, life history, and phylogeny. J Mamm Evol 23:251–263.

Veitschegger K, et al. 2018. Resurrecting Darwin's Niata—anatomical, biomechanical, genetic, and morphometric studies of morphological novelty in cattle. Sci Rep 8:9129.

Vella CM, et al. 1999. Robinson's Genetics for Cat Breeders and Veterinarians. Butterworth-Heinemann.

Velthuis HHW, Van Doorn A. 2006. A century of advances in bumblebee domestication and the economic and environmental aspects of its commercialization for pollination. Apidologie 37:421–451.

Venkataraman VV, et al. 2015. Solitary Ethiopian wolves increase predation success on rodents when among grazing gelada monkey herds. J Mammal 96:129–137.

Verano JW, Ubelaker DH eds. 1992. Disease and demography in the Americas. Smithsonian Institution Press.

Verdiglione R, Rizzi C. 2018. A morphometrical study on the skull of Padovana chicken. Italian J Anim Sci 17:785–796.

Verein für Deutsche Schäferhunde 1922. Körbuch für Deutsche Schäferhunde. Kämpfe.

Vickaryous MK, Hall BK. 2006. Human cell type diversity, evolution, development, and classification with special reference to cells derived from the neural crest. Biol Rev 81:425–455.

Vickrey AI, et al. 2015. Convergent evolution of head crests in two domesticated columbids is associated with different missense mutations in EphB2. Mol Biol Evol 32:2657–2664.

Vickrey AI, et al. 2018. Introgression of regulatory alleles and a missense coding mutation drive plumage pattern diversity in the rock pigeon. eLife 7:e34803.

Vignal A, et al. 2019. A guinea fowl genome assembly provides new evidence on evolution following domestication and selection in Galliformes. Mol Ecol Resour 19:997–1014.

Vigne J-D. 2011. The origins of animal domestication and husbandry: a major change in the history of humanity and the biosphere. Compte Rendu Biol 334:171–181.

Vigne J-D, Helmer D, Peters J. 2005a. New archaeological approaches to trace the first steps of animal domestication: general presentation, reflections and proposals. In: The First Steps of Animal Domestication: New Archaeozoological Approaches (Vigne J-D, Peters J, Helmer D eds). Oxbow Books, 1–16.

Vigne J-D, Peters J, Helmer D eds. 2005b. The First Steps of Animal Domestication: New Archaeozoological approaches (Proceedings of the 9th ICAZ Conference). Oxbow Books.

Vigne J-D, et al. 2009. Pre-Neolithic wild boar management and introduction to Cyprus more than 11,400 years ago. PNAS 106:16135–16138

Vigne J-D, et al. 2012. First wave of cultivators spread to Cyprus at least 10,600 y ago. PNAS 109:8445–8449.

Vigne J-D, et al. 2016. Earliest "domestic" cats in China identified as leopard cat (*Prionailurus bengalensis*). PLoS ONE 11:e0147295.

Vilà C, et al. 1997. Multiple and ancient origins of the domestic dog. Science 276:1687–1689.

Vilà C, et al. 2001. Widespread origins of domestic horse lineages. Science 291:474–477.

Voit M. 1909. Das Primordialcranium des Kaninchens. Anat Hefte 38:425–616.

von Lengerken H. 1955. Ur, Hausrind und Mensch: Versuch eines Überblicks. Deutsche Akademie der Landwirtschaftswissenschaften zu Berlin.

von Stephanitz MEF. 1921. Der deutsche Schäferhund in Wort und Bild. Druck von A. Kämpfe.

von Holdt BM, et al. 2017. Structural variants in genes associated with human Williams-Beuren syndrome underlie stereotypical hypersociability in domestic dogs. Sci Adv 3:e1700398.

Voroshilova S. 1974. Growth and development of guinea-fowl broilers (in Russian). Ptitsevodstvo 24:20.

Wada N, Nohno T, Kuratani S. 2011. Dual origins of the prechordal cranium in the chicken embryo. Dev Biol 356:529–540.

Waddington C. 1930. Developmental mechanics of chicken and duck embryos. Nature 125:924–925.

Waddington CH. 1932. Experiments on the development of chick and duck embryos, cultivated in vitro. Philos Trans R Soc Lond 221:179–230.

Wagner GP, Booth G, Bagheri-Chaichian H. 1997. A population genetic theory of canalization. Evolution 51:329–347.

Wagner GP, Pavlicev M, Cheverud JM. 2007. The road to modularity. Nat Rev Genet 8:921–931.

Wagner F, Ruf I. 2020. "Forever young"—Postnatal growth inhibition of the turbinal skeleton in brachycephalic dog breeds (*Canis lupus familiaris*). Anat Rec 304:154–189.

Wanek N, et al. 1989. A staging system for mouse limb development. J Exp Zool 249:41–49.

Wang Y, et al. 2012. The crest phenotype in chicken is associated with ectopic expression of HOXC8 in cranial skin. PLoS ONE 7:e34012.

Wang N, et al. 2013. Assessing phylogenetic relationships among Galliformes: a multigene phylogeny with expanded taxon sampling in Phasianidae. PLoS ONE 8:e64312.

Wang J, et al. 2014a. Evidence for the evolutionary origin of goldfish derived from the distant crossing of red crucian carp × common carp. BMC Genet 15:33.

Wang G-D, et al. 2016. Out of southern East Asia: the natural history of domestic dogs across the world. Cell Res 26:21–33.

Wang Y, et al. 2018. The formation of the goldfish-like fish derived from hybridization of female koi carp × male blunt snout bream. Front Genet 9:437.

Wang M-S, et al. 2020. 863 genomes reveal the origin and domestication of chicken. Cell Res 30:693–701.

Wanzala W, et al. 2005. Ethnoveterinary medicine: a critical review of its evolution, perception, understanding and the way forward. Livestock Research for Rural Development 17.

Warinner C, et al. 2015. Direct evidence of milk consumption from ancient human dental calculus. Sci Rep 4:7104.

Washburn SL. 1960. Tools and human evolution. Sci Am 203:62–75.

Watase S. 1887. On the caudal and anal fins of goldfishes. Journal of the College of Science, Imperial University of Tokyo, Japan 1:247–267.

Waters MR, Stafford Jr T. 2013. The first Americans: a review of the evidence for the Late-Pleistocene peopling of the Americas. In: Paleoamerican Odyssey (Graf KE, Ketron CV, Waters MR eds). Texas A&M University Press, 543–562.

Watson JM. 1939. The development of the Weberian ossicles and anterior vertebrae in the goldfish. Proc R Soc Lond B 127:452–472.

Watson JPN. 1975. Domestication and bone structure in sheep and goats. J Archaeol Sci 2:375–383.

Wayne RK. 1986. Cranial morphology of domestic and wild canids: the influence of development on morphological change. Evolution 40:243–261.

Weber M. 1927. Die Säugetiere. Fischer Verlag.

Wehrend A. 2013. Neonatologie beim Hund: Von der Geburt bis zum Absetzen. Schlütersche Verlagsgesellschaft.

Weintraub K. 2016. 20 years after Dolly the sheep led the way—where is cloning now. Sci Am.

Weisbecker V, Goswami A. 2011. Neonatal maturity as the key to understanding brain size evolution in homeothermic vertebrates. Bioessays 33:155–158.

Weisbecker V, et al. 2015. The evolution of relative brain size in marsupials is energetically constrained but not driven by behavioral complexity. Brain Behav Evol 85:125–135.

Weissbrod L, et al. 2017. Commensal origins of the house mouse. PNAS 114:4099–4104.

Weitzel EM, Codding BF. 2016. Population growth as a driver of initial domestication in Eastern North America. R Soc Open Sci 3:160319.

Werneburg I. 2009. A standard system to study vertebrate embryos. PLoS ONE 4:e5887.

Werneburg I, Geiger M. 2017. Ontogeny of domestic dogs and the developmental foundations of carnivoran domestication. J Mamm Evol 24:323–343.

Werneburg I, Sánchez-Villagra MR. 2011. The early development of the echidna, *Tachyglossus aculeatus* (Mammalia: Monotremata), and patterns of mammalian development. Acta Zool 92:75–88.

Werneburg I, Spiekman SNF. 2018. Mammalian embryology and organogenesis. From Gametes to Weaning In: Mammalia Series: Handbook of Zoology / Handbuch der Zoologie (Zachos F, Asher RJ eds). De Gruyter.

Werneburg I, Yaryhin O. 2019. Character definition and tempus optimum in comparative chondro-cranial research. Acta Zool 100:376–388.

Werneburg I, et al. 2013. Development and embryonic staging in non-model organisms: the case of an afrotherian mammal. J Anat 222:2–18.

Werneburg I, et al. 2016. Evolution of organogenesis and the origin of altriciality in mammals. Evol Dev 18:229–244.

Werth E. 1944. Die primitiven Hunde und die Abstammungsfrage des Haushundes. Z Tierzuecht Zuechtungsbiol 56:213–260.

West-Eberhard MJ. 2003. Developmental plasticity and evolution. Oxford University Press.

West-Eberhard MJ. 2005. Phenotypic accommodation: adaptive innovation due to developmental plasticity. J Exp Zool Part B 304:610–618.

Weston EM. 2003. Evolution of ontogeny in the hippopotamus skull: using allometry to dissect developmental change. Biol J Linn Soc 80:625–638.

Wheeler JC. 2012. South American camelids: past, present and future. J Camelid Sci 5:1–24.

Whewell W. 1840. The Philosophy of the Inductive Sciences. John W. Parker.

Whitehead H, et al. 2019. The reach of gene-culture coevolution in animals. Nat Commun 10:2405.

Whitman BD. 2004. Domestic Rabbits & their Histories: Breeds of the World. Leathers Publishing.

Whittemore K, et al. 2019. Telomere shortening rate predicts species life span. PNAS 116:15122–15127.

Widowski TM, Lo Fo Wong DMA, Duncan IJH. 1998. Rearing with males accelerates onset of sexual maturity in female domestic fowl. Poult Sci 77:150–155.

Wiener G, Han J, Long R. 2003. The Yak. 2nd ed. Regional Office for Asia and the Pacific of the Food and Agriculture Organization of the United Nations.

Wiener P, Wilkinson S. 2011. Deciphering the genetic basis of animal domestication. Proc R Soc Biol Sci Ser B 278:3161–3170.

Wigger A, et al. 2009. Lumbosakraler Übergangswirbel beim Deutschen Schäferhund: Häufigkeit, Formen, Genetik und Korrelation zur Hüftgelenksdysplasie. Tierarztl Prax 200:7–13.

Wilfert L, et al. 2016. Deformed wing virus is a recent global epidemic in honeybees driven by *Varroa* mites. Science 351:594–597.

Wilkins NP, et al. 1995. Fluctuating asymmetry in Atlantic salmon, European trout and their hybrids, including triploids. Aquaculture 137:77–85.

Wilkins AS, Wrangham RW, Fitch WT. 2014. The "domestication syndrome" in mammals: a unified explanation based on neural crest cell behavior and genetics. Genetics 197:795–808.

Wilkins A. 2017a. Revisiting two hypotheses on the "domestication syndrome" in light of genomic data. Vavilov J Genet Breeding 21:435–442.

Wilkins AS. 2017b. Making Faces. The Evolutionary Origins of the Human Face. Harvard University Press.

Wilkins AS. 2019. A striking example of developmental bias in an evolutionary process: the "domestication syndrome". Evol Dev 22:43–153.

Wilkins A, Wrangham RW, Fitch WT. 2021. The neural crest domestication syndrome hypothesis, explained: Reply to Johnsson, Henriksen, and Wright. Genetics, iyab098, https://doi.org/10.1093/genetics/iyab098.

Willerslev R, Vitebsky P, Alekseyev A. 2015. Sacrifice as the ideal hunt: a cosmological explanation for the origin of reindeer domestication. J R Anthropol Inst 21:1–23.

Williams GC. 1957. Pleiotropy, natural selection, and the evolution of senescence. Evolution 11:398–411.

Wilson PN. 1958. The effect of plane of nutrition on the growth and development of the East African dwarf goat. Part II. Age changes in the carcass composition of female kids. J Agric Sci 51:4–21.

Wilson EO. 1984. Biophilia. The Human Bond with Other Species. Harvard University Press.

Wilson DE, Reeder DM eds. 2005. Mammal Species of the World: A Taxonomic and Geographic Reference. Johns Hopkins University Press.

Wilson LA. 2013. The contribution of developmental palaeontology to extensions of evolutionary theory. Acta Zool 94:254–260.

Wilson LA. 2018. The evolution of ontogenetic allometric trajectories in mammalian domestication. Evolution 72:867–877.

Wilson A, Wilson D, Robin L. 2017. The ought-ecology of ferals: an emerging dialogue in invasion biology and animal studies. Aust Zool 39:85–102.

Wilson LAB, et al. 2021. Modularity patterns in mammalian domestication and developmental hypotheses for diversification. Evol Lett 5:385–396.

Witt KE, et al. 2015. DNA analysis of ancient dogs of the Americas: Identifying possible founding haplotypes and reconstructing population histories. J Hum Evol 79:105–118.

Witten PE, et al. 2006. Vertebrae fusion in Atlantic salmon (*Salmo salar*): development, aggravation and pathways of containment. Aquaculture 258:164–172.

Witten PE, Huysseune A, Hall BK. 2010. A practical approach for the identification of the many cartilaginous tissues in teleost fish. J Appl Ichthyol 26:257–262.

Witten P, Hall BK. 2015. Teleost skeletal plasticity: modulation, adaptation, and remodelling. Copeia 103:727–739.

Witten P, Cancela M. 2018. What aquaculture does for taxonomy, evo-devo, palaeontology, biomechanics and biomedical research. J Appl Ichthyol 34:429–430.

Wolf P, Kamphues J. 1996. Untersuchungen zu Fütterungseinflüssen auf die Entwicklung der Incisivi bei Kaninchen, Chinchilla und Ratte. Kleintierpraxis 41:723–732.

Wolpert L. 1994. The evolutionary origin of development: cycles, patterning, privilege and continuity. Development 1994:79–84.

Woods GL, et al. 2003. A mule cloned from fetal cells by nuclear transfer. Science 301:1063.

Wrangham RW. 2018. Two types of aggression in human evolution. PNAS 115:245–253.

Wrangham R. 2019. The goodness paradox: the strange relationship between virtue and violence in human evolution. Pantheon.

Wright S. 1934. An analysis of variability in number of digits in an inbred strain of guinea pigs. Genetics 19:506.

Wright S, Wagner K. 1934. Types of subnormal development of the head from inbred strains of guinea pigs and their bearing on the classification and interpretation of vertebrate monsters. Am J Anat 54:383–447.

Wright TF, et al. 2004. Sex-linked inheritance of hearing and song in the Belgian Waterslager canary. Proc R Soc Lond B 271:S409-S412.

Wright D, et al. 2009. Copy number variation in intron 1 of SOX5 causes the Pea-comb phenotype in chickens. PLoS Genet 5:e1000512.

Wright D. 2015. The genetic architecture of domestication in animals. Bioinform Biol Insights 9:11–20.

Wu H, et al. 2014. Camelid genomes reveal evolution and adaptation to desert environments. Nat Commun 5:5188.

Xiang H, et al. 2018. The evolutionary road from wild moth to domestic silkworm. Nat Ecol Evol 2:1268–1279.

Xie J, et al. 2011. Ecological mechanisms underlying the sustainability of the agricultural heritage rice–fish co-culture system. PNAS 108:E1381-E1387.

Xu P, et al. 2014. Genome sequence and genetic diversity of the common carp, *Cyprinus carpio*. Nat Genet 46:1212–1219.

Yamasaki M, Tonosaki A. 1988. Developmental stages of the society finch, *Lonchura striata var. dornestica*. Dev Growth Differ 30:515–542.

Yang DY, et al. 2008. Wild or domesticated: DNA analysis of ancient water buffalo remains from north China. J Archaeol Sci 35:2778–2785.

Yang Y, et al. 2013. Proteomic analysis of cow, yak, buffalo, goat and camel milk whey proteins: quantitative differential expression patterns. J Proteome Res 12:1660–1667.

Yilmaz O, Wilson RT. 2012. The domestic livestock resources of Turkey: status, use and some physical characteristics of mules. Journal of Equine Science 23:47–52.

Yindee M, et al. 2010. Y-chromosomal variation confirms independent domestications of swamp and river buffalo. Anim Genet 41:433–435.

Yom-Tov Y, Yom-Tov S, Baagøe H. 2003. Increase of skull size in the red fox (*Vulpes vulpes*) and Eurasian badger (*Meles meles*) in Denmark during the twentieth century: an effect of improved diet? Evol Ecol Res 5:1037–1048.

Yoshimura K, et al. 2012. Inheritance and developmental pattern of cerebral hernia in the crested Polish chicken. J Exp Zool Part B 318:613–620.

Young NM, et al. 2014. Embryonic bauplans and the developmental origins of facial diversity and constraint. Development 141:1059–1063.

Young NM, et al. 2017. Craniofacial diversification in the domestic pigeon and the evolution of the avian skull. Nat Ecol Evol 1:95.

Zachos F, Hartl G, Suchentrunk F. 2007. Fluctuating asymmetry and genetic variability in the roe deer (*Capreolus capreolus*): a test of the developmental stability hypothesis in mammals using neutral molecular markers. Heredity 98:392–400.

Zachos FE. 2016. Species Concepts in Biology. Historical Development, Theoretical Foundations and Practical Relevance. Springer International.

Zanatta DB, et al. 2009. Evaluation of economically important traits from sixteen parental strains of the silkworm *Bombyx mori* L (Lepidoptera: Bombycidae). Neotrop Entomol 38:327–331.

Zanella M, et al. 2019. Dosage analysis of the 7q11.23 Williams region identifies BAZ1B as a major human gene patterning the modern human face and underlying self-domestication. Sci Adv 5:eaaw7908.

Zedda M, et al. 2008. Comparative bone histology of adult horses (*Equus caballus*) and cows (*Bos taurus*). Anat Histol Embryol 37:442–445.

Zedda M, et al. 2020. A first comparison of bone histomorphometry in extant domestic horses (*Equus caballus*) and a Pleistocene Indian wild horse (*Equus namadicus*). Integr Zool 15:448–460.

Zeder MA. 1994. Of kings and shepherds: specialized animal economy in Ur III Mesopotamia. In: Chiefdoms and Early States in the Near East (Stein G, Rothman M eds). Prehistory Press, 175–191.

Zeder MA. 2006. Central questions in the domestication of plants and animals. Evol Anthropol 15:105–117.

Zeder MA. 2012a. Pathways to animal domestication. In: Biodiversity in Agriculture: Domestication, Evolution, and Sustainability (Gepts P, et al. eds). Cambridge University Press, 227–259.

Zeder MA. 2012b. The domestication of animals. J Anthropol Res 68:161–190.

Zeder MA. 2012c. The broad spectrum revolution at 40: resource diversity, intensification, and an alternative to optimal foraging explanations. J Anthropol Archaeol 31:241–264.

Zeder MA. 2015. Core questions in domestication research. PNAS 112:3191–3198.

Zeder MA. 2017. Domestication as a model system for the extended evolutionary synthesis. Interface Focus 7:20160133.

Zeder MA. 2018. Why evolutionary biology needs anthropology: evaluating core assumptions of the extended evolutionary synthesis. Evol Anthropol 27:267–284.

Zeder MA. 2020. Straw foxes: domestication syndrome evaluation comes up short. Trends Ecol Evol 35:647–649.

Zeder MA, et al. eds. 2006. Documenting Domestication: New Genetic and Archaeological Paradigms. University of California Press.

Zeder MA, Lemoine X, Payne S. 2015. A new system for computing long-bone fusion age profiles in *Sus scrofa*. J Archaeol Sci 55:135–150.

Zelditch ML, Lundrigan BL, Garland Jr T. 2004. Developmental regulation of skull morphology. I. Ontogenetic dynamics of variance. Evol Dev 6:194–206.

Zelditch ML, Calamari ZT, Swiderski DL. 2016. Disparate postnatal ontogenies do not add to the shape disparity of infants. Evol Biol 43:188–207.

Zeller U, Göttert T. 2019. The relations between evolution and domestication reconsidered—Implications for systematics, ecology, and nature conservation. Glob Ecol Conserv 20:e00756.

Zeuner FE. 1963. A History of Domesticated Animals. Harper & Row.

Zhan S, et al. 2020. Genomic landscape and genetic manipulation of the black soldier fly *Hermetia illucens*, a natural waste recycler. Cell Res 30:50–60.

Zhang H, et al. 2013a. Morphological and genetic evidence for early Holocene cattle management in northeastern China. Nat Commun 4:2755.

Zhang Y, et al. 2013b. Primary genome scan for complex body shape-related traits in the common carp *Cyprinus carpio*. J Fish Biol 82:125–140.

Zhang Z, et al. 2018. Whole-genome resequencing reveals signatures of selection and timing of duck domestication. GigaScience 7:giy027.

Zhang H, et al. 2020a. A unique feature of swine ANP32A provides susceptibility to avian influenza virus infection in pigs. PLoS Path 16:e1008330.

Zhang S-j, et al. 2020b. Genomic regions under selection in the feralization of the dingoes. Nat Commun 11:1–10.

Zhou JF, et al. 2003. Genetic divergence between *Cyprinus carpio carpio* and *Cyprinus carpio haematopterus* as assessed by mitochondrial DNA analysis, with emphasis on origin of European domestic carp. Genetica 119:93–97.

Zhou Z, et al. 2018. An intercross population study reveals genes associated with body size and plumage color in ducks. Nat Commun 9:1–10.

Zhu L, et al. 2014. A morphometric study on the skull of donkey (*Equus asinus*). Int J Morphol 32:1306–1310.

Zimmer C, Emlen DJ. 2013. Evolution: Making Sense of Life. Roberts and Co.

Zimmermann W. 1961. Zur Domestikation der Chinchillas. Z Tierzuecht Zuechtungsbiol 76:343–348.

Zollikofer CPE, Ponce de León MS. 2010. The evolution of hominin ontogenies. In: Semin Cell Dev Biol. Elsevier, 441–452.

Zuidhof MJ, et al. 2014. Growth, efficiency, and yield of commercial broilers from 1957, 1978, and 2005. Poult Sci 93:2970–2982.

Index

Page numbers in italics indicate figures and tables.

African sharptooth catfish (*Clarias gariepinus*), 224

Agassiz, Louis, on Darwin, 140

Aldrovandi, Ulisse, 60

algae (*Polysiphonia* sp.), 5

alpacas, domesticated mammals, *51*, 51–52

American catfish (*Ictalurus punctatus*), 224

American Cavy Breeders Association, 57

American Kennel Club, 28

American Poultry Association, 31, 60

American Standard of Perfection, 31, 60

Americas: biogeographic diversity of, 32–33; dogs in, 39–40; human occupation, 25; landraces and breeds in, 31

amphibian model *Xenopus*, 101

Anas platyrhynchos, development and staging, *95*

ancient DNA studies, 15; dogs, 37

animals: categories of landraces and breeds for domestic, 28; culturesphere of, 247–48; feralization and wild phenotype, 205–6; nomenclature separation of wild and domestic, 26–27; selective breeding and welfare of, 218

Anna Karenina Principle, domestication, 31

aquaculture, 234; marine and freshwater species, 224–25; salmonids, 223–24

archaeological record, 11; dogs in, 36; Neolithic transition, 22

Aristotle, study of domestic animals, 183

artificial selection, 3, *3*

Atlantic salmon (*Salmo salar*), 224, 233; heads of male adults, *226*; life history and metabolic changes, *232*; sexual selection in, 226; vertebrae fusion in, *230*; wild populations, 233. *See also* fish domestication

avian domesticates, 8–9, 11. *See also* domesticated birds

axial skeleton: body shape, 136, *137*; developmental repatterning in mammals, 136–38

Bactrian camels, 49–50, *51*, 101

barn mice, longitudinal study of, 211–12

Bateson, William, on Mendelian inheritance, 67

bees, domestication of, 238–40. *See also* insect domestication

Belyaev, D. K., 209; silver fox experiment, 81

Bengalese finch, 162, 208, *208*

bighorn sheep *Ovis canadensis*, 193

biological features, domestication criteria, 31–32

biological web-of-life, 247

biomedical research, model organisms for, 97

biophilia, concept of, 42

biosphere, census of biomass composition, 23, *23*

birds: avian hybrids, 72; estimated time line of domestication, *12*; modularity in skull of, 127, 129–30; potential domesticate, *34*. *See also* domesticated birds

birds and life history, 164; altricial Muscovy duckling, *162*; clutch size variation, 164–65; condition at hatching, 162, 164; evolvability in, 164; fecundity, 164–65; growth in chickens, 165–66; longevity in chickens, 166; precocial pigeon, *162*; qualitative scoring framework of hatchlings, 162, *163*; selective breeding in broiler chickens, *166*. *See also* domesticated birds; life history and growth

black soldier fly (*Hermetia illucens*), 237

bluefin tuna (*Thunnus thynnus*), 225

body plan: anatomical variation within, 97; mammalian, 136

body shape, developmental repatterning in mammals, 136–38

bone histology, life history and, 161

bone microstructure, study of, 161

bonobos *Pan paniscus*, 142

Bosch, Hieronymus, creatures of paintings of, 172, *172*

bovines: development and staging *Bos taurus*, *93*; domestication of, 46–49; examples of wild and domestic bovids, *48*; horns in bovids and caprids, 192–94, *193*. *See also* cattle

brachycephaly: diversity in domesticated mammals, *196*; skull variation, 195, 197

brain anatomy and size changes: domesticated birds, 186–88; domesticated fish, 230–31; domesticated mammals, 182, *182*–86; feral populations, 205–6; selective breeding, 201–2 breed(s): in Americas, 31; bovines, 46–49, *48*; cats, 41, *41*; cavy (*Cavia porcellus*), *57*; chondrodysplasia in dogs and cattle, 118; conventions for defining, 28; definition, 28; dog litter size variation by, 158, *159*; dogs, 37, *38*, 85, *85*; dogs in Americas, 39–40; dogs' litter size variation, 158; donkey, *54*; ducks, *64*; extinction of local, 29–31; geese, *64*, 65; goats, 43, *44*; horse, *54*; pigeon (*Columba livia*), 59; pigs, 46, *47*; selective breeding and size changes, 177; sheep, 45, *45*

Buffon: network of dog breeds, 183; theory of transmutation, 71

Busch, Wilhelm: studies of, 190

Call of the Wild (film), 207

camels: Bactrian, 49–50, *51*, 101; dromedary, 49–50, *51*, 101; South American camelids, *51*, 51–52

canary (*Serinus canaria*): domesticated, 162, 209; domestication, 25, 66, 171; hybridization, 70; illustration, *62*

Canis lupus. *See* wolves

Capra hircus. *See* goats

carbon isotopes, captive animal rearing, 20

carp (*Cyprinus carpio*): domestication of, 170, 220–22; fish farming for, 222; relation to goldfish, 223; variations in body shape and squamation, *222*. *See also* fish domestication

cat(s), 40–41; average crown-rump length (CRL) of embryos by age, *131*; brachycephaly, 195, *196*; breeds, 41, *41*; caecum length in, 154, *154*; coat coloration, *105*; developmental plasticity of coats, 104–6; development and staging, *93*; facial length in, 103; feral, 203; genetic mechanism, 79; genetic variation analysis, 41; Hemingway mutation, 117; misplaced biophilia, 42; polydactyly in, 117; protein-coding gene *BRAF* in, 118

cat breeds: Abyssinian, *41*; Bengal, *41*; British Longhair, *41*; British Shorthair, *41*; Burmese, 105–6; Japanese Bobtail, 138; Manx, *41*, 117; Neva Masquerade, *41*; Norwegian Forest, *41*; Siamese, *41*, 105, 105–6; Sphynx, *41*

cattle: average crown-rump length (CRL) of embryos by age, *131*; changes in body proportions, 178, *179*; chondrodysplasia in, 118; digestive system, 197; domestication of bovine species, 46–49; double-muscling in breeds of, 86–87, *87*; Dove's "fabulous unicorn", 104, *104*; gestation length, 149, *150*; head shape variation, 181, *180*; horns in bovids, 192–93, *193*; nomenclature, 28; ontogenetic tooth changes, 106, *108*; respiratory and digestive types of, *179*; selective breeding and size changes, 177; skull of vaca ñata *vs.* standard, *30*

cattle breeds, 195; Aberdeen-Angus, 117, *179*; Ankole-watusi, 192, *193*; Austrian Blondvieh, *193*; Bali cattle, 46; banteng, 46; Belgian Blue bull, *87*; Cape buffalo, *193*; Chiania bull, 177; common cattle (*Bos taurus*), 27, 48–49; Dexter, 118, 177; gayal or mithan, 46; Holstein-Friesian, 177, *179*;

Poll Hereford, 117; Red Poll, 117; Texas Longhorn, 192, *193*; vaca ñata, 28, *30*; water buffalo, 46–47; wild and domestic bovids, *48*; yak, 47–48

cavy (*Cavia porcellus*): breeds of, *57*; digit number in, 113, *113*; domestication, 56–57; malformation, 118

Chagas disease, *Trypanosoma cruzi* and, 244

chicken(s): bone microstructure, 161; brains of, 188–89, *188*; breeds of, 61; changes in size and body proportions, 181; comb types, *123*; cranial vault of skull, 200; Darwin on, 6; domestication, 60–61, *61*; domestication centers, 60; epigenetic changes in, 109–10; epistasis, 80; experimental generation of polydactyly in, 115, *116*; fluctuating asymmetry in, 111; genetic mechanism, 79; genetic studies of domestication, 14–15; growth in, 165–66; interactions of brain-skull interaction, 111, *112*; introgression, 70; isolating mechanisms and assortative mating, 74; late embryonic stage skull, *132*, 133; movement and temperature variation during embryonic life, 109–10; neural crest, 99–101, *100*; phylotypic stage, 97, *98*; red junglefowl and White Leghorn experiments, 213–14; sagittal sections of skulls, *189*; selective breeding, 23, 144, *166*; species delimitation, 73; structural genetic changes, 77; yellow skin phenotype of, 84; zone of polarizing activity (ZPA) of wing, 115, *116*

chicken breeds: Araucana, *61*; Barbu d'Anvers, *61*; Brahma, *61*; Hubbard, *61*; Lohmann Selected Leghorns, 74; Polish, *61*; red junglefowl, *61*, 164, 188, *188*, 213–14; Red Leghorns, 74; Shamo, *61*; Silkie, *61*, 115, 116; White Crested Polish, 74, 189–90, *189*; White Leghorn, *61*, 110, 213–14, 215

Childe, V. Gordon, on domestication and environment, 22

chimpanzees *Pan troglodytes*, 142

chinchillas, domestication, 171

Chinook salmon (*Oncorhynchus tshawytscha*), 230

chondrodysplasia: brachycephaly, 195; in dogs and cattle, 118

climate curve, weather patterns, 21–22, *22*

Clutton-Brock, J. A.: on breed, 28; on domestication, 32

cobia (*Rachytcentron canadum*), 225

cognitive buffer hypothesis, 184

coho salmon (*Oncorhynchus kisutch*), 223, 225, 232

Columba livia. *See* pigeons

commensal pathway, domestication, 4, *4*

comparative embryology, domesticated animals in, 92, 94, 96

computed tomography (CT), X-ray, 174–75

conservation, feralization and, 203

Coppinger, Raymond, pedomorphosis hypothesis, 134, *134*

Corded Ware Culture, 15

Coturnix. *See* quail

cows. *See* cattle

Crampe, Hugo, rat breeding experiments, 67

cranial changes: in domesticated birds, 199–201; morphology of pigs, 16, *17*
cranial evolutionary allometric (CREA) hypothesis, 200
Cretan dwarf deer *Candiacervus*, 185
crucian carp (*Carassius carassius*), 223, 234
cultural evolution, domestication, 21
culturesphere, domesticated and not domesticated animals, 247–48

Dactylopius coccus (cochineal insect): domesticated insect, 241–42; illustration of, *237*
damselfish (*Stegastes nigricans*), 5
Danish Institute of Agricultural Sciences, 212
Darwin, Charles: on chicken breeds, 60; *Descent of Man*, 140, *140*, 244; diversity of domesticated species, 139, 183; diversity of pigeons, 199; on domestication, 6; domestication and, 139–40; on "lop-eared" rabbit, 81–82; *The Variation of Animals and Plants under Domestication*, 139, 183
Darwinian Demons, 144
Daubenton, L. J. M., Buffon and, 71
Deep Ecology movement, Naess and, 167
dental changes: dental traits of hairless dogs, 67, *68*, *88*, 88–89; developmental plasticity in tooth growth, 106–7, *108*; in domesticated mammals, 197–98; domestication in dogs, 15–16; domestication in pigs, 15–16; growth in dogs, 158–60; growth in mammals, 151–53
de Saint-Exupéry, Antoine, 29
The Descent of Man (Darwin), 140, *140*, 244
developmental biology, evolutionary, 91–92
developmental evolutionary biology, genotype-phenotype map (GPM), *76*, 76–77
developmental genetics, in domestication, 74–77
developmental plasticity: case of cat and rabbit coats, 104–6, *105*; Slijper's two-legged goat, 106, *107*; in tooth growth, 106–7, *108*
developmental repatterning: body shape, axial skeleton and, in mammals, 136–38; chicken comb types, *123*; domestic mammal skull growth, 133–36; evolution schematic diagram, *123*; heterochrony, *123*, 124–25, 132; heterometry, 122, *123*; heterotopy, 122, *123*, 135; heterotypy, 122, *123*, 124; sequence heterochrony, 124–25; 3D geometric morphometrics approach, 134–35; variation in, 143. *See also* ontogenetic change
development process, changes in, *91*
digital imaging, morphospace studies, 174–76
dingoes: feral, 206; reproduction of, 157. *See also* dog(s)
directed pathway, domestication, *4*, 4–5
diseases: insects providing drugs for, 238; zoonoses and, 244–45
disparity, term, 169
diversity, domesticated mammals and birds, 6. *See also* morphological diversification
DNA studies: bovines, 47, 49; domestication, 11, 13, 15
doctor fish (*Garra rufa*), 225

dog(s), 36–37; in the Americas, 39–40; ancient breeds, 37; ancient DNA studies, 37; average crown-rump length (CRL) of embryos by age, *131*; beginnings of domestication, 38–39; bony labyrinth location in wolf skull, *18*; brachycephaly, 195; breeds of, 37, *38*; cervical ribs in Pug dogs, 138; changes in body proportions, 178, *179*, *180*; chondrodysplasia in, 118; coat color and type in breeds, 85, *85*; combinations of mutations in three genes, *85*; Coppinger's simplification of pedomorphosis hypothesis, 134, *134*; dental signs of domestication, 15–16; dental variation and case of hairless, 67, *68*, *88*, 88–89; dingo as feral, 206; diversity in Americas, 40; double-clawed condition, 113; facial length in, 102; genetic mechanism, 79; hairless breeds, 67, *68*, *88*, 88–89; head shape variation, *180*, 181; hybridization of, 40; hypersociability of, 84–85; inner ears, 17, 19; myostatin gene mutation, 86–87, *87*; network of genealogical relationships, *71*; old network of dog breeds, 71–72; osteological signs of domestication, 16, *18*, 19; pedomorphosis hypothesis, *134*, 134–35; phenotypic variation in polydactyly in, 113–15; postcranial functional anatomy and locomotion in, 198–99; prenatal changes in skull shape among breeds, *130*; prenatal development, 130, *131*, 133; sample of breeds, *38*; selective breeding and size changes, 177–78; sexual size dimorphism (SSD), 178; skeletal variation in hind feet of, *114*; whole-genome resequencing of, 84
dog breeds, 39–40; Airdale Terrier, *38*, *85*; Australian Terrier, *38*, *85*; Basset Hound, *38*, 85, *85*, 118; Bearded Collie, *38*, *85*; Border Collie, *85*; brachycephalic syndrome (BS), 195, *196*; Bull Terrier, *217*; "Canis mexicana", *87*; Chihuahuas, 177; Chinese crested, *88*, 89; Chow-Chow, 158; Collies, 28; Corgi, 85, 118; Dachshund, 85, 118, 120; Dynastic Egyptian dogs, *179*; French Bulldog, 195, *196*, 199; German Shepherd, 138, 178, *180*; Giant Poodles, 158; Golden Retriever, *38*, *85*; Great Danes, 158, 177; Great Pyrenes, 28, 113–15; Greyhound, 199; hairless, 67, *68*, 88–89; Italian Maremmas, 28; Mastiffs, 113, 177; Newfoundlands, 113; Norwegian Lundehund, 114; number of newborns by, *159*; Old Welsh Grey, 28; Patagonian Sheepdog, 28, *29*; Pekingese, 113, 177; Pomeranians, 158; poodles, 113, *155*, 138, 195; Pug, *38*, 138, 195; St. Bernard, 113, 114, 217, *217*; Sardinian dog *Cynotherium sardous*, 185; Staffordshire Bull Terrier, *38*; Turkish Anatolian Shepherds, 28; Whippet, *38*, 87
dogs and life history: brain growth in, 155, *155*; caecum length, *154*; gastrointestinal tract of, *154*; gestation length, *150*; litter size variation, 158, *159*; longevity, 160; seasonality of reproduction and fecundity, 157–58; sequence of dental, skeletal and sexual maturity, 158–60. *See also* life history and growth
"Dolly" (cloned sheep), 67

domesticated animals, 8; coat variation in, 176; comparative embryology of, 92, 94, 96; culturesphere of, 247–48; divergence from the wild, 169–70; evo-devo approach to study of, 91–92; evolutionary veterinary medicine, 120–21; innovation or novelty, 129; misplaced biophilia and, 42; morphology studies, 183; novelties, 129; patterns of morphological or genetic diversification, *171*; production of, 167; tempo and mode of morphological diversification, 170–71; welfare of, 218. *See also* morphological diversification

domesticated birds, 8–9, 11; brain anatomy and size changes, 186–88; canary (*Serinus canaria*), *62*, 66; changes in size and body proportions, 181–82; chicken (*Gallus gallus*), 60–61, *61*; condition at hatching, 162, 164; cranial changes in, 199–201; development and staging, 95; diversity of, 6; domestic and wild forms, *10*; duck (*Anas platyrhynchos*), 63, *64*; estimated time line for, *12*; evolutionary relationships among species, *10*; geese (*Anser*), 63, 64, 65; growth in, 165–66; guinea fowl (*Nimida meleagris*), 61, *62*, 63; Japanese quail (*Coturnix cuturnix*), *62*, 65–66; munia (*Lonchura striata*), 208, *208*; origin of human language and, 208–9; pigeon (*Columba livia*), 58–59, *59*; primary works on development and staging of species, 95; sites of domestication, 25–26, *26*; turkey (*Meleagris gallopavo*), *62*, 65; wild and domestic species, *62*. *See also* birds and life history; chicken(s); ducks

domesticated mammals: body shape, axial skeleton and developmental repatterning in, 136–38; body size changes, 176–78; bovines, 46–49, *48*; brachycephaly case, 195, *196*, 197; brain anatomy and size changes, 182, 182–86; caecum length in dogs and cats, *154*; cats, 40–41, *41*; cavy (*Cavia porcellus*), 56–57, *57*; cranial changes in, 191–92; developmental repatterning in skull growth of, 133–36; diversity of, 6; dogs, 36–37, *38*; domestic and wild ancestors, *7*; embryological series of domestic pig, *96*; estimated time line for, *12*; European rabbit (*Oryctolagus cuniculus*), 57–58, *58*; evolutionary relationships among species of, *9*; ferrets, 43, *43*; goats, 43–44, *44*; growth in, 153–55; horse and donkey (Perissodactyla), 54–56; llamas and alpacas, *51*, 51–52; Old World camelids, 49–50, *51*; pigs, 46, *47*; primary works on development and staging of species, 93–94; rabbits, 57–58, *58*; reindeer, 52–53, *53*; sheep, *45*, 45–46; sites of domestication, 25, *26*; South American camelids, *51*, 51–52; wild ancestors of, *7*

domesticated phenotype, term, 81

domesticates: avian, 8–9, 11; avian hybrids among, 72–73; experimental embryology of, 103–4; mammalian, 6–8; mammalian hybrids among, 72–73; potential, 33–34, *34*; self-domestication hypothesis, 142; variation in morphological

diversification of, 201–2. *See also* domesticated birds; domesticated mammals

domestication: beginnings and antiquity of, 11, 13–15; comparing *sensu stricto* vs. artificial selection, 3; conceptualization, 2; cultural evolution of, 21; cultural influences of, 31–33; Darwin and, 139–40; definition, 1–2; evo-devo approach to, 90; experimental studies of, 206–7, 215; genetics of, 67–69; geography of, 25–26; human self-, 141–43; isotopic markers of, 19–20; morphological innovation or novelty, 129; pathways to, 3, *4*, 4–5; phenotypes, 86; rates of evolution in, 216–17; reconstructing history of, 21; selective breeding, 144; silver foxes, 209–10; species delimitation and, 73; variation in process of, 35. *See also* experimental studies; fish domestication; insect domestication

domestication *sensu stricto*, 3

domestication syndrome, 8, 79, 169; brain reduction in, 182; correlation of traits and, 81–82, 201; neural crest and, 101–2; phenotypes and, 79; term, 81

domestic populations, genetic diversity changes, 69

domestic productions, breeding of, 139

donkey: breeds, *54*; domestication of, 54–56; mule and hinny as hybrids, 72–73

Dove, Franklin, 104; fabulous unicorn, *104*

dromedary camels, 49–50, *51*, 101

ducks, 8; avian hybrids, 72; bone microstructure, 161; brains of, 190; changes in size and body proportions, 182; clutch size variation, 164–65; domestic, from mallard, 63, *64*; domestication of, 162; domestic breeds, *64*

Duerst, J. U., domesticated animal studies, 183

ecosystem engineering, 2

elephant (*Loxodonta africana*), 156

embryo, Darwin comparing human and dog, *140*

embryology: development and staging of domesticated birds, 95; development and staging of domesticated mammals, 93–94; domesticated animals in comparative, 92, 94, 96; domestic pig series, *96*; experimental, of domesticates, 103–4

embryonic development: fluctuating asymmetry, 110–11; movement and temperature variation, 109–10; neural crest, 99–101; phylotypic stage, 97, *98*; timing of change in, 97–98

embryonic morphospace, 98

environment, domesticated and wild animals, 23–24

environmental appreciation, decline of, 184

environmental crisis, Neolithic transition and, 22–24

epigenetics, genetic mechanism, 79, *80*

epistasis, genetic mechanism, 79, *80*

Equus. See horses

Escherichia coli, 245

European or wels catfish (*Silurus glanis*), 224

European sea bass (*Dicentrarchus labrax*), 224, 229

evo-devo. *See* evolutionary developmental biology

evolution: experimental, 3; rates in domestication process, 216–17; variation in rates, 219 evolutionary development: digit number in cavies, dogs and chickens, 112–17; domestication and malformations, 117–18, 120; epigenetic changes in chickens, 109–10; interactions of developing tissues (brain and skull), 111, *112*; neural crest, 99–101; neural crest, faces and transcription factor *Runx2*, 102–3; neural crest and domestication syndrome, 101–2; variation in, 121. *See also* developmental plasticity

evolutionary developmental biology (evo-devo), 90; model organisms, 97; research program, *92*, 121; studying domesticated animals, 91–92

evolutionary relationships: domesticated birds, *10*; domesticated fish, *221*; domesticated mammals, *9*

evolutionary veterinary medicine, 120–21

evolution diagram, developmental repatterning in, *123*

experimental studies: behavioral selection in mice and minks, 212–13; developmental plasticity in pigs, 214; domestication, 206–7, 215; longitudinal study of barn mice, 211–12; quantifications of rate change, 215; red junglefowl and White Leghorn chicken experiments, 213–14; silver fox experiment, 209–10; tamed and laboratory rat experiments, 210–11; urban fauna, 215–16

extinction, local breeds, 29–31

farming, Neolithic transition and, 24–25

fecundity: of dogs, 157–58; mammals, 147, 148–49

Fédération Cynologique Internationale, 28, 71

Felis catus. See cat(s)

feral, term, 203

feralization, 203–4; conservation, 203; contamination by, 204; genetics and, 204–5; variation in, 219; wild phenotype and, 205–6

ferrets: domestication of, 43; examples of, *43*; global trade, 43

Fertile Crescent, 13, 25; goats diversity, 43; pigs in, 46

fish domestication: aquaculture of salmonids, 223–24; Atlantic salmon (*Salmo salar*), 224, 226, 233, 234; brain size changes, 230–31; carp (*Cyprinus carpio*), 170, 220–22; coho salmon (*Oncorhynchus kisutch*), 223, 225, 232; crucian carp (*Carassius carassius*), 223, 234; evolutionary relationships among species, *221*; fluctuating asymmetry, 233; gene duplication and, 233–34; goldfish (*Carassius auratus*), 223, 226, 227, *227*, 228, 234; grass carp (*Ctenopharyngodon idella*), 222; guppies (*Poecilia reticulata*), 230; Indian carp (*Catla catla*), 222; life history and reproductive changes, 231–33, *232*; marine and freshwater aquaculture species, 224–25; morphological changes with aquaculture, 225, 226–30; Nile tilapia (*Oreochromis niloticus*), 224, 230; panga (*Pangasius hyophthalmus*), 224; silver carp (*Hypophthalmichthys molitrix*), 222; swordfish (*Chrysophrys aurata*), 224; trout (*Salmo trutta*), 223, 233; turbot (*Scophthalmus maximus*), 224; variation in, 234–35; westslope cutthroat trout *Oncorhynchus clarkii lewisi*, 233

"Fitch" ferrets, 43

fluctuating asymmetry: embryonic development, 110–11; fish domestication, 233

Food and Agriculture Organization, 57

foxes (*Vulpes vulpes*), 216; experimenting with, 79

Friedrich von Stephanitz, Max Emil, on German shepherd, 72

Fromm, Erich, on biophilia, 42

fruit fly Drosophila, 207, 238; domesticated insect, 242–43; illustration of, *237*

'Gait keeper' mutation, horses, 78

Galapagos finches, 139

Gallus gallus (chicken), 6: development and staging, 95; domestication, 60–61, *61*, 162. *See also* chicken(s)

Galton, Francis: on domestication, 31; on tamed seal, 14; on wolf domestication, 39

Gambian giant rat: detecting land mines, *33*; potential domesticate, 33–34

geese (*Anser*): avian hybrids, 72; breeds, 181; changes in size and body proportions, 181; clutch size variation, 165; domestication of, 63, 65, 162; wild and domestic, *64*

gene-culture coevolution, 247

gene duplication, fish domestication and, 233–34

gene flow, domestic and wild populations, 13

The Genesis of Species (Mivart), 140

genetic analysis, 13; cats, 41; chickens, 60; dogs, 37; feralization and, 204–5; geese, 63

genetics of domestication, 67–69; architecture and diversity of traits, 79–80; assortative mating, 73, 74; avian hybrids among domesticates, 72–73; chicken domestication, 14–15; correlation of traits, 81–82; dental variation and case of hairless dogs, 88–89; developmental genetics, 74–77; effects of trait selection on genes, 83–85; epigenetics, 79, *80*; epistasis, 79, *80*; genotype-phenotype map (GPM), *76*, 76–77; hybridization and, 69–70; isolating mechanisms, 74; linkage, 79, *80*; locomotion in horses, 78; mammalian hybrids among domesticates, 72–73; mutations, 74–77; myostatin gene mutation, 86–87; old network of dog breeds, 71–72; pleiotropy, 79, *80*; population size and geography, 82–83; quantitative trait loci (QTL) mapping, 80; species delimitation, 73; structural genetic changes, 77; variation of, 89

genetic variation, population size and geography, 82–83

genomic imprinting, 73

genomic tools, tracing neural crest cells, 102

Geoffroy Saint-Hilaire, Isidore, diversity of domesticated mammals, 183

geography of domestication, 25–26; genetics and, 82–83; map showing sites, *26*

gerbils, wild (*Meriones unguiculatus*), 185

German Mammalogical Society, 134

gestation length, mammals, 149, *150*

global trade, ferrets, 43

goat(s): average crown-rump length (CRL) of embryos by age, *131*; brachycephaly, 195, *196*; developmental plasticity of Slijper's two-legged, 106, *107*; development and staging, *93*; domestication of, 43–44; examples of domesticated, *44*; feral, 204; genetic diversity, 44; horns in caprids, 193–94, *194*; osteological signs of domestication, 16; selective breeding and size changes, 177; sexual size dimorphism (SSD), 194; structural genetic changes, 77

goat breeds, 43, *44*, 177, 193

golden hamster, domestication, 171

goldfish (*Carassius auratus*), 234; caudal and anal fins of, *228*; creation and development of, 223; CT images of skull of four varieties, *227*; variations of, 226

goose. *See* geese (*Anser*)

Gray, Asa, *140*

growth differentiation factor-8 (GDF-8), muscle development, 86

guinea fowl (*Numida meleagris*): clutch size variation, 165; development and staging, *95*; domestication, 61, *62*, 63; domestication of, 25

guinea pigs. *see* cavy

Guinness World Records, "Matilda" as World's Oldest Living Chicken, 166

Haeckel, Ernst: on breeding experiments, 67; on Porto-Santo rabbit, 204

hair anatomy, domestication, 15

Haustiere-zoologisch gesehen (Herre and Röhrs), 183

Hemingway, Ernest, on polydactyl cats, 117

heterochrony: observed pattern of, 132; repatterning, *123*, 124–25

heterometry, repatterning, 122, *123*

heterotopy, repatterning, 122, *123*

heterotypy, repatterning, 122, *123*, 124

Hilzheimer, Max, 134

hinny, mammalian hybrid, 72–73

His, Wilhelm, on heredity, 109

Histoire naturelle, générale et particulière (Buffon), 71, 183

Homo floresiensis, 185

Homo sapiens, 21; evolution of, 142–43, 182; self-domestication of, 142–43

honey bee *Apis mellifera*: domesticated insect, 238–40; illustration, *237*. *See also* insect domestication

honey hunting, *240*

horse(s): average crown-rump length (CRL) of embryos by age, *131*; coat variations, 170; dental variation in, 170, 197–98; development and staging, *93*; digit number in, 112; domestication of, 54–56; gestation length, 149, *150*; head shape variation, *180*, 181; locomotion in, 78; modularity in adult skulls, *128*, 129–30; morphometric studies of skull growth, 135; mule and hinny as hybrids, 72–73; ontogenetic tooth changes, 106, *108*; pace and gait, 78; pedomorphism hypothesis, 135;

protein-coding gene *BRAF* in, 118; skull of domesticated breeds, *128*; skull shape variation of breeds, 135; structural genetic changes, 77

horse breeds, *54*, 138, 177; Andalusian, *54*; Appaloosa, *54*; Bay, *54*; Belgian, *54*; Colombian Paso, 28; Comtois, *54*; Falabella, *54*, *128*; Percheron, *54*; Shetland pony, *54*

human-animal interactions: earliest domestication phase, 13–14; Ego, Eco and Evo views of, *3*; geography of domestication, 25–26; intensity of, *4*; pet keeping and, 32

human dental calculus, dairy consumption, 15

human health, Neolithic transition and, 24–25

human language, domesticated birds and origin of, 208–9

human variation, human self-domestication and, 141–43

human-wasp relationship, 236

hybridization, 6; adaptive introgression, 69; avian hybrids, 72; bovines, 46, 48; of dogs, 40; ferrets, 43; geese, 63; genetic variation and, 69–70; mammalian hybrids, 72–73; turkeys, 65

hypersociability, dogs, 84–85

influenza, ferret and, 43

innovation, term, 129

insect domestication, 236; beetle petting zoo, 238; cochineal insect *Dactylopius coccus*, *237*, 241–42; cultural perception and insects as pets, 238; diseases and zoonoses, 244–45; fruit fly *Drosophila melanogaster*, *237*, 242–43; giant water bug (*Lethocerus indicus*), *237*; honey bee (*Apis mellifera*), *237*, 238–40; human-insect interaction, 236; insects as food, 236–37; mosquitoes, 243; palm weevil (*Rhynchophorus ferrugineus*), *237*; silkworm (*Bombyx mori*), *237*, 241; uses for medicine, 238; variation in, 246

Institute for Cytology and Genetics, Novosibirsk (Russia), 209

intensive breeding, 1, 3, *4*

International Cat Association, 41

introgression, by hybridization, 69–70

Iron Age, 197

Japanese amberjack (*Seriola quiqueradiata*), 224

Japanese ayu (*Plecoglossus altivelis*), 224

Japanese quail (*Coturnix coturnix*): domestication of, 65–66, 162, 164; illustration, *62*

juvenilization, 134, *134*, 142

Kiel school, domestic animal studies, 183

King, Helen Dean, 210–11, *211*

Klatt, Berthold, domestic animal studies, 183

Kuo, Zing-Yang, on experimentalists, 110

Lamarck, Jean-Baptiste, 71

landraces: in Americas, 31; definition, 28

L'Enfant sauvage (film), 207

life history and growth, 144–45; birds, 162–66; bone microstructure and, 161; dogs, 157–60; domestic animal production, 167; mammals, 145–57; variation in evolution, 168; Western World Social Paradigm, 167. *See also* birds and life history; dogs and life history; mammals and life history
linkage: clusters, 80; genetic mechanism, 79, *80*
Linnaeus, Carolus, breeds of domestic pigeon, 73
llamas, domesticated mammals, *51*, 51–52
Lonchura striata. *See* domesticated birds
London, Jack, *Call of the Wild*, 207
longevity: in chickens, 166; dogs, 160; of mammals, 155–57; mutation accumulation theory, 156
Lorenz, Konrad: notion of "Verhausschweinung", 141; studies of ducks and birds, 74

magnetic resonance imaging (MRI), 174
malformations: cavy mutants, 118; chickens, 117; newborn lambs with, *119*, 120
mallard duck (*Anas platyrhynchos*), domestication, 63, *64*. *See also* ducks
mammalian domesticates, 6–8; domestic and wild forms, *7*. *See also* domesticated mammals
mammalian skull: dermal and endochondral bones, *126*; modularity during development, 125–26; modularity in horse skull, *128*, 129; modularity of, 127, 129–30; prenatal changes in shape among dog breeds, *130*; schematic, *126*
mammals: body shape, axial skeleton in, 136–38; changes in body proportions in, 178, *179, 180,* 181; changes in body size in, 176–78; coat variation in, 176; developmental repatterning in, 136–38; estimated time line of domestication, *12*; modularity in skull of, 127, *128*, 129–30; ontogenetic trajectories of shape before birth, 130–33; potential domesticate, 34, *34*
mammals and life history: altricial/precocial continuum, 145–46; brain development at birth, 145–46; condition at birth, 145–46; dental maturity, 151–53; fecundity, 147, 148–49; gestation length, 149, *150*; growth in domesticated, 153–55; labor and birth, 149–51; litter size variation, 148–49; longevity of, 155–57; metabolic rate and size, 156; parturition in, 149–51; seasonality of reproduction and fecundity, 147–48; sexual maturity in, 151; size, modularity, disparity and altricial/precocial continuum, 146–47; skeletal maturity, 151–53; telomere shortening in aging, 156–57. *See also* life history and growth
Mammal Species of the World, mammalian taxonomy, 26
marsupials, 8; brain size, 184; neural crest in, 102
material culture, domestication, 15
"Matilda", World's Oldest Living Chicken, 166
medicine, evolutionary veterinary, 120–21
Meleagris gallopavo. *See* turkeys
Melopsittacus undulatus. *See* domesticated birds
Mesocricetus auratus. *See* domesticated birds
mesolithic rock art, honey hunting, *240*

mice: artificial selection in, 215; behavioral selection, 212–13; development and staging, *93*; experimental generation of polydactyly in, *116*; interactions of brain-skull interaction, *112*; *Mus musculus*, 40, *93*, 156, 211–12; *Peromyscus polionotus*, 204
migrations, dogs in Americas, 40
milk, domestication research, 15
mink *Neovison vison*, 148, 198, 212
Mivart, George, on Darwin, 140
mixed-species association, 5
model organisms, 97
modularity: horse skull, *128*; relation to ontogeny, 125–26; in skull of mammals and birds, 127, 129–30; skull shape of mammals, 146
molecular biology, experimental embryology and, 103
Morgan, Thomas Hunt, founder of genetics, 67
morphogenesis, 91
morphological diversification, 169; body proportion changes in mammals, 178, *179, 180,* 181; body size changes in mammals, 176–78; brachycephaly, 195, *196,* 197; brain of chickens, 188–89; brain of ducks, 190; changes in body size and proportions in birds, 181–82; changes in brain size and anatomy in birds, 186–88; changes in brain size and anatomy in mammals, 182, 184–86; coat variation in mammals, 176; cranial changes in birds, 199–201; cranial changes in mammals, 191–92; dental changes in mammals, 197–98; divergence from wild, 169–70; historical studies, 183; horns in bovids and caprids, 192–94, *193, 194*; sexual size dimorphism in sheep and goats, 194; skull variation, 195, 197; tempo and mode of patterns in domesticated animals, 170–71; variation of domesticates, 201–2
morphological traits, of domestication, 14
morphospace: digital imaging approaches, 174–76; human and dog skulls in Cartesian grid, *174*; quantification of, 173–74; universe of potential phenotypes, 172; unoccupied, 172, *173*
mosquitoes (*Anopheles* species), 243. *See also* insect domestication
mouse. *See* mice
mule, mammalian hybrid, 72–73
Muscovy duck (*Cairina moschata*), 25, 62, 162, 164. *See also* ducks
muscularity, extreme, myostatin gene mutation, 86–87
Mus musculus. *See* mice
mutagenesis screen, *3*
mutation accumulation theory, 156
mutations, in domestication, 74–77
Muybridge, Eadweard, on locomotion in animals, 78
Mycobacterium bovis, bovine tuberculosis agent, 244
myostatin gene mutation, extreme muscularity in domesticated forms, 86–87
Myotragus, 152, 185
myxomatosis, 245
Myxoma virus, 245

Naess, Arne, Deep Ecology movement, 167
Native Americans, adoption of European domesti-
    cates, 32
Natural History Museum Bern (NMBE), 160
Neolithic revolution, 21
Neolithic transition, 2, 21–22; environmental crisis and,
    22–24; farming communities, 239; mismatches and
    human health, 24–25; weather patterns, 21–22, *22*
neural crest: cells of cranial, in chicken embryo, *100*;
    development of, 99–101; development of face in
    mammal with, *100*; development programs, 101;
    domestication syndrome and, 101–2; regulatory
    changes to, 99
niche construction, 2
nitrogen isotopes, captive animal rearing, 20
nomenclature: cattle, 28; domesticated species,
    breeds and landraces, 26–28
novelty, term, 129
Numida meleagris. See guinea fowl

obstetrical dilemma, mammals, 150
Old World camelids: domestication of, 49–50;
    examples of, *51*
Oliver, Jamie, on animals eaten, 42
On Growth and Form (Thompson), *174*
On the Origin of Species (Darwin), 139, 140
ontogenetic change: average size of embryos by age,
    131; body shape and axial skeleton, 136–38;
    developmental repatterning, 122, 124–25, 136–38;
    domestic repatterning in domestic mammal skull
    growth, 133–36; modularity, 125–26; modularity
    in mammal and bird skulls, 127, 129–30; morpho-
    logical innovation or novelty, 129; shape trajectories
    before birth in mammals, 130–33; skull shapes in
    dog breeds, *130*; skull shapes in equids and horses,
    128; 3D geometric morphometrics approach,
    134–35. *See also* developmental repatterning
Oryctolagus cuniculus. See rabbit(s)
osteological signs of domestication: dogs, 16, *18*, 19;
    goats, 16; pigs, 16, *17*; sheep, 16, 19–20
ostrich (*Struthio camelus*), 8, 25
Ovis aries. See sheep
Owen, Richard, vaca ñata in writings, 28
oxygen isotopes, captive animal rearing, 20

parasitic mite *Varroa destructor*, 240
parturition, in mammals, 149–51
pathogens, Pleistocene Megafauna Extinctions, *245*
pathways to domestication, *4*, 4–5; commensal, *4*, *4*;
    directed, *4*, 4–5; prey, *4*, *4*
pedomorphosis hypothesis: dogs, *134*, 134–35;
    horses, 135; pigs, 135
Perissodactyla (horse and donkey): breeds, *54*;
    domestication of, 54–56
pet keeping: domestication and, 32; insects as pets, 238
Phyle, term, 141
pig(s): average crown-rump length (CRL) of
    embryos by age, *131*; body shape changes and

variation, 136, *137*; brachycephaly, 195, *196*;
    cranial morphology, 16, *17*; dental signs of
    domestication, 15–16; developmental plasticity in,
    214; development and staging, *94*; digit number
    in, 112; embryological series of domestic pig, *96*;
    feral, 203–4, 205–6; gestation length, 149, *150*;
    isotopic markers of domestication, 20; litter size,
    148–49; osteological signs of domestication, 16,
    17; prenatal development, 130; selective breeding
    and size changes, 177; sexual maturity, 151; skull
    of domestic, and wild boar, *17*; Vietnamese Pot-
    bellied, 151, *151*; zoonoses and, 244
pig breeds, 46, *47*, 195, *196*; Angel Saddleback, 46;
    Bentheim Black Pied, 46; Berkshire, 46; Chato
    Murciano, 46; Duroc, 46; Kune Kune, 46; Large
    White, 46; Middle White, *137*; Piétrain, 46;
    Swabian-Hall, 46; Vietnamese Pot-bellied, 46, 151,
    151; Woolly, 46
pigeons: African speckled, 70, 74; breeds, *59*, *200*;
    Darwin on, 6, 199; development and staging, *95*;
    domestication of, 58–59, 162; domestic forms of,
    73; foot feathers, 75; head crests in, 75, *75*
plants: domesticated animals and, 23–24;
    domestication of, 13; geography of tomato
    domestication, 26
pleiotropy, genetic mechanism, 79, *80*
Pleistocene Megafauna Extinctions, *245*
poisson mandarin (*Perca* sp.), 224
polecat, ferret from, 43
polydactyly: digit variation in dogs, 113–15, *114*;
    Silkie chickens, 115
population densities, Neolithic transition, 24
population size, genetics and, 82–83
postcranial functional anatomy, dogs, 198–99
prey pathway, domestication, *4*, *4*
production animals: gestation length, *150*; Western
    World Social Paradigm, 167

quail (*Coturnix coturnix*), 162, 164, 175; develop-
    ment and staging, *95*; domestication of Japanese,
    62, 65–66
quantitative trait loci (QTL) mapping, 80

R2OBBIE-3D, digitalization, 176
rabbit(s): coat coloration, *105*; cottontail *Sylvilagus*,
    245; developmental plasticity of coats, 104–6;
    domestication of, 57–58; feral, 204; litter size of,
    148; myxoma virus in, *245*; *Oryctolagus cuniculus*,
    26, *82*, *93*, 245; skull variation in domesticated and
    lop-eared, 81–82, *82*; structural genetic changes, 77
rabbit breeds, 58, *105*; dwarf black, *58*; English Angora,
    58; European (*Oryctolagus cuniculus*), 26, 57–58,
    58; French Lop, *58*; Himalayan, *105*, 105–6;
    Holland Lop, *58*; Mini Lop, *58*; Mini Rex, *58*; New
    Zealand, *58*; Porto-Santo rabbit, 204
race, term, 141
rainbow trout (*Oncorhynchus mykiss*), 225, 231, 232.
    See also fish domestication

rats (*Rattus norvegicus*): development and staging, *94*; litter size, 148; tamed and laboratory experiments, 210–11

reindeer: domesticated mammals, 52–53; herd of, *53*

Remane, Adolf, 141

reproduction: seasonality of dogs, 157–58; seasonality of mammals, 147–48

Rütimeyer, L., domesticated animal studies, 183

Saudi Arabia, mixed-species association in, 5

selective breeding, 4, 144; animal welfare and, 218; brain size reduction after, 201–2; chickens, 23, 144, *166*; domesticated fish, 232–33; domestication and, 13; domestic dogs, 177–78, 186; geese, 181; mismatches of humans with environment, 24–25; within species, *3*

self-domestication: concept of, 143; human, 141–43; hypothesis, 38

severe acute respiratory syndrome (SARS)-associated corona virus, ferret and, 43

sexual maturity, 144; acceleration of, 151; in chickens, 164; in dogs, 158–60; in foxes, 209; in guinea fowl, 165; late, in mammals, 152; neural crest and, 101, 142; in rats, 148

sexual size dimorphism (SSD): dog breeds, 178; sheep and goats, 194

sheep: average crown-rump length (CRL) of embryos by age, *131*; body shape changes and variation, 136, *137*; development and staging, *94*; "Dolly" clone, 67; domestication of, 45–46; gestation length, 149, *150*; horns in caprids, 193–94; isotopic markers of domestication, 20; osteological signs of domestication, 16, 19–20; selective breeding and size changes, 177; sexual size dimorphism (SSD), 194; skulls of newborn lambs, *119*, 120

sheep breeds, 45, 195; ancient, 45; Arles Merino, 45; Crossbreed, 45; European mufflon, 45; Hissar, 45; South Down, 138; Suffolk, 45, 136, *137*

Sicilian pygmy elephant *Elephas falconeri*, 185

Signor-Lipps effect, 11

silkmoth *Bombyx mori*: domesticated insect, 241; illustration of, *237*; morphological difference between wild and domesticated, *242*. See also insect domestication

silver fox experiment, 209–10

skeletal growth: dogs, 158–60; in mammals, 151–53

skulls: case of brachycephaly, 195, 197; chondrocranium of mammals and birds, 132–33; cranial changes in domesticated mammals, 191–92; developmental repatterning in domestic mammal growth, 133–36; modularity in, mammals and birds, 127, 129–30; prenatal changes in shape of, in dog breeds, *130*; vaca ñata *vs.* standard cattle, 28, *30*

Slijper, E. J., two-legged goat, 106, *107*

Smellie, William, translation of Buffon's work, 72

South America: camelids, llamas and alpacas, *51*, 51–52; potential domesticates, *34*

sparrows (*Passer domesticus*), 84

"Species and Race", Remane, 141

species delimitation, domestication and, 73

Studer, Theophil, 133

Sus domesticus. See pig(s)

symbiosis, 5

SYNBREED project, chicken diversity panel, 60

tameness, 1; Belyaev on selection for, 81; pleiotropy and, 79; self-domestication and, 142; wolf domestication, 38

Temnick, Caenraad, on chicken types, 73

Teng, Camilla S., on skull roof bones, *112*

Thompson, D'Arcy: geometric transformations, 173, *174*; human and dog skulls in Cartesian grid, *174*

tilapias, 224. See also fish domestication

time line, domestication (estimated): of bird species, *12*; of mammalian species, *12*

time management, Neolithic transition, 24–25

tissue transplantation, 103

tomato, geography of domestication, 26

tree shrews (*Tupaia*), brain weight, 185

Truffaut, François, *L'Enfant sauvage*, 207

Trypanosoma cruzi, Chagas disease, 244

turkeys (*Meleagris gallopavo*): clutch size variation, 165; development and staging, *95*; domestication of, 25, 65, 162; illustration, *62*

turtle, potential domesticate, 34, *34*

University of Maine, 104

University of Tübingen, 133

Unnatural Selection (van Grouw), 182

urban ecology, 42

USA National Pork Producers Council Terminal Sire Line National Evaluation Program, 154

vaca ñata, 28; skull of, *30*

van Grouw's *Unnatural Selection*, 182

The Variation of Animals and Plants under Domestication (Darwin), 139, 183

Varro, on rabbits, 57

veterinary medical database (VMDB), 160

veterinary medicine, evolutionary, 120–21

von Baer, Karl, embryonic development, 183

warble flies (*Hypoderma*), 237

weather patterns, Neolithic transition, 21–22

Western World Social Paradigm, 167

Whewell, William, on theory's truth, 139

White Fang (film), 207

Williams-Beuren syndrome (WBS), 84

Wilson, E. O., on biophilia, 42

The Wistar Institute, 210

wolves: bony labyrinth location in skull, *18*; brain to body weight, *155*; *Canis lupus*, *38*; development and staging, *93*; dogs descending from, 36; domestication of, 38–39; inner ears, 17, 19; reproduction, 157; sample of dog breeds and, *38*. *See also* dog(s)

World Canine Organization, 71

World Health Organization, "One Health" approach, 167

Wright, Sewall: founder of genetics, 67; on polydactyly, 113

zebra finch (*Taeniopygia guttata*), 8, *95*, 162

zebrafish (*Danio rerio*), 220; brain-skull interaction, *112*

zone of polarizing activity (ZPA), chicken wing, 115, *116*

zooarchaeological record: chickens in, 60; dental signs of domestication, 16; digital documentation, 175

zooarchaeology, domestication, 11, 13–15

Zoobiquity Conferences, 120

zoological knowledge, misplaced biophilia and, 42

zoonoses: definition, 244; diseases and, 244–45

zoos, longevity of mammals in, 155–56

Image Credits

1.1. Artwork by Gabriel Aguirre; silhouettes courtesy of phylopic and the following contributors: Michael Keesey (human female and male), Steven Traver (goat, rabbit, cow, and pig), Carlos Cano (carp), Ryan Cupo (rat), David Orr (cat), Cristopher Silva (water buffalo and Bali cattle), Mercedes Yrayz and Michael Keesey (horse), Madeleine Price (mouse), Sharon Wegner (duck), Lukasiniho (camel), uncredited/public domain (ferret), Matthew Stewart (ass), Birgit Lang (turkey), Mystica (reindeer), Rececca Groom (goose), an ignorant atheist (dog).

1.2. Diego Aguirre, modified and adapted from Stern (2011).

1.3. Adapted from Larson and Burger (2013).

1.4. Gabriel Aguirre using the "phylogeny subsets" tool downloaded from http://vertlife.org/. The tree is based on Upham et al. (2019).

1.5. Gabriel Aguirre using the "phylogeny subsets" tool downloaded from http://vertlife.org/. (Jetz et al. 2012).

1.6. Gabriel Aguirre, dates based on: Larson and Fuller (2014) for dog, goat, sheep, taurine cattle, llama, vicugna, indicine cattle, donkey, horse, reindeer, dromedary, water buffalo, cavy, yak, Bactrian camel, rabbit, banteng, and gaur; Lord et al. (2020a) for cavy; Qiu et al. (2015) for yak; Zeller and Göttert (2019) for dog, pig, and ferret; Ottoni et al. (2017) for cat. Silhouettes courtesy of phylopic and the following contributors: Michael Keesey (human female and male), Steven Traver (goat, rabbit, cow, pig, and sheep), Carlos Cano (carp), Ryan Cupo (rat), David Orr (cat), Cristopher Silva (water buffalo and Bali cattle), Mercedes Yrayz and Michael Keesey (horse), Madeleine Price (mouse), Lukasiniho (camel), uncredited/public domain (ferret), Matthew Stewart (ass), Mystica (reindeer), an ignorant atheist (dog), Flappieh/Wikimedia commons (Cavy), treehill/Wikimedia commons (yak).

1.7. Gabriel Aguirre, dates based on Larson and Fuller (2014) for pigeon, Muscovy duck, duck, and goose; Gilbert and Shapiro (2014) for pigeon; Peters et al. (2016) for chicken; Sharpe et al. (2018) for turkey; and Vignal et al. (2019) for guinea fowl. Silhouettes courtesy of phylopic and the following contributors: Birgit Lang (turkey), Maija Karala (Muscovy duck), Rececca Groom (goose), Sharon Wegner (duck), Ferran Sayol (pigeon), Steven Traver (chicken).

1.8. Artwork by Jaime Chirinos.

1.9. Parts a and b courtesy of Laura Wilson; c and d modified from Schweizer et al. (2017).

1.10. Modified from Platt et al. (2017, fig. 5a).

1.11. Reproduced from Yinon M. Bar-On et al. (2018, fig. S5).

1.12. Drawn by Diego Aguirre, adapted and modified from Zeder (2018).

1.13. Reproduced with permission from Barrios et al. (2016).

1.14. Courtesy of Alejandro Daniel Mosquera.

1.15. Adi Haririe/Shutterstock.com.

2.1. Images sourced from Shutterstock, produced by: Iakov Filimonov (wolf), Kazlouski Siarhei (Bearded Collie), Dora Zett (Whippet), cynoclub (Staffordshire Bull Terrier),

Dmitry Kalinskovsky (Airdale Terrier), Andreas Gradin (Australian Terrier), Billion Photos (Basset Hound), Jagodka (Pug), Jenson (Golden Retriever).

2.2   Images sourced from Shutterstock, produced by (*left to right, top to bottom*): Konovalov (Manx cat), Eric Isselee (Neva Masquerade), Matt Gore (Siamese cat), cynoclub (Bengal cat), Eric Isselee (Sphynx cat), digitalienspb (Norwegian Forest cat), PHOTOCREO Michal Bednarek (British Shorthair), Viorel Sima (British Longhair), likuzia (Abyssinian).

2.3   Images sourced from Shutterstock, produced by: vandycan (polecat-colored ferret), Couperfield (silver color ferret).

2.4   Images sourced from Shutterstock, produced by (*left to right, top to bottom*): Eric Isselee (Toggenburger goat), prapann (brown goat), a_v_d (black-white goat), Eric Isselee (Rove), Farinoza (goat), JackF (long haired goat), Eric Isselee (Angora goat), Iakov Filimonov (black goat), Belikin (white goat).

2.5   Images sourced from Shutterstock, produced by (*left to right, top to bottom*): volkova natalia (European mufflon), Eric Isselee (crossbreed), photomaster (Hissar), Eric Isselee (Ouessant), Eric Isselee (Suffolk), Eric Isselee (Arles Merino).

2.6   Images sourced from Shutterstock, produced by (*left to right, top to bottom*): Eric Isselee (wild boar), photomaster (domestic pig), New Africa (black mini pig), Eric Isselee (domestic pig), photomaster (Vietnamese Pot-bellied pig), Eric Isselee (Göttingen minipig), Eric Isselee (Kounini pig), Eric Isselee (Mangalitsa), Eric Isselee (Oxford Sandy and Black).

2.7   Images sourced from Shutterstock, produced by (*left to right, top to bottom*): Worraket (Banteng), wanchai (gaur), Eric Isselee (domestic Asian water buffalo), Eric Isselee (yak), Daniel Prudek (domestic yak), Dennis Jacobsen (aurochs skeleton), Vander Wolf (cattle).

2.8   Images sourced from Shutterstock, produced by (*left to right, top to bottom*): photomaster (camel), lara-sh (dromedary), Iakov Filimonov (guanaco), mariait (llama), Eric Isselee (alpaca), Alessandro Pinto (vicuña).

2.9   Longtaildog/Shutterstock.

2.10   Images sourced from Shutterstock, produced by (*left to right, top to bottom*): Anton Starikov (donkey), Alexia Khruscheva (Bay horse), Eric Isselee (Comtois), Eric Isselee (Belgian), Eric Isselee (Appaloosa), cynoclub (Falabella), Eric Isselee (Andalusian), Eric Isselee (Shetland Pony), Eric Isselee (Percheron).

2.11   Images sourced from Shutterstock, produced by (*left to right, top to bottom*): Domestic breeds of the cavy: Eric Isselee (long-hair cavy, hairless cavy), cynoclub (Peruvian guinea), Katya Nikitina (Abyssinian), Galyna Syngaievska (cavy unknown breed), Photok.dk (cavy unknown breed).

2.12. Images sourced from Shutterstock, produced by (*left to right, top to bottom*): Eric Isselee (European rabbit *Oryctolagus cuniculus*), Eric Isselee (French Lop), photographer (New Zealand), Eric Isselee (Mini Lop), ravl (dwarf black), Eric Isselee (English Angora), Piyathep (Holland Lop), Jiang Hongyan (white rabbit), cynoclub (Mini Rex).

2.13   Images sourced from Shutterstock, produced by (*left to right, top to bottom*): stockphoto mania (Homing pigeon), photomaster (Fantail), Eric Isselee (Tippler), Eric Isselee (Old German Owl pigeon), Rosa Jay (rock dove *Columba livia*), Eric Isselee (Jacobin), Eric Isselee (Tumbler), photomaster (pigeon), photomaster (German Modena).

2.14   Images sourced from Shutterstock, produced by (*left to right, top to bottom*): Narupon Nimpaiboon (Red junglefowl), photomaster (Brahma), Eric Isselee (Silkie), yevgeniy11 (White Leghorn), Sutham (Shamo), Valentina_S (Hubbard), KinoMasterskaya (Polish), lunamarina (Araucana), cynoclub (Barbu d'Anvers).

2.15   Images sourced from Shutterstock, produced by (*left to right, top to bottom*): photomaster (wild quail *Coturnix japonica*), Roman Teteruk (domestic Japanese quail), veleknez (wild turkey *Meleagris gallopavo*), photomaster (domestic turkey), Eric Isselee (wild helmeted guinea fowl), Eric Isselee (domestic helmeted guinea fowl *Numida meleagris*), Eric Isselee (wild canary), dule (domestic canary).

2.16   Images sourced from Shutterstock, produced by (*left to right, top to bottom*): Fekete Tibor (Mallard *Anas platyrhynchos*), and domestic forms: photomaster (crested duck), Olhastock (white domestic duck), Shutterstock (duck), Eric Isselee (Indian Runner), Margo Harrison (Muscovy duck).

2.17   Images sourced from Shutterstock, produced by (*left to right, top to bottom*): E.O. (Greylag goose *Anser anser*), Iakov Filimonov (swan goose *Anser cygnoides*), TR STOCK (domestic goose), Kaiskynet Studio (heavy goose), Eric Isselee (domestic goose).

3.1   Courtesy of Martin Fischer, Jena.

3.2.   Drawn by Diego Aguirre, modified from Frantz and Larson (2019).

3.3.   Buffon and Daubenton (1755), from the French National Library. gallica.bnf.fr/AU.

3.4.   Drawn by Jaime Chirinos.

3.5   Drawn by Diego Aguirre, modified from Green et al. (2017).

3.6   Drawn by Jaime Chirinos, modified from Nickel et al. (1986).

3.7   Drawn by Diego Aguirre, modified from Agnvall et al. (2018).

3.8   From Darwin (1868).

3.9   Images sourced from Shutterstock, credited to (*left to right, top to bottom*): billion photos, Kazlouski Siarhei, Andreas Gradin, Jenson, Dmitry Kalinovsky, and Csanad Kiss.

3.10   Eric Isselee/Shutterstock.

3.11   From Hernández (1651).

3.12   Drawn by Vera Primavera based on photos in Kupczik et al. (2017).

4.1   Drawn by Diego Aguirre.

4.2   Drawn by Diego Aguirre.

4.3   Adapted from Keibel (1897).

4.4   Images by Daniel Núñez-León, embryo procured by Hiroshi Nagashima, Niigata, Japan.

4.5   Drawn by Vera Primavera and Diego Aguirre.

4.6   Reproduced from Dove (1936).

4.7   Drawn by Vera Primavera.

4.8   Modified from Slijper (1942). Artwork by Vera Primavera.

4.9   Drawn by Vera Primavera, silhouettes from phylopic by Mercedes Yrayz (horse) and Steven Traver (cow).

4.10   Adapted from Teng et al. (2019).

4.11   Reproduced from Wright (1934).

4.12   Modified and adapted from Kadletz (1932).

4.13   Drawn by Vera Primavera, based on Lange and Müller (2017) and work by Li et al. (2006).

4.14   Drawn by Vera Primavera, based on Diogo et al. (2019).

5.1   Drawn by Diego Aguirre, adapted from Richardson et al (2009).

5.2   Chicken heads by Jaime Chirinos; cross histological section by Vera Primavera.

5.3   Left image from Weber (1927), right image courtesy of Daisuke Koyabu, modified from Koyabu et al. (2014).

5.4   Modified from Heck et al. (2018b).

5.5   Drawn by Vera Primavera.

5.6   Reproduced from Gaupp (1906, fig. 393).

5.7   Drawn by Vera Primavera.

5.8   Drawn by Jaime Chirinos.

5.9   Reproduced from Darwin (1871).

6.1   Reproduced from Heck et al. (2018a).

6.2   Photomaster/Shutterstock.

6.3   (a) Drawn by Gabriel Aguirre based on data from McGrosky et al. (2016) and Snipes and Snipes (1997); (b) drawn by Vera Primavera.

6.4   Drawn by Gabriel Aguirre using data from Schleifenbaum (1973).

6.5   Pigeon by Jeanne Menjoulet; Muscovy duckling by materialscientist/Wikimedia commons.

6.6   Reproduced from Zuidhof et al. (2014) with permission.
7.1   Artwork by Diego Aguirre.
7.2   Reproductions of drawings of Hieronymus Bosch by Vera Primavera.
7.3   Drawn by Jaime Chirinos.
7.4   Drawn by Vera Primavera, modified from Thompson (1917).
7.5   Reproduced from Duerst (1931).
7.6   Artwork by Vera Primavera.
7.7   Artwork by Vera Primavera.
7.8   Artwork by Vera Primavera.
7.9   Plotted by Gabriel Aguirre, based on the data and analyses in Kawabe et al. (2017).
7.10   Reproduced from Darwin (1868).
7.11   (a–h) Drawn by Vera Primavera based on Riedel (1993); (i) Anan Kaewkhammul/
       Shutterstock; (j) Charles Lemar Brown/Shutterstock; (k) drawn by Jaime Chirinos.
7.12   Drawn by Jaime Chirinos.
7.13   Drawn by Vera Primavera, adapted from Ravn-Mølby et al. (2019).
7.14   (a–j) by Kristoff Veitschegger and Tímea Bodogán, from Veitschegger et al. (2018); (k)
       by Madeleine Geiger.
7.15   Drawn by Vera Primavera, adapted from Darwin (1868) and Young et al. (2017).
8.1   Drawings by Vera Primavera.
8.2   Courtesy of The Wistar Institute.
8.3   Reproduced from Geiger and Sánchez-Villagra (2018).
9.1   Plotted by Gabriel Aguirre using the fishtree package in R (Chang et al. 2019; Rabosky
       et al. 2018).
9.2   Drawings by Vera Primavera, (a–b) based on Antipa (1909).
9.3   Drawings by Vera Primavera, adapted from Perry et al. (2019).
9.4   Photos by Jorge Carrillo; 3D models and their snapshots by Stephan Spiekman.
9.5   Photos by Vangert and r.classen/Shutterstock; drawings by Vera Primavera based on
       Bateson (1894).
9.6   Reproduced with permission from Witten et al. (2006).
10.1   Images sourced from Shutterstock, produced by (*left to right, top to bottom*): Protasov
       AN, Vinicius Tupinamba, Eric Isselee, Daniel Prudek, Protasov AN.
10.2   Redrawn by Achillea/Wikimedia commons.
10.3   Drawings by Jaime Chirinos, based on imaging in Takai et al. (2018).
10.4   Modified by Jaime Chirinos from Doughty et al. (2020).